I0051042

Electrically Conductive Membrane Materials and Systems

Electrically Conductive Membrane Materials and Systems offers in-depth insight into the transformative role of electrically conductive materials in membrane separation processes for desalination and water treatment. The book focuses on the intelligent design of conductive membranes and systems, fouling and related phenomena, fouling control using electrically conductive materials, and electrically tunable membrane systems for microfiltration, ultrafiltration, nanofiltration, reverse osmosis, and membrane distillation.

With rising concerns around inaccessibility to freshwater and the ever increasing threats of population growth, climate change, and urban development, the book brings electrically conducting materials to the forefront of membrane separation technology with an emphasis on their role in the mitigation of fouling and related phenomena. Electrically conducting materials expand the versatility of membrane technology and ultimately improve access to safe water.

The book is important reading for scientists, engineers, entrepreneurs, and enthusiasts from the water industry who seek to familiarize themselves with a groundbreaking area of study within modern desalination and water treatment.

- Explores novel membrane materials and systems from preparation methods, materials selection, and their application in monitoring, fouling control, and performance enhancement.
- Examines the mechanism of fouling prevention and cleaning in various electrically conductive materials.
- Evaluates the scalability of antifouling materials and coatings, as well as electrically enhanced processes for monitoring and control in membrane separation technology.
- Assesses advantages and limitations of applying electrically conductive membrane systems to fouling control for specific water treatment applications.
- Provides a critical review of scientific literature in the specialized area of electrical conductive materials and systems for membrane technology.

Electrically Conductive Membrane Materials and Systems

Fouling Mitigation for Desalination and Water Treatment

Farah Ahmed, Raed Hashaikeh, and Nidal Hilal

CRC Press
Taylor & Francis Group
Boca Raton   London   New York

CRC Press is an imprint of the
Taylor & Francis Group, an **informa** business

First edition published 2023
by CRC Press
6000 Broken Sound Parkway NW, Suite 300, Boca Raton, FL 33487–2742

and by CRC Press
4 Park Square, Milton Park, Abingdon, Oxon, OX14 4RN

CRC Press is an imprint of Taylor & Francis Group, LLC

© 2023 Farah Ahmed, Raed Hashaikeh, and Nidal Hilal

Reasonable efforts have been made to publish reliable data and information, but the author and publisher cannot assume responsibility for the validity of all materials or the consequences of their use. The authors and publishers have attempted to trace the copyright holders of all material reproduced in this publication and apologize to copyright holders if permission to publish in this form has not been obtained. If any copyright material has not been acknowledged please write and let us know so we may rectify in any future reprint.

Except as permitted under U.S. Copyright Law, no part of this book may be reprinted, reproduced, transmitted, or utilized in any form by any electronic, mechanical, or other means, now known or hereafter invented, including photocopying, microfilming, and recording, or in any information storage or retrieval system, without written permission from the publishers.

For permission to photocopy or use material electronically from this work, access www.copyright.com or contact the Copyright Clearance Center, Inc. (CCC), 222 Rosewood Drive, Danvers, MA 01923, 978–750–8400. For works that are not available on CCC please contact mpkbookspermissions@tandf.co.uk

Trademark notice: Product or corporate names may be trademarks or registered trademarks and are used only for identification and explanation without intent to infringe.

Library of Congress Cataloging-in-Publication Data
Names: Ahmed, Farah (Farah Ejaz), author. | Hashaikeh, Raed, author. |
    Hilal, Nidal, author.
Title: Electrically conductive membrane materials and systems and systems for
    fouling mitigation / Farah Ahmed, Raed Hashaikeh, Nidal Hilal.
Description: Seventh edition. | Boca Raton : CRC Press, 2023. | Includes
    bibliographical references.
Identifiers: LCCN 2022045961 (print) | LCCN 2022045962 (ebook) |
    ISBN 9780367702069 (hardback) | ISBN 9780367702113 (paperback) |
    ISBN 9781003144991 (ebook)
Subjects: LCSH: Water—Purification—Membrane filtration. | Fouling. |
    Materials—Electric properties.
Classification: LCC TD442.5 .A34 2023 (print) | LCC TD442.5 (ebook) |
    DDC 628.1/64—dc23/eng/20221207
LC record available at https://lccn.loc.gov/2022045961
LC ebook record available at https://lccn.loc.gov/2022045962

ISBN: 978-0-367-70206-9 (hbk)
ISBN: 978-0-367-70211-3 (pbk)
ISBN: 978-1-003-14499-1 (ebk)

DOI: 10.1201/9781003144991

Typeset in Times
by Apex CoVantage, LLC

Contents

Preface and Acknowledgments

As inaccessibility to freshwater becomes an escalating global concern, membrane-based separation provides a versatile option for water treatment and desalination. With the ability to separate substances from water over a wide range from monovalent salts to micron-sized suspended particles, membrane development has seen tremendous growth in the last six decades. It therefore comes as little surprise that many large-scale facilities based on membrane separation technology have sprung during this time. While much of the scientific world's attention is on polymer materials for both membrane and turbulence promoters, metals and ceramics have also attracted attention and have formed niche markets in water purification. Nevertheless, all membrane processes continue to endure setbacks, as they suffer from yet unresolved operational challenges. Of these, fouling of the membrane and spacer represents perhaps the greatest obstacle in further lowering the cost of established processes and expanding emerging processes toward full-scale use. The deposition of unwanted feed constituents blocks the effective surface area of the membrane and spacer, deteriorates membrane performance and increases energy costs. While fouling alone is problematic, it does not help that it is often accompanied by an array of other process-specific challenges. Effectively tackling fouling and related phenomena in membrane processes consists of two stages:

1. Real-time non-invasive reliable monitoring
2. Timely and effective non-invasive control strategies

Electrically conductive materials have the potential to simultaneously undertake both these roles and are already proving to be transformational in new-generation materials for membrane processes. In fact, they can go further in improving performance through in situ pore size tuning and electrically enhanced heating. A major benefit of applying electrically conducting materials is the choice of virtually endless possibilities of fabrication methods and material selection, which can be tailored to both the membrane and the spacer, to yield intelligent electrically conductive membrane systems for fouling mitigation.

The exciting ongoing journey in this field of research was initiated with our groundbreaking work on the first ever use of electrically conducting membranes for in situ electrolytic self-cleaning, the findings of which were submitted to the *Journal of Membrane Science* in 2014. This sparked the interest of researchers worldwide, and there has been no looking back ever since.

In the two introductory chapters of this text, we convey the significance of membrane separation technology by introducing the historical context and fundamentals of membrane processes, as relevant to water purification; we describe the various types of fouling in membrane processes, categorizing them into high pressure processes, low pressure processes and membrane distillation.

In Chapter 3, we examine existing industrial practices and recent research developments for monitoring and control in membrane processes, with emphasis on fouling and related phenomena. Here, we identify the limitations of some of these techniques and pay special attention to electrochemical impedance spectroscopy as a powerful tool for real-time characterization.

In the following chapter, we contemplate mechanisms for electrical conductivity and electrochemical activity in selected classes of materials including conductive polymers, metals, polymer composites, and carbon-based nanomaterials. The goal of this chapter is for the reader to elucidate the relationship between structure and related electrical/electrochemical properties and to describe common membrane fabrication methods using these materials. It provides the foundation for the discussion that ensues in the three subsequent chapters. These chapters review the progress in electrically conducting membranes (Chapter 5) and spacers (Chapter 6) for the mitigation of fouling and related phenomena. Chapter 7 focuses exclusively on membrane distillation, highlighting studies in conductive membranes for in situ detection of pore wetting and fouling. In Chapter 8, we extend our inquiry to other functions of electrically conductive membranes and spacers, i.e. electrically tunable pore size and electrical membrane heaters for performance enhancement.

The book ends by discussing future prospects in electrically conducting membrane systems, more specifically the simplified desalination plant, process optimization, and integration of these materials into existing membrane modules for sustainable desalination and water treatment.

We believe that this book provides an integrated state of the art on an area that is fresh, exciting and, if we may add, revolutionary to membrane separation technology.

Farah Ahmed
Raed Hashaikeh
Nidal Hilal
Abu Dhabi, United Arab Emirates
August 2022

Author Biographies

Dr. Farah Ahmed is a research fellow at the Water Research Center–New York University Abu Dhabi. She is an affiliate member at the Mohammed bin Rashid Academy of Scientists in the United Arab Emirates. Dr. Ahmed received her doctorate in interdisciplinary engineering from Khalifa University of Science and Technology in 2018. Her research expertise lies in the area of advanced membrane materials for desalination and water treatment, with a focus on emerging low-energy technologies. She has authored many articles in prestigious international journals and book chapters and has delivered invited lectures at various international conferences and institutions around the world. Dr. Ahmed serves on the editorial and advisory boards of a number of journals including *Membranes* and *Separation Technologies*, a specialty section of *Frontiers in Environmental Chemistry*.

Raed Hashaikeh is Tenured Professor of Mechanical Engineering at New York University-Abu Dhabi and a member of the Mohammed bin Rashid Academy of Scientists in the UAE. He received his MSc and PhD in Materials Engineering from McGill University in 2000 and 2005, respectively. In 2008, He joined the Mechanical and Materials Engineering Department at the Masdar Institute in Abu Dhabi as Assistant Professor and went through the ranks to become a Full Professor in 2016. Between 2017 and 2019, he was Professor at the Chemical Engineering Department, Khalifa University. Before moving to UAE, he spent several years (2006–2008) at FPInnovations-Paprican Division, Canada, as a scientist. He was awarded the Natural Sciences and Engineering Research Council of Canada (NSERC) Industrial Research and Development Fellow in 2006. He was also a visiting scholar at MIT between 2008 and 2009. He has built core strengths in materials processing, characterization, and applications. The objectives of his research in electrically conductive membranes is to develop high-performing membrane materials for specific water treatment and desalination applications and to apply these multifunctional membranes to fouling control and performance enhancement.

Professor Nidal Hilal is a Chartered Engineer in the United Kingdom, a registered European Engineer, an elected Fellow of both the Institution of Chemical Engineers and the Learned Society of Wales. He received his bachelor's degree in chemical engineering in 1981 followed by a master's degree in advanced chemical engineering from Swansea University in 1986. He received his PhD degree from Swansea University in 1988. In 2005, he was awarded a doctor of science degree (DSc) from the University of Wales in recognition of an outstanding research contribution in the fields of water processing, including desalination and membrane science and technology. He was also awarded the prestigious Kuwait Prize (Kuwait Medal) of Applied Science for the year 2005 by the Emir of Kuwait and the Menelaus Medal 2020 by the Learned Society of Wales for excellence in engineering and technology.

His research interests lie broadly in the identification of innovative and cost-effective solutions within the fields of nanowater, membrane technology, water treatment, desalination, and colloid engineering. He has published eight handbooks and around 600 articles in refereed scientific literature. He has chaired and delivered lectures at numerous international conferences and prestigious organizations around the world.

Professor Hilal sits on the editorial boards of a number of international journals, is an advisory board member of several multinational organizations, and has served in or consulted for industry, government departments, research councils, and universities on an international basis.

Abbreviations

$\Delta\Pi$	osmotic pressure
ΔP	pressure drop
ϵ_0	permittivity of free space
ε_k	dielectric constant of the kth layer
ε_m	dielectric constant of the membrane material
δ	membrane thickness
A	membrane area
AA	ascorbic acid
AA2	alginic acid
AC	alternating current
AFM	atomic force microscopy
AGMD	air gap membrane distillation
ALD	atomic layer deposition
AOM	algogenic organic matter
ASTM	American Society for Testing and Materials
ATR	attenuated total reflectance
BB	building blocks
BC	bacterial cellulose
BP	biopolymer
BPA	bisphenol A
BSA	bovine serum albumin
C	membrane active layer capacitance
CA	cellulose acetate
CAGR	compound annual growth rate
CC	carbon cloth
CDI	capacitive deionization
CECP	cake-enhanced concentration polarization
CFD	computational fluid dynamics
CFS-MFIUF	Crossflow Sampler Modified Fouling Index Ultrafiltration
CIP	cleaning-in-place
ck	capacitance of the kth layer
CLSM	confocal laser scanning microscopy
CMC	critical micelle concentration
CNS	carbon nanostructures
CNT	carbon nanotube
CP	concentration polarization
CTAB	cetrimonium bromide
CVD	chemical vapor deposition
d	thickness of the active layer
DAF	dissolved air flotation
DC	direct current
DCMD	direct contact membrane distillation

DDTI	detection of dissolved tracer intrusion
DFT	density functional theory
DI	deionized
d_k	thickness of the kth layer
DLVO	theory named after Derjaguin, Landau, Verwey, and Overbeek
DOTM	direct observation through the membrane
DP	diffusion polarization
ECPNC	electrically conductive polymer nanocomposite
ED	electrodialysis
EDTA	ethylene diamine tetraacetic acid
EEM	excitation emission matrix
EIS	electrochemical impedance spectroscopy
EPS	extracellular polymeric substances
ES	electrospun/electrospinning
ESR	electron spin resonance
FA	fulvic acid
FDM	fused deposition modeling
FFM	feed fouling monitor
FO	forward osmosis
FORUS	fiber-optic reflectance UV-vis spectrometry
FTIR	Fourier transform infrared
G	conductance
GNP	graphene nanoplates
GO	graphene oxide
GOR	gain output ratio
HA	humic acid
HDPE	high density polyethylene
HEMA	hydroxyethylmethacrylate
HER	hydrogen evolution reaction
HET	heterogeneous electron transfer
HLB	hydrophilic-lipophilic balance
HS	humic substances
HAS	human serum albumin
ICP	intrinsically conducting polymer
IDA	iminodiacetic acid
IoT	internet of things
IP	interfacial polymerization
J_c	critical flux
K	permeability constant
LC-OCD	liquid chromatography with organic carbon detection
LDPE	low density polyethylene
LEP	liquid entry pressure
LIG	laser-induced graphene
LMH	liters per square meter per hour
LMW	low molecular weight organics
MB	methylene blue

MBR	membrane bioreactor
MD	membrane distillation
MED	multieffect desalination
MENA	Middle East and North America
MF	microfiltration
MFI	membrane fouling index
MPM	multiphoton microscopy
MRI	magnetic resonance imaging
MS	multiple sclerosis
MSF	multistage flash
MTS	methyltrichlorosilane
MWCNT	multiwalled carbon nanotube
MWCO	molecular weight cut-off
NASA	National Aeronautics and Space Administration
NC	networked cellulose
NF	nanofiltration
NIPS	non-solvent-induced phase separation
NMP	N-methyl-2-pyrrolidone
NOM	natural organic matter
OCT	optical coherence tomography
OER	oxygen evolution reaction
OPEX	operating expenses
OSW	Office of Saline Water
p	porosity
PA	polyamide
PAA	polyacrylic acid
PAI	polyamide-imide
PAM	polyacrylamide
PAMPSA	2-acrylamido-2-methyl-1-propanesulfonic acid
PAN	polyacylonitrile
PANI	polyaniline
PAO1	*P. aeruginsa*
PBI	poly[2,2'-(m-phenylene)-5,5'-dibenzimidazole]
PC	polycarbonate
PDA	polydopamine
PDMS	polydimethylsiloxane
PDT	pressure decay test
PEG	polyethylene glycol
PEO	polyethylene oxide
PES	polyethersulfone
PET	polyethylene terephthalate
PH, PVDF-co-HFP	poly(vinylidene fluoride-co-hexafluoropropylene)
PI	polyimide
PLA	polylactic acid
POME	palm oil mill effluent

PP	polypropylene
PPSU	poly(phenylene sulfone)
PPy	polypyrrole
pQA	polymeric quaternary ammonium
PRO	pressure retarded osmosis
pSBMA	poly(sulfobetaine methacrylate)
PSf	polysulfone
PSS	polystyrenesulfonate
PTFE	polytetrafluoroethylene
PTS	phthalocyanine tetrasulfonic acid
PVA	polyvinyl alcohol
PVC	polyvinyl chloride
PVD	physical vapor deposition
PVDF	polyvinylidene fluoride
PVF	polyvinyl formal
PVP	polyvinylpyrrolidone
QCM-D	quartz crystal microbalance with dissipation monitoring
R	universal gas constant
RO	reverse osmosis
ROS	reactive oxygen species
RSM	response surface methodology
SA	sodium alginate
SANS	small-angle neuron scattering
SAXS	small-angle X-ray scattering
SCE	Spacer Configuration Efficacy
SDBS	sodium dodecyl benzene sulfonate
SDI	silt density index
SDS	sodium dodecyl sulfate
SEC	specific energy consumption
SEM	scanning electron microscopy
SFE	surface free energy
SGMD	sweeping gas membrane distillation
SLA	stereolithography
SLS	selective laser sintering
SPVDF/QPVA	polyvinylidene fluoride/quaternary ammonium polyvinyl alcohol
SR	salt rejection
SR-μCT	synchrotron-based X-ray microtomography
SRFA	Suwannee River fulvic acid
STEC	specific thermal energy consumption
SWCNT	single-walled carbon nanotube
SWM	spiral wound module
SWRO	seawater reverse osmosis
T	absolute temperature
tCLP	Transverse crossed layer of parallel
TDS	total dissolved solids
TEM	transmission electron microscopy

TEP	transparent exopolymer particles
TFC	thin film composite
TMC	trimesoyl chloride
TMP	transmembrane pressure
TMPS	trimethoxy(propyl)silane
TOC	total organic carbon
TOM	total organic matter
TP	temperature polarization
TPC	temperature polarization coefficient
TPMS	triply periodic minimal surfaces
TRL	technology readiness level
UF	ultrafiltration
UNESCO	United Nations Educational, Scientific and Cultural Organization
UTDR	ultrasonic time-domain reflectometry
UV	ultraviolet
VF	vacuum filtration
VMD	vacuum membrane distillation
VUV	vacuum ultraviolet

1 Introduction to Membrane Separation Processes

1.1 INTRODUCTION

A rampant global issue, water scarcity in arid regions continues to worsen on a daily basis. Although our planet is made up mostly of water, clean water is a valuable resource to which many have little to no access. According to the World Health Organization, more than 2 billion people around the world lack access to safe and clean water (Organization, 2017).

By 2050, it is expected that up to 5.7 billion people could be affected by water scarcity (WWAP, 2018). Freshwater reserves are being depleted at astounding rates, much faster than they can be renewed. Growing stress on freshwater reserves has long-lasting impacts on global health, sanitation, and economy. There are three major threats to freshwater availability: population growth, climate change, and urban development. Rapid urban growth has accelerated water stress in urban areas (Flörke et al., 2018). Increasing industrial wastes has also polluted freshwater sources. Agricultural and textile industries are the top two largest polluters of water on earth. Industrial wastewater is the aqueous discard of substances either dissolved or suspended in water during an industrial manufacturing process (Woodard & Curran, 2006). In addition, climate change presents one of the largest threats to water availability, with increased risk of drought (Diffenbaugh et al., 2015; Haddeland et al., 2014; Konapala et al., 2020; Padrón et al., 2020; Yuan et al., 2019). Protecting water availability is also necessary to maintain agricultural production, especially as the global population continues to soar.

Diversifying water supply through sustainable technologies is a necessity in the face of rising water stress. Sustainable desalination and water treatment allow us to tap into contaminated streams and seawater to extract freshwater to meet our needs. Water may be around us, but too much of it is unsuitable for direct use due to its high content of suspended solids, bacteria, viruses, salts, and/or minerals. Many sources of water such as industrial waste from oil and gas, also known as produced water, and groundwater reserves also contain varying concentration of salts and contaminants that need to be treated to produce safe, potable water.

Although the global installed capacity has grown to just under 100 million m^3 per day, desalination still only accounts for roughly 1% of the earth's drinking water (Angelakis et al., 2021). Extraordinary water stress has driven coastal communities to incorporate desalination and wastewater treatment in national water security strategies as a means of diversifying water supply (Eke et al., 2020). Water-intensive

DOI: 10.1201/9781003144991-1

industries, including oil and gas, electronics, and chemical manufacturing, have also directed attention to water treatment as they require water with different levels of purity for processing (Frost & Hua, 2019). The upward trajectory of desalination capacity in particular means that the market size is likely to reach USD 32 billion by as early as 2025 (*Water Desalination Equipment Market Size, Share & Trends Analysis Report By Technology (RO, MSF, MED), By Source (Seawater, River Water), By Application (Municipal, Industrial), By Region, And Segment Forecasts, 2020– 2027*, 2020). Global water and wastewater treatment is seeing an even more rapid surge and is expected to grow to USD 500 billion by 2028 (Snehal Mohite, 2022). Unsurprisingly, water-scarce regions such as the Middle East and North Africa are currently leaders in seawater desalination, while water treatment is applied globally. However, many of the methods employed remain energy-intensive and unsustainable, with a plethora of economic and environmental challenges.

Membrane technology has transformed separation and purification. It is a versatile field with endless possibilities of materials, configurations, and applications with noteworthy contributions to desalination and water treatment. Membrane technology is already playing a role in moving toward sustainable desalination and water treatment. According to Singh (Rajindar Singh, 2015), the development of advanced membrane materials with high efficiencies and tunable morphology at a large scale have contributed extensively to the growth of membrane technology. Advances in membranes with superior transport properties have opened up exciting new applications within water purification. High separation efficiencies, relatively low costs, and ease of operation make membrane technology a viable approach toward sustainable desalination and water treatment. The increasing need for purification of water demands improvements in membrane science and engineering. The purpose of this chapter is to introduce the reader to membrane processes, including a historical overview of synthetic separation membranes and the fundamentals of membrane separation.

1.2 HISTORICAL EVOLUTION OF MEMBRANE PROCESSES

Once an area confined exclusively to biological research, membrane transport and membrane separation have witnessed unprecedented growth in the late 20th and early 21st centuries. Today, membrane technology encapsulates an entire field in the form of synthetic membranes, from materials and fabrication techniques to applications and system design. While desalination and water purification have existed for centuries, the first known investigation of osmotic transport through animal membranes was reported in 1748 by Abbé Jean Antoine Nollet (J.-A. Nollet, 1752; J. A. Nollet, 1764), who recognized the relation between a semipermeable membrane, osmotic pressure, and flux.

Interest in biological materials led to the 1845 discovery of an asymmetric pore structure in animal membranes (Hendricks, 2010). In the same year, the accidental synthesis of microporous nitrocellulose, also known as collodion, led to further interest in cellulose-based materials (Hendricks, 2010) for membranes. During this time, scientists ardently carried out studies on osmotic phenomena and the theory of solutions (Hammel & Scholander, 1976), using membranes merely to assist their work. Along with J. van't Hoff's groundbreaking findings on osmotic pressure and chemical equilibrium in 1887, these fundamental studies laid the foundation for reverse osmosis (RO) and other membrane separation processes.

Simultaneous interest in porous materials and solution theory gave rise to early membranes, using different types of diaphragm and later nitrocellulose. In 1907, early microfiltration (MF) membranes emerged when Bechhold prepared nitrocellulose membranes of graded pore size (Bechhold, 1907). Other authors built on Bechhold's work, and microfiltration collodion membranes became commercially available by the 1930s (Elford, 1930). Bechhold also invented the term "ultrafiltration", which took on its modern form decades later. While other materials such as cellulose acetate were explored, development in membrane technology remained slow until the end of World War II, when filters were required to test water quality, and much of the drinking water supply across Europe had collapsed (Hardenbergh, 1946; Lowe, 2012). Microporous membranes were used mainly to remove microorganisms and particulate matter from fluid streams or for estimating the size and shape of particles and macromolecules. Growth in the area of membrane technology was complemented by rapid developments in polymer science following the war (Glater, 1998) when materials other than cellulose were explored. In fact, condensation polymers used in high performance RO and nanofiltration (NF) membranes today can be traced back to the development of nylon, the first synthetic polyamide, by Carothers (Hermes, 1996).

During the war, German scientists developed membrane filters to culture microorganisms in drinking water and examine the bacterial quality of water (Hamilton et al., 1983; Ismail & Goh, 2015). A filter was used to remove bacteria from a water sample, and then bacteria was allowed to grow on the filter surface for 24 hours. This concept was further developed through research sponsored by the U.S. Army Chemical Corps, and by 1956, use of the membrane filter method for bacteriological examination of water was permitted by the U.S. federal government (Services & Center, 1960). Millipore Corporation became the first to commercially produce microfiltration membranes (Porter, 1986). Thus began the modern era of membrane technology. Through the 1950s and 1960s, the Office of Saline Water (OSW), established by the U.S. Department of the Interior, funded the initial development of desalination technology (Krishna, 2004; Matchette, 1998). Between 1950 and 1980, dramatic growth in RO membrane materials development through OSW funding led to the publication of approximately 1200 federal government desalination reports. Transport mechanisms and models for different membrane systems were also developed and widely used (Kucera). It was in this period that the foundation for most of today's membrane-based technologies was built. Although Reid and Berton revealed the potential of high salt rejection membranes (Saleh & Gupta, 2016), the high thickness of these membranes meant that the flux was impractically low. Perhaps the most transformational discovery in membrane separation technology was the Loeb–Sourirajan membrane. Until this point, membranes were considered slow, expensive, and inefficient (Baker, 2012). Loeb and Sourirajan were the first to develop high-flux membranes suitable for industrial desalination through phase inversion (Loeb, 1960; Matsuura, 2020). Their work revealed that cellulose acetate (CA) membranes prepared via phase inversion have an asymmetric structure with a thin, porous, active layer of CA and an integrally supported porous CA substrate.

Over the next two decades, a series of new polymeric materials were applied to membrane separation including polyvinyl chloride, polyamide, polysulfone, polycarbonate, and cellulose triacetate (Kucera). Following the work of Loeb and Sourirajan, other membrane fabrication techniques were developed, and it became possible to make ultrathin selective layers with high permeability (da Silva Biron et al., 2018). It was during this time that start-ups targeting seawater desalination began to spring up, especially in Southern California (Hendricks, 2010).

Simultaneously, various packaging methods for membrane modules were also developed, including spiral wound (Vrouwenvelder & Kruithof, 2011), hollow-fiber (Cohen et al., 1972), and plate-and-frame (da Silva Biron et al., 2018). Plate-and-frame modules consist of flat-sheet membranes stacked one over another in a casket with spacers individual compartments forming the module. The packing density of plate-and-frame modules ranges from 100 to 400 m^2 m^{-3}. Spiral wounds are similar in that they use flat-sheet membranes, but these are wrapped around a permeate collection tube in a spiral manner, separated by spacers. Compared to plate-and-frame, spiral wound modules are more space-efficient and are therefore the most widely used module, especially for RO desalination systems. In spiral wound modules, the feed flows parallel to the central tube. The packing density of spiral wound modules ranges is much higher than that of plate-and-frame modules and ranges from 300 to 1000 m^2 m^{-3}. Hollow fiber modules consist of an array of narrow hollow fibers as long as 1–2 m, each with a dense skin layer and a porous support layer. A hollow fiber membrane module is constructed by packing together thousands of fibers, which increases packing density. The feed can pass through the bore of the hollow fiber, and the permeate is collected on the outside; this arrangement is known as "inside feed" (Balster, 2016). The alternate, where the feed enters the module on the outside and the permeate is collected from the fiber bore, is also possible and is known as "outside feed". The outside diameter (OD) of hollow fibers ranges from 0.65 to 2 mm, with a wall thickness between 0.1 and 0.6 mm (Howe et al., 2012). Hollow fiber modules are commonly used in ultrafiltration (UF) systems, although optimization for specific applications is a factor in fiber diameter and flow direction. Hollow fiber membrane modules can be packed with densities as high as 30,000 m^2 m^{-3} (Scholz et al., 2011). Figure 1.1 shows a scanning electron microscope (SEM) image of the cross section of a PVDF hollow fiber membrane (Tan et al., 2006).

FIGURE 1.1   SEM image of a PVDF hollow fiber membrane.

Source: Tan, X., Tan, S., Teo, W. K., & Li, K. (2006). Polyvinylidene fluoride (PVDF) hollow fibre membranes for ammonia removal from water. *Journal of Membrane Science, 271*(1–2), 59–68. Reprinted with permission.

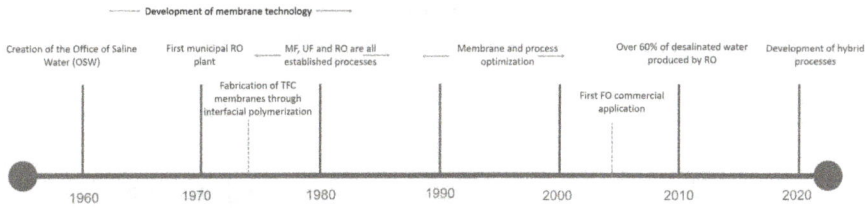

FIGURE 1.2   Development of membrane separation technology.

These developments in polymeric materials, membrane fabrication, and packaging quickly moved membrane processes such as MF, ultrafiltration (UF), RO, and electrodialysis (ED) from the lab to commercial large-scale plants. MF/UF membranes then started to be used for the treatment of municipal water resources and alongside bioreactors for sewage treatment plants. Figure 1.2 shows the development of the membrane separation industry between 1960 and 2020.

It can be argued that membrane technology is a direct product of "strategic science and technology", i.e. the combined efforts and activity of science administrators, industrialists, policy makers, public groups, and movements (Van Lente & Rip, 1998).

There is little doubt that membrane technology has strengthened our capabilities to transform production processes, enable new technologies, and prioritize environment and public health. Today, the use of membrane technology extends to wide-ranging industrial applications, such as desalination, gas separation, pharmaceutical drug release, energy conversion, waste treatment, dairy purification, tissue engineering, to name a few (Siekierka et al., 2021). Despite decades of progress in materials with superior properties, enhanced permeability, and tunable selectivity, membrane technology is still in need of gradual progress toward sustainable operation. To introduce these concepts, we define selectivity and permeability here. *Selectivity* is the extent to which desired molecules are separated from the rest, whereas *permeability* is how fast molecules pass through the membrane material.

1.3 FUNDAMENTALS OF MEMBRANE SEPARATION

John A. Howell defines a membrane as "a thin barrier between two fluids which restricts the movement of one or more components of one or both fluids across the barrier". A membrane can be considered a semipermeable barrier between two phases that allows selective transport from one side to the other under the effect of a driving force. The driving force can be a gradient in concentration, pressure, electric field, or vapor pressure. In any membrane process, there are three streams: feed, product or permeate, and retentate (Figure 1.3).

Membranes are often classified by transport, as porous or non-porous (Mulder, 1996); this classification largely determines the transport mechanism through the membrane, as shown in Figure 1.4. Other classifications are by membrane material type: inorganic, organic, or hybrid membranes exist for water treatment and desalination. Membranes also differ by structure depending on the homogeneity of the cross section, which makes the membrane isotropic or anisotropic. Anisotropic membranes can be formed with the

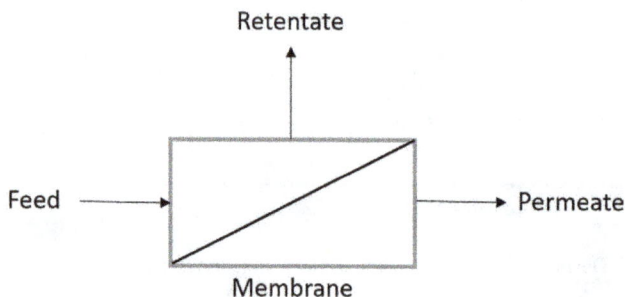

FIGURE 1.3   Schematic of a membrane process; the diagonal line represents the membrane.

same material or with different layers, as is the case with thin film composite (TFC) membranes. Membranes can also be classified according to their configuration in the module: flat-sheet, hollow fiber, spiral wound, etc., as discussed in Section 1.2. Figure 1.5 shows the classification of membranes in different ways.

1.3.1 POROUS MEMBRANES

In porous membranes, separation is based on size-exclusion, or sieving, as with MF and UF. For these pressure-driven processes, separation efficiency is determined by pore size and distribution within the membrane as well as parameters such as pore geometry and tortuosity. Particles smaller than the pore size permeate through the membrane by convective flow, and larger substances are retained by the membrane. Parameters such as number of layers and pore tortuosity also affect the efficiency of filtration in porous media. Tortuosity is the ratio of pore length to membrane thickness; low tortuosity or short transport distance contributes to high permeability in porous membranes.

1.3.2 NON-POROUS MEMBRANES

Non-porous membranes rely on selective diffusion through the membrane material for separation. This solution-diffusion mechanism relies on dissolution in the membrane matrix and subsequent diffusion through the membrane (Vrentas & Vrentas, 2002). Such highly dense membranes are commonly used in desalination processes such as reverse osmosis (RO) and more recently, forward osmosis (FO). Nanofiltration (NF) membranes have pores in the range of 1–10 Å (Otero et al., 2008), and while they are used for desalination, their structure is somewhat "loose" compared to that of RO membranes. This makes transport through NF membranes slightly more complex as it is a combination of solution-diffusion and charge effects (Schaep et al., 1998), as we will see in Section 1.3.3. Most dense barriers exist as thin films with thicknesses of 100–200 nm on porous support layers. The support layer is mechanically, thermally, and chemically stable (Kaur et al., 2012). The thinner active layer helps retain high fluxes. Support and barrier layers

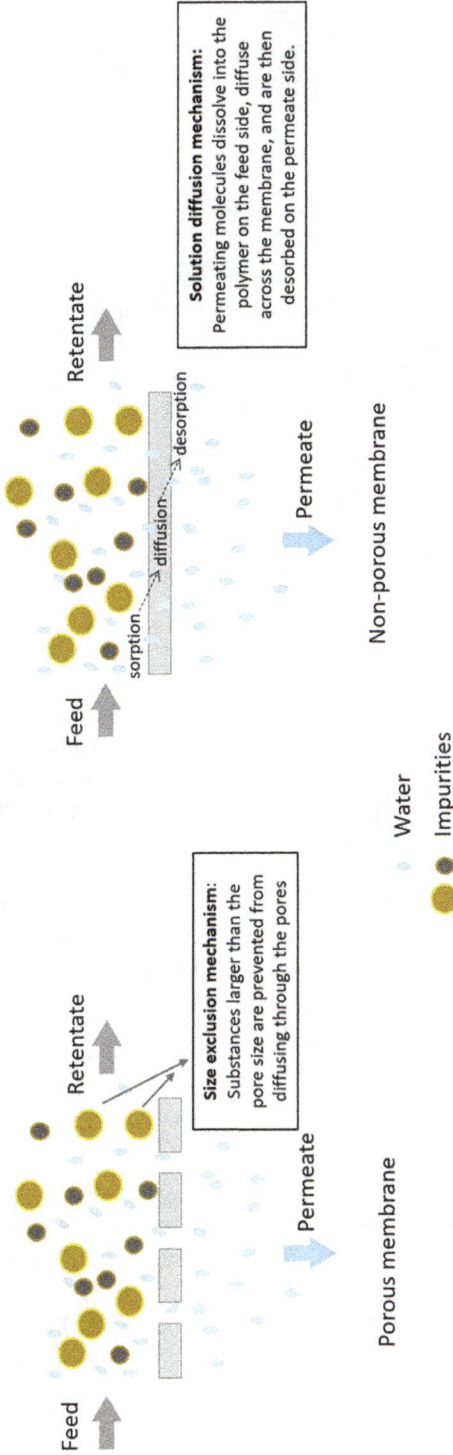

Solution diffusion mechanism: Permeating molecules dissolve into the polymer on the feed side, diffuse across the membrane, and are then desorbed on the permeate side.

Size exclusion mechanism: Substances larger than the pore size are prevented from diffusing through the pores

Retentate

Feed

Permeate

Non-porous membrane

Porous membrane

Water

Impurities

FIGURE 1.4 Transport mechanisms in porous and non-porous membranes.

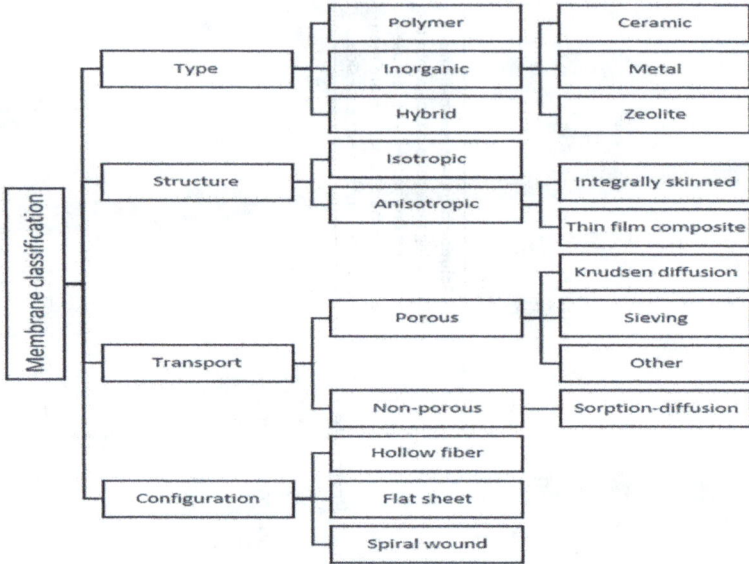

FIGURE 1.5   Classification of membranes by type, structure, transport, and configuration.

in TFC membranes are commonly prepared using the phase inversion technique, although novel techniques such as electrospinning are now replacing traditional methods with improved performance as support (Yoon et al., 2006). The dense layer is commonly prepared on top of the support layer using the interfacial polymerization process.

The amount of driving force required is determined in part by the potential gradient across the membrane. Membrane processes are equilibrium processes in which permeate flow through the membrane will eventually reach steady-state if the driving force is kept constant (Rajindar Singh, 2015). In general, for pressure driven processes, mass transport is expressed by:

$$J = K \cdot \frac{\Delta P}{\delta} \qquad \text{Equation 1.1}$$

where J is the transmembrane flux, K the membrane permeability constant, ΔP the pressure difference across the membrane, and δ is the membrane thickness. In the case of RO where there is an osmotic pressure difference $\Delta\Pi$ between both sides of the membrane, the driving force is $\Delta P - \Delta\Pi$.

Physicochemical properties of the membrane determine separation performance. Separation depends on the size, shape, chemical properties, and/or electrical properties of the substance that needs to be separated. Apart from permeability, the rejection coefficient R is a primary indicator of the performance of a membrane process.

$$\% \text{ Rejection} = \frac{\text{Feed solute concentration-permeate solute concentration}}{\text{Feed solute concentration}} \times 100 \quad \text{Equation 1.2}$$

The trade-off between membrane productivity and selectivity is a known phenomenon and a critical problem in membrane processes (H. B. Park et al., 2017). All membranes exhibit this trade-off, which was reported by Robeson in 1991 (Robeson, 1991) primarily for gas separation. He analyzed selectivity vs. permeability graphs for various gas separation membranes and found an upper bound on the trade-off between membrane permeability and selectivity that limits the performance of polymeric membranes. When first developed, most membranes were unable to break past the Robeson upper bound; however, over time novel membrane fabrication technologies and materials with better separation performance resulted in membranes that exceeded this upper bound, causing Robeson to revise his original correlation in 2008 (Robeson, 2008). Figure 1.6 shows Robeson's original and revised upper bound

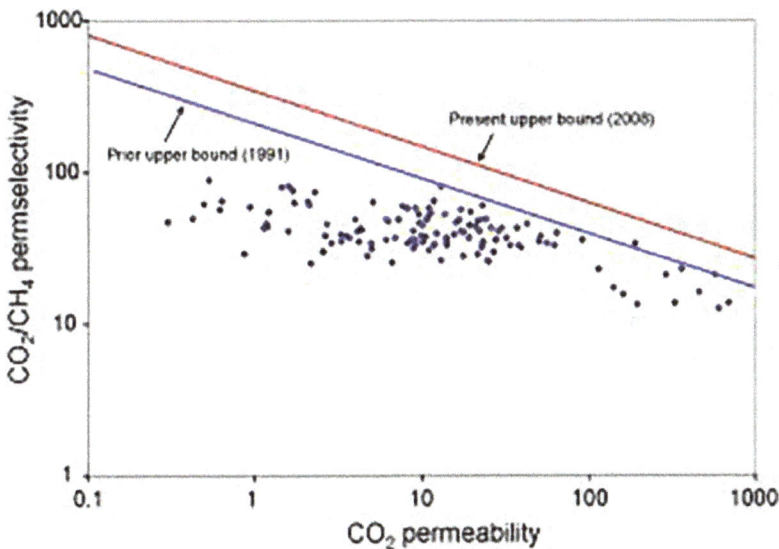

FIGURE 1.6 Robeson's 1991 and 2008 upper bound curves representing trade-off effect of polymeric membranes during gas separation.

Source: Xiao, Y., Low, B. T., Hosseini, S. S., Chung, T. S., & Paul, D. R. (2009). The strategies of molecular architecture and modification of polyimide-based membranes for CO_2 removal from natural gas—A review. *Progress in Polymer Science, 34*(6), 561–580. With permission. Reprinted with permission.

curves indicating selectivity and permeability trade-off for gas separation membranes. The revised analysis proved that advanced membrane preparation strategies can adjust the trade-off between membrane permeability and selectivity.

The concept of this trade-off was later extended to desalination membranes (Geise et al., 2011) and was attributed to the trade-off between the water diffusion coefficient and water/salt diffusion selectivity. Yang et al. further studied the upper bound relationship for TFC polyamide membranes used in desalination, surveying >300 published articles (Z. Yang et al., 2019). Their work confirmed the trade-off between water permeance and water/NaCl selectivity for polyamide-based desalination membranes. Fabrication conditions and modification methods greatly affect separation properties.

1.4 MEMBRANE PROCESSES

Desalination and water treatment using membrane technology is implemented as a supply-side strategy for improving freshwater supplies. Together with seawater desalination, increasing volumes of wastewater have prompted the development of membrane processes in past half century. Ample research in polymeric materials resulting from an aggressive desire to establish energy-efficient seawater desalination has also benefitted other water purification applications, as polymers tailored for microfiltration, ultrafiltration, and nanofiltration resulted as an unintended consequence of desalination research, later erupting into their own areas.

Figure 1.7 shows the range of particle size rejected in various membrane processes. Anything that can filter larger particles than the range shown is considered a conventional filter.

Porous membranes with a pore size of 0.1–1 µm are also typical for membrane distillation (MD), which, as we note, is a thermally driven process in which volatile compounds vaporize and are transported through the pores.

Membrane processes are classified loosely by the types of matter rejected, operating pressures, and nominal pore dimensions (Howe et al., 2012). For example, a "tight" UF membrane may have similar separation characteristics as a "loose" NF membrane.

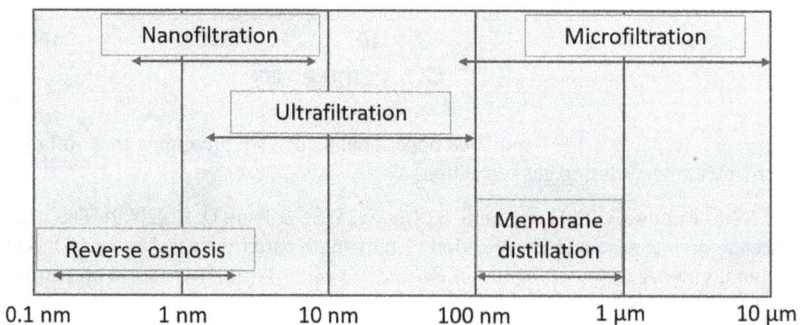

FIGURE 1.7   Size of particles rejected in various membrane processes.

Today, microfiltration, ultrafiltration and nanofiltration are all established processes for water treatment. In this section, we provide an overview of some of the major membrane processes used for desalination and water treatment.

1.4.1 MICROFILTRATION

Microfiltration is a low-pressure separation process that is used to concentrate, purify, or separate macromolecules, colloids, and suspended particles. With nominal pore sizes in the range of 0.1–10 μm, MF membranes are widely applied to wastewater treatment, including biological treatment when combined with bioreactors, as well as pretreatment to desalination. The wide range of pore sizes means that MF membranes are able to separate colloids, bacteria, aerosols, and various macromolecules from fluids. MF is widely used in large-scale municipal wastewater treatment and treatment of toxic industrial waste (Khan & Boddu, 2021). MF is also used as pretreatment to UF, NF and RO. Common applications for MF are in food and dairy, biotechnology, pharmaceuticals, as well as latex emulsions treatment. The typical operation involves low transmembrane pressures (TMPs) of less than 2 bars in cross-flow configuration (Ray et al., 2020). Today, its wide range of applications make the MF membrane market one of the largest in the world, second only to hemodialysis membranes (P. Pal, 2020). However, when a higher degree of separation is needed and cost can be compromised, UF membranes replace MF as they can achieve a higher degree of purification than low-cost MF membranes. Research in MF has focused increasingly on aspects of wastewater treatment, membrane fabrication, and modification and fouling studies, whereas modeling and simulation have received decreasing attention between 2009 and 2018 (Anis et al., 2019).

1.4.1.1 Mode of Operation

Membrane processes can operate in either dead-end or cross-flow filtration modes. The first MF systems operated on the dead-end filtration mode, which is still commonly employed in lab-scale studies. Industrial operation favors cross-flow filtration due to low fouling propensity. In dead end filtration, the feed is pressurized perpendicular to the membrane surface (Bhave, 2014), as shown in Figure 1.8a. In MF,

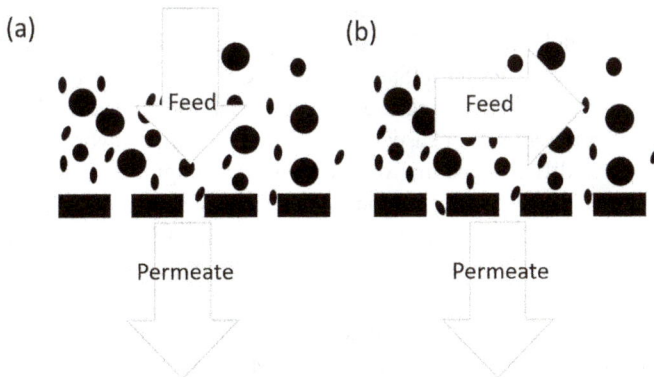

FIGURE 1.8   Schematic representation of (a) dead-end filtration and (b) cross-flow filtration.

separated solids form a filter cake on the surface of the membrane, which increases mass transfer resistance and affects permeability over time. Dead-end filtration is a batch process in which unwanted particulates are accumulated on the membrane until filtration can no longer process, at which point the process should be stopped, and the filter needs to be cleaned or replaced.

In cross-flow filtration, also known as tangential flow filtration, the feed suspension flows parallel to the membrane while permeate flows perpendicular to the membrane (Figure 1.8b). The retentate stream is the remainder of particles that continue to flow across the membrane after a pressure differential causes permeate to transport through the membrane. Unlike dead-end, the parallel flow in cross-flow systems limits the deposition of particles on the membrane surface as deposited particles are often swept by the high velocity fluid toward the retentate (X. Li & Li, 2016). As a result, cross-flow filtration is theoretically a continuous mode of operation although, practically, not all fouling is prevented. Nevertheless, cross-flow filtration is popular for industrial processes due to low membrane clogging and concentration polarization. These phenomena are elaborated in Chapter 2. The performance of cross-flow membranes is affected by the nature of the membrane material, pore geometry, surface properties, and membrane thickness (Bhave, 2014). Cross-flow systems still suffer from fouling, even if to less of an extent than dead-end filtration systems. Viscosity limits of the pump affect the amount of suspended solids that can be handled by a cross-flow MF system.

1.4.1.2 Membrane Materials

MF membranes are prepared from various organic materials including PVDF, PTFE, PE, PP, cellulose, PSf, polycarbonate, PA, etc. Ceramic MF membranes are gaining attention in industrial applications as more manufacturers join the market. This is due to their superior chemical, thermal, and mechanical stabilities as compared to polymer membranes. However, the lack of control over pore size has largely limited the use of ceramic membranes to low pressure applications. In addition, most ceramic membranes involve expensive raw materials and fabrication methods. Low-cost alternatives have been identified as promising substitutes for expensive raw materials: these include kaolin clay (Rekik et al., 2017), ball clay (Abd Aziz et al., 2019; Mohamed Bazin et al., 2019), bentonite clay (Liangxiong et al., 2003; Saja et al., 2020), and dolomite (Malik et al., 2020), all of which have been used in the synthesis of ceramic membranes (Sandhya Rani & Kumar, 2021). Low-cost ceramic membranes are increasingly being used for oily wastewater treatment, treatment of textile effluent, and removal of suspended solids. Techniques involved for preparing ceramic supports differ depending on the element configuration and often involve paste casting, dry pressing, or extrusion (Sandhya Rani & Kumar, 2021). Although a longer operational lifetime is characteristic of ceramic MF membranes due to their ability to withstand high temperatures, a wide pH range, and transmembrane pressures, the capital cost of installing ceramic systems is typically 80–90% higher than that for polymer systems per square meter of membrane area (Wagner, 2001).

Morphological and permeation properties of MF membranes are commonly characterized using thermogravimetric analysis, X-ray diffraction, Fourier transform infrared spectroscopy, scanning electron microscopy, liquid–liquid displacement porometry, hydraulic permeability and pure water permeation (Randeep Singh & Purkait, 2019).

1.4.1.3 Applications

Wastewater treatment currently constitutes the largest industrial-scale application for MF. MF is often combined with other methods such as biological treatment (Campos et al., 2002; Pillay et al., 1994; Shimizu et al., 1996) and electrical/electrochemical processes (Juang et al., 2013; Nataraj et al., 2007; G. C. C. Yang et al., 2003). Zuo et al. coupled a commercial hollow fiber MF membrane with a biocathode microbial desalination cell to exploit both conventional anaerobic biological treatment with electrodeionization simultaneously (Zuo et al., 2018). Park et al. combined electrochemical treatment with MF to enhance removal of organics and simultaneously generate hydrogen fuel in the presence of wastewater organic colloids (H. Park et al., 2013).

Anis's review of research trends indicates that the use of MF is rising in the food industry but has not changed significantly for medicine and desalination (Anis et al., 2019). In the food industry, MF is applied to the filtration of milk for removal of spores, bacteria, etc. However, in recent years, the tighter pore sizes of UF and NF membranes have lowered demand for MF membranes. In desalination, MF can be used to remove larger suspended solids during the pretreatment step to avoid fouling.

1.4.2 ULTRAFILTRATION

UF is also a pressure-driven process that, like MF, operates at relatively low pressures. UF membranes are membranes whose pore size falls between 0.1 and 0.01 μm. As compared to MF, UF membranes have lower permeability and operate at slightly higher transmembrane pressures of up to 5 bars. As in MF, separation in UF is based on size exclusion. UF membranes are able to separate macromolecules, colloids, proteins, and bacteria. Low-molecular-weight or hydrated ions can be transmitted through UF membranes. Size cut-offs in UF membranes are expressed in terms of molecular weight rather than micrometers (Spivakov & Shkinev, 2005). Usually, suspended solids and solutes with molecular weights greater than 300 kDa are retained by UF membranes (Shah & Rodriguez-Couto, 2021). A membrane with a molecular weight cut-off (MWCO) of 100,000 means that most solutes with this molecular weight or more will be rejected by the membrane. UF is commonly used to remove pathogens (bacteria, viruses, etc.) and suspended solids from aqueous streams.

Due to the reliability of UF to produce high-quality permeate and its moderate operational and capital costs compared to conventional water treatment processes, UF has simplified water treatment by replacing a series of conventional methods often applied to water treatment. Granular filters, which rely on depth filtration, often require a series of conventional methods such as coagulation, flocculation, and sedimentation facilities for effective purification; these are all replaced by employing UF membranes. Compared to granular filtration, the flux through UF membranes is roughly two orders of magnitude higher (Howe et al., 2012). A major advantage of UF membranes is that the quality of permeate water is independent of fluctuations in raw water quality. As a result, the UF market is briskly expanding and is currently estimated at USD 2.1 billion.

1.4.2.1 Applications

There are many applications of UF within desalination and water treatment. UF is used in large-scale water treatment for removal of viruses and bacteria and for separation of oil and water. It is also preferred as pretreatment for RO desalination when higher-quality feed is required. Both MF and UF employed as treatment of secondary treated wastewater before RO and UV radiation to generate drinking water (Rajindar Singh & Hankins, 2016). UF membranes are applied to various aspects to water purification, waste effluent treatment, food and dairy industry, toxic heavy metal removal from industrial waste streams, recovery of precious elements, and production of drinkable water (Christensen & Plaumann, 1981; Khan & Boddu, 2021).

Apart from wastewater treatment and desalination, another UF application that concerns human health is in the medical industry. UF membranes are often employed in dialysis for the treatment of failing kidneys through the removal of toxins from the bloodstream (Ficheux et al., 2013). UF is also widely employed in implants, diagnostic assays (Beda-Maluga et al., 2015), dialysis, and drug delivery (Boyd, 2003) systems due to the ability of UF membranes to retain microorganisms of various sizes. The pore size range of UF membranes also makes them suitable for pharmaceutical applications in the manufacture of drugs and for protein separation. Other areas where UF is relevant are the textile and petroleum industries. In the textile industry, they are used mainly for concentrating dye and treating effluent. In the petroleum industry, they are used to treat oily wastewater.

1.4.2.2 Materials

Over the past couple of decades, ceramic, carbon, and metallic membranes have become available in the UF pore size range and have found niche applications. However, special module housings and membrane geometries discourage replacement of polymeric UF membranes for water treatment. Benefits of UF include low operating pressure, high flux, low energy consumption as compared to NF and RO, and high-quality permeate as compared to MF, and UF provides disinfection through removal of bacteria and viruses. UF membranes are porous and have an asymmetric cross-sectional structure with a more dense skin layer with thicknesses between 1 and 3 µm (Mulder & Mulder, 1996). UF membranes exhibit extremely high removal of bacteria and suspended solids but low to moderate rejection of organic compounds. If the feedwater is not high in total organic compounds (TOC) (Xia et al., 2004), UF may be used as a standalone technology (Laîné et al., 2000). For surface water treatment, pretreatment methods such as coagulation (Yu et al., 2019), adsorption (K. Li et al., 2014), peroxidation (M. Li et al., 2019), or biological treatment (Huang et al., 2011) can improve permeate quality and reduce fouling of UF membranes.

1.4.2.3 Research Trends

Much of the research surrounding UF today focuses on developing new materials with high flux and/or antifouling property. A statistical review by Al Aani et al. shows that membrane fouling, modeling, and wastewater are the three most dominant research subfields within ultrafiltration, accounting for more than half of the total scientific articles published between 2009 and 2018 (Al Aani et al., 2020). China and the United States lead the research in UF, with over 2000 articles each, between the first mention of UF in 1907 and 2022.

Wang et al. recently developed an ultralow-pressure UF membrane by incorporating polyamic acid into PES via in situ polycondensation (K. Wang et al., 2022). Polyamic acid is compatible with PES, and uniform distribution led to improved hydrophilicity and refined pores. The water flux of the composite membrane was nearly double that of PES alone under 0.3 bars, with a BSA rejection ≥90%. The membrane was also used for the removal of oil from water with ≥92% efficiency and improved anti-oil fouling properties. Lu et al. developed UF membranes with antibacterial properties from PVDF/quaternary ammonium polyvinyl alcohol (SPVDF/QPVA) (Lu et al., 2022). The membrane was able to remove 95% of proteins, 100% of phenolphthalein, and 87% of ß-naphthol from synthetic wastewater. In another study, high strength sulfonated PVDF membranes were compared to PVDF membranes (Ayyaru & Ahn, 2022); it was found that sulfonation enhanced pure water flux and antifouling properties, with high rejection for BSA and humic acid. To introduce a charge-based mechanism into UF, Nieminen and coworkers employed TEMPO-mediated oxidation to enhance the surface charge of a commercial regenerated CA UF membrane, allowing Na_2SO_4 rejection of 36% as compared to 4% in the unmodified membrane (Nieminen et al., 2022). Low exposure induces surface charge, while high exposure serves to open the membrane pores.

Li and coworkers introduced micronano bubbles to the UF feed to improve permeate flux by reducing apparent viscosity of substances present in natural organic matter (NOM) (X.-X. Li et al., 2022).

1.4.3 NANOFILTRATION

Nanofiltration is a relatively new technique that falls between UF and RO. In fact, some of the early names given to NF were "ultralow pressure RO", "charged UF", "loose RO", "tight UF", "open RO" (Schäfer et al., 2005). Researchers agree that NF is a somewhat complex process to define (Van der Bruggen et al., 2008). NF is known for high rejection of multivalent ions and low molecular weight organic compounds and for moderate rejection of ions at lower pressures than RO. NF could replace RO in applications where high rejections are not a priority but high fluxes are necessary (Geise et al., 2010).

Although the steric pore model has been used to determine the pore size of NF membranes, pores in the range of 1 nm were first characterized using atomic force microscopy (AFM) by Hilal's group (Bowen et al., 1997; N. Hilal et al., 2005). The typical pore size of 1 nm found in NF membranes corresponds to an MWCO of 300–500 Da (Wu et al., 2017). The unique combination of solute transport via steric, Donnan, dielectric transport effects led to niche applications in several areas including water and wastewater treatment, pharmaceuticals, and food processing (Wu et al., 2017). In many cases, its lower pressure compared to RO became advantageous in applications where complete rejection of ions was not priority.

As pretreatment to RO, NF can provide high-purity feed as it retains multivalent ions and low molecular weight organics. Notably, NF membranes are suited for the removal of solutes whose molecular weight is between 200and 1000 g mol^{-1} from aqueous streams (Peeva et al., 2010). NF is often applied to water softening, in which larger ions such as Ca^{2+} and Mg^{2+} need to be removed. NF differs from both RO and UF in that, in addition to sorption diffusion and size exclusion, transport through NF membranes is also a result of Donnan exclusion or electrostatic interactions between solutes and the charged membrane. Separation in NF membranes is a combination of surface charge and sieving mechanisms. The former can be described by an electrostatic and steric-hindrance model in which "solute retention is a function of the ratio of charge density of the membrane to ionic concentration of the solvent and proportion of solute dimension in the solvent to pore size of the membrane" (Madaeni, 2015). As such, solution characteristics such as pH and ionic strength strongly affect the retention of trace organics by NF membranes. It is important to note that negatively charged species have greater rejection rates compared to uncharged molecules due to electrostatic repulsion between the negatively charged membrane and the molecules (Childress & Elimelech, 2000). Adsorption is also recognized as a step in the transport mechanism of NF (Madaeni, 2015).

Molecules with high dipole moment have lower retentions compared to non-polar molecules due to electrostatic interactions that direct the dipole toward the membrane (Van der Bruggen et al., 1999). Furthermore, charge effects are more dominant when the pores of the NF membrane are larger. Yoon and coworkers found that a porous hydrophobic NF membrane retains many hydrophobic pharmaceutical compounds due to a combination of hydrophobic adsorption and size exclusion, although the transport phenomenon depends on feedwater chemistry and membrane material. pH and salt conditions of the feed greatly affect NF performance. The presence of salt and high pH can increase membrane pore size and membrane permeability (J. Luo & Wan, 2013); however, if the salt content is increased, the salt-induced viscosity increase in the solution prevents improvement of permeate flux. Many mechanisms contribute to the effect of pH on NF membranes: membrane swelling, charge variation, electrostatic effect, co-ions competition, charge balance, etc. (J. Luo & Wan, 2013).

The preparation of NF membranes aims to control membrane structure and optimize performance. Hence NF membranes are often manufactured using phase inversion and interfacial polymerization. Another possible technique is to modify existing UF and RO membranes via functionalization methods, i.e. coating, self-assembly, plasma treatment, chemical treatment, etc., and/or through embedding nanofillers (Zhao et al., 2013). Over the years, novel and improved methods for NF fabrication have allowed development of membranes with enhanced selectivity, flux, and antifouling property. Interfacial polymerization (IP) remains the process of choice to fabricate polymeric NF membranes.

In comparison with ultrafiltration and reverse osmosis, nanofiltration has always been a difficult process to define and describe (Van der Bruggen et al., 2008; Geens et al., 2006). The specific features of nanofiltration membranes are mainly the combination of very high rejections for multivalent ions (>99%) with low to moderate rejections for monovalent ions (< about 70%) and high rejection (>90%) of organic compounds with molecular weight above the molecular weight cut-off of the membrane. The mechanism of the mass transport depends strongly on the membrane structure, on the interactions between the membrane and transported molecules. The separation efficiency can be governed by the sieving effect (when the size of the nanopores and that of the solute molecules have the main effect) or by the solution and diffusion properties of the solute molecules. In the case of charged molecules, the electrical field has a determined role in the transport. Three parameters are crucial for operation of a nanofiltration unit: solvent permeability or flux through the membrane, rejection of solutes, and yield or recovery.

1.4.3.1 Research Trends

Research in NF seeks to overcome some of these drawbacks with the technology. While low-pressure separation at the cost of loss of selectivity has traditionally been a selling point for NF, there has been a rising need to improve pollutant rejection using NF membranes, even those present at very low concentrations. For example, the removal of monovalent ions such as nitrates is recognized as important with a maximum limit of 50 ppm set by the EU to prevent possible health impacts on infants (Van der Bruggen et al., 2008). NF membranes, however, only show moderate nitrate rejection and are not appropriate for complete removal.

Traditionally, separation efficiency in membranes made via interfacial polymerization (IP), such as those used for NF and RO, is dictated by the molecular design of the monomers involved and reaction conditions (Wu et al., 2017). More recently, enhancing separation efficiency has extended to mixed matrix membranes using organic or inorganic fillers. Many materials have been investigated as fillers for NF membranes including zeolite (Ahmadi et al., 2022; Dong et al., 2016; Kong et al., 2021; Priyadarshini et al., 2018), carbon nanotubes (CNTs) (Alshahrani et al., 2020; Ganji et al., 2022; Manorma et al., 2021; Vatanpour et al., 2014; Vatanpour et al., 2011), MoS_2 (S. Li et al., 2022; X. Wang et al., 2021; Xie et al., 2022), TiO_2 (Batool et al., 2021; Gholami et al., 2022; Wei et al., 2020), and graphene (Balaji et al., 2021; Lai et al., 2016; Shao et al., 2014). Johnson and Hilal recently published an extensive review on nanocomposite membranes for NF (Johnson & Hilal, 2022).

Other challenges that need the attention of researchers today include membrane fouling, long-term chemical stability, and the need to simulate real systems. For many applications, the amount of literature available shows that steps have already been taken to address these issues.

1.4.4 Reverse Osmosis

RO is the most established membrane-based desalination process on an industrial scale (Nidal Hilal et al., 2015) for both seawater and brackish water desalination. It is a high pressure process that is capable of rejecting low-molecular-weight organics, monovalent and multivalent salts and of allowing water molecules to diffuse through (Demeuse, 2009). Due to several advantages such as low energy costs, modularity, and compactness, RO desalination capacity has surpassed thermal in recent decades. The energy consumption of RO as a desalination process is low as no phase change is involved.

Osmosis is the spontaneous process by which water molecules move spontaneously from a solution of low solute concentration to a solution of high solute concentration across a semipermeable membrane that allows selective passage of water molecules to be transported (Baumgarten & Feher, 2012). Osmotic pressure is the pressure that needs to be applied to stop this flow of water molecules across the membrane. RO relies on applying a hydraulic pressure greater than the osmotic pressure of the feed solution to reverse the spontaneous phenomenon of osmosis, thus forcing water to move from a high-solute region to a low-solute region through a non-porous semipermeable membrane. Depending on the osmotic pressure of the feed solution, RO units are operated at TMPs of 20–80 bars. Existing commercial membranes cannot operate at higher pressures due to material limitations, which is another reason research in inorganic membranes is picking up again. Additionally, for highly saline feed solutions, the pressure requirements and therefore operational costs are high. The efficiency of the RO process depends on membrane material, feedwater characteristics, and operational parameters.

For an ideal solution, the osmotic pressure π is estimated via the van't Hoff equation:

$$\pi = CRT$$ Equation 1.3

where C is the molar concentration of a non-permeable solute in mol/L, R is the universal gas constant (= 0.08206 L atm mol K^{-1}), and T is the absolute temperature in K. Osmotic pressure ranges from 0.04 to 0.08 bar for every 100 ppm of total dissolved solids (TDS) in the solution ("Reverse Osmosis Principles," 2010). To treat seawater at 35,000 ppm TDS, operating pressures as high as 100 bars may be required.

RO systems are characterized by product flow rate. Recovery is the percentage of feedwater recovered as permeate. It is given by:

$$\% \text{ Recovery} = \frac{\text{Permeate flow rate}}{\text{Feed flow rate}} \times 100$$ Equation 1.4

The design of the RO system involves choosing the appropriate recovery. While high recovery may seem desirable, it is accompanied by a plethora of other challenges such as compromised permeate water quality and shorter membrane lifetime due to increased fouling. Recovery rates depend greatly on feed characteristics including salinity, pretreatment, and design configuration.

Various techniques are used to characterize RO membranes. SEM is widely used to study morphology, while transmission electron microscopy (TEM) helps extract elemental and structural information. Other tools used to characterize the physico-chemical properties of RO membranes include electron spin resonance (ESR) and atomic force microscopy (AFM). The reader is referred to a recent review by Qasim et al. on the reverse osmosis process, including membranes, pretreatment steps, and challenges (Qasim et al., 2019).

While RO is mostly applied for desalination of brackish and seawater, it is also used to treat industrial effluents from the food, chemical, petroleum, pharma, and textile industries (Khan & Chandra, 2021). Most NF and RO membranes in industry are of the spiral wound configuration. Today, over 60% of the world's desalinated capacity is from RO. High pressure pumps and energy recovery devices have significantly helped lower energy costs associated with RO. For example, a modern seawater RO plant can operate at 3–5 kWh m^{-3}. Yet integration with renewable energy and hybrid desalination technologies that enable brine recovery and improvements in membrane materials addressing fouling, chlorine resistance, and removal of specific contaminants offer further potential in bringing energy consumption closer to the minimum thermodynamic limit. For typical seawater with a concentration of 35,000 mg L^{-1} at ambient temperature, the thermodynamic limit to separate water is 0.78 kWh m^{-3}.

1.4.5 MEMBRANE DISTILLATION

Membrane distillation (MD) is a thermally driven separation process that combines thermal distillation with membrane technology. It is a promising technique for saltwater desalination and has several benefits over conventional methods, especially when waste heat or renewable energy is available and if the feed's constituents make it difficult for treatment via RO. MD is described in more detail in Chapter 7, so we provide only a brief overview here.

MD separates volatile compounds from an aqueous stream through a vapor pressure gradient resulting from a temperature difference across the two sides of the membrane. The vapor pressure gradient causes pure water vapor to form on the feed side and to transport through the pores of the hydrophobic porous membrane, where it is condensed and collected as permeate water. MD offers several advantages over conventional thermal methods such as multistage flash (MSF). These include relatively low operating temperatures (40–80 °C), as well as the compactness and modularity that come with membrane processes. Compared to other membrane-based desalination processes, MD is less prone to fouling, operates at lower hydraulic pressures, demonstrates 100% rejection of nonvolatile compounds, and is not sensitive to feedwater salinity.

Although the first patent was filed in the early 1960s, progress in MD has been low due to lack of suitable membrane materials, low flux, low energy efficiency, and the tendency of the membrane to undergo so-called pore wetting during long-term operation. Nonetheless, many new developers have prompted research into pilot-scale MD systems. At the same time, its use has been explored in domains other than desalination, such as removal of heavy metals, recovery of valuable minerals from desalination brine, and clarification of juices (Tai et al., 2019).

1.4.6 FORWARD OSMOSIS

As we have seen, osmosis is the spontaneous and selective passage of solvent from a dilute solution to a concentrated solution through a semipermeable membrane separating the two (Feher, 2017; Ozer & Brazy, 2009). The osmotic pressure is the pressure that needs to be applied to prevent this spontaneous motion and to maintain equilibrium. Transport in FO is guided by an osmotic pressure gradient across a semipermeable membrane. As such, no applied hydraulic pressure needs to be applied. Instead, FO relies on a concentrated draw solution with high osmotic pressure on one side of the membrane that draws in water molecules from the feed side, which is at a lower osmotic pressure (Eyvaz et al., 2018). The now diluted draw solution can then be further processed to separate draw solutes from purified water. Without the need for high pressures, FO is less susceptible to fouling and scaling;

however, water purification with FO is a two-step process, and its energy consumption and feasibility depend on the efficiency of the process used to recover the draw
solution. Some methods that have been applied to draw solution recovery after FO
include MD, RO, NF, and UF (H. Luo et al., 2014). Performance of an FO system
depends on feed characteristics and membrane properties but also on the choice of
draw solution (Amjad et al., 2018). An appropriate draw solution must:

1. Exhibit a high osmotic pressure to ensure sufficient osmotic gradient between
   feed and draw solution; lower molecular weight and viscosity in water is desirable to reduce the effects of concentration polarization (Johnson et al., 2018).
2. Be readily regenerated using low-cost methods to keep overall costs low
   (McCutcheon et al., 2006).

Figure 1.9 shows how osmotic pressure varies with solution concentration for different draw solutions (Cath et al., 2006).

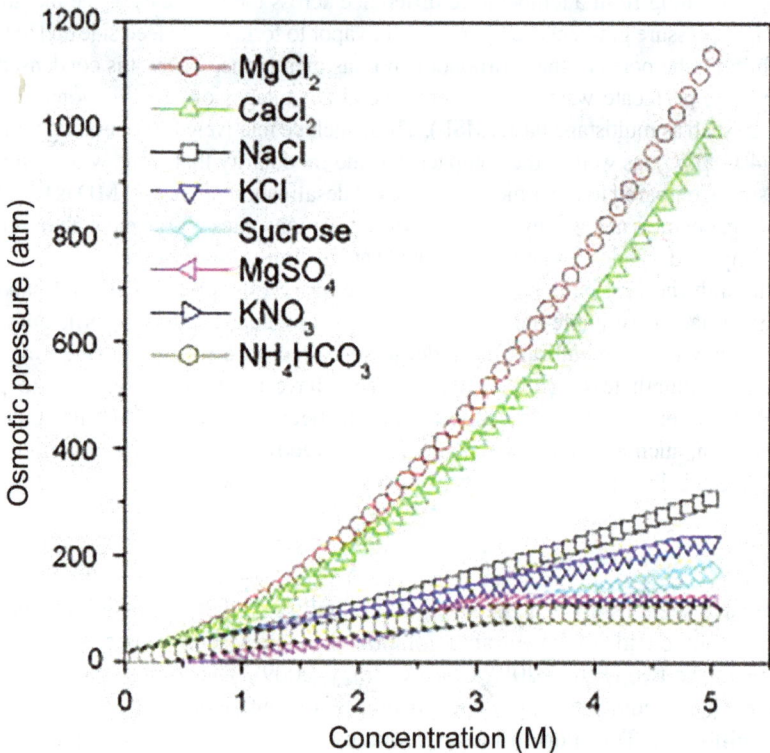

FIGURE 1.9   Osmotic pressure as a function of solution concentration at 25 °C for various
potential draw solutions.

Source: Cath, T. Y., Childress, A. E., & Elimelech, M. (2006). Forward osmosis: Principles,
applications, and recent developments. *Journal of Membrane Science, 281*(1), 70–87. Reprinted with permission.

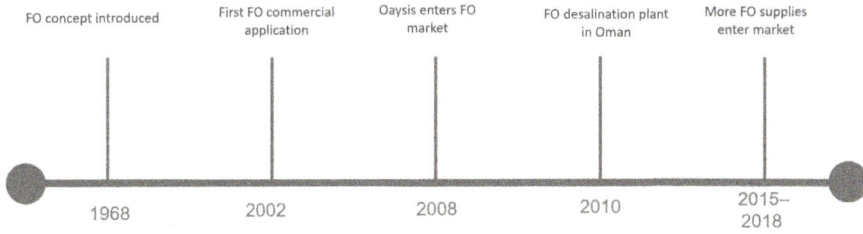

FIGURE 1.10   Milestones in the development of FO desalination.

Figure 1.10 shows the timeline of FO development since the concept was first suggested in 1968.

1.5 MEMBRANE MATERIALS

Polymers remain the predominant class of materials for membrane preparation, largely due to ease of processability, high chemical resistance, ability to operate under a wide pH range, and tailorability (M. PURKAIT & Singh, 2020). Polymeric materials are available in a variety of structures. Properties of polymeric membranes are dictated by many polymer characteristics such as molecular weight, chain flexibility, crystallinity, change in pH, change in temperature, density, melting point, glass transition temperature, and surface charge (Lalia et al., 2013; M. PURKAIT & Singh, 2020). The current generation of membrane materials most commonly employs CA (H. Lonsdale et al., 1965), PSf (Mamah et al., 2021), PES (Ahmad et al., 2013; Zhao et al., 2013), PAN (Lohokare et al., 2008), PVDF (Kang & Cao, 2014), PP (Ariono & Wardani, 2017), PVA (Katz & Wydeven Jr, 1982), and PTFE (Zhang et al., 2016).

Most early work surrounding membrane technology focused almost entirely on RO desalination. Membrane materials with simultaneous high selectivity and high permeability are desirable for RO. The very first high performance RO membranes were prepared in 1959 with CA membranes of thicknesses between 3.7 and 22 μ. The thicker membranes could reject salt with an efficiency of 99%, at 41.4 bars using 1 M aqueous NaCl as feed. While this points to the trade-off between membrane permeability and selectivity, the concept was not elucidated in liquid separation membranes before the early 21st century. The breakthrough work by Loeb and Sourirajan was monumental in advancing membrane technology. They used a dry-wet phase inversion technique to fabricate CA membranes that were an order of magnitude thinner than other films enabling high fluxes, at the same salt rejection (Loeb, 1981). Phase inversion is a widely used technique for fabricating polymeric membranes in which a homogeneous polymer solution is precipitated into a gel or thin film (Kesting, 1985). Non-solvent-induced phase separation, also known as the Loeb–Sourirajan process (Chung & Feng, 2021), is a type of phase inversion process in which a casted polymer solution is immersed in a coagulation bath containing a non-solvent (Baig et al., 2020; Zare & Kargari, 2018). An exchange between the solvent and non-solvent drives the polymer solution to turn into a polymer-rich phase and a liquid-rich phase. The polymer-rich phase or the membrane is formed with an asymmetric

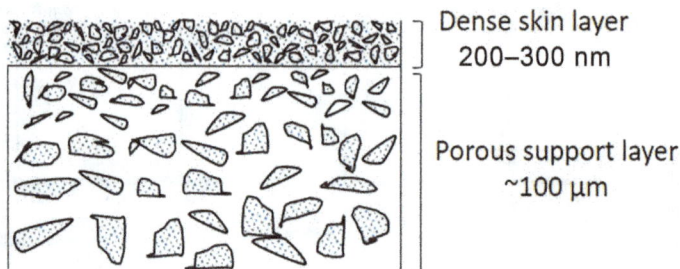

FIGURE 1.11   Cross-sectional structure of an asymmetric membrane.

cross-sectional structure resulting from different rates of solvent exchange on the surface of the cast film and that under the skin layer. This results in asymmetric membranes with a porous sublayer with a dense skin layer (Figure 1.11), which made Loeb and Sourirajan's cellulose acetate membranes feasible for commercial application (H. K. Lonsdale, 1982). Membrane morphology is controlled through polymer type, concentration, and choice of solvent.

For cellulosic membranes, an annealing step is often employed to eliminate micropores and improve salt rejection (Baker, 2012). The annealing temperature determines membrane properties and end performance. These membranes had relatively poor thermal stability and could only be operated under 35 °C and within a pH range of 3–7 (D. Pal, 2003).

In the 1960s, cellulosic esters such as cellulose acetate butyrate were explored in an attempt to improve membrane properties. In addition, a deeper understanding of NIPS ensued as the effect of variables such as solution composition, evaporation time, precipitation temperature and annealing temperature were investigated (Baker, 2012). Typical cellulose acetate membranes used today reject NaCl with about 96% efficiency (Baker, 2012). Interest has expanded to novel cellulose morphologies such as cellulose nanofibers extracted from natural sources. These membranes have shown superior performance in terms of flux for UF (Ma et al., 2014).

Soon after, interest in exploring non-cellulosic materials for membranes increased with the aim of developing economically feasible membranes with high flux and rejection. The next breakthrough in membrane materials came about when Cadotte prepared a thin film composite (TFC) membrane consisting of a thin film of polyamide (PA) polymerized on a porous support layer (Rozelle et al., 1970). Selection of different materials for the support and active layers allowed each layer to be optimized individually. The first TFC PA membrane was fabricated via a polycondensation reaction of branched poly(ethylene imine) and 2,4-diisocyanate on a porous polysulfone (PS) membrane. In 1981, a patent for the design of three-layer TFC membranes was issued, incorporating a cross-linked polyamide active layer, a microporous support, and a polyester fabric (Cadotte, 1981). This was significantly different from integrally skinned asymmetric CA membranes where both the support and active layer are made of the same material. TFC membranes prepared via interfacial polymerization (IP) revolutionized RO desalination and have been the standard for RO desalination for more than five decades now (Kumar et al., 2017). Interfacial polyamide membranes surpassed CA membranes in terms of flux and rejection (Kucera). By 1971, Dupont made TFC PA membranes in flat-sheet and the newer hollow fiber configurations (Dipak et al., 2015). To increase the chlorine resistance of TFC membranes, approximately 10% of linear aromatic PA rings were substituted by sulfonic acid groups. Early TFC membranes (NS200, PEC 1000) had a chlorine resistance of a few hundred ppm h, which indicates the approximate time, in hours, of exposure to 1 ppm of free chlorine after which eventual degradation of the material may occur (Baker, 2012). Decades later, chlorine resistance continues to be a challenge with PA membranes, although several publications and patents have been issued in recent years for PA membranes withstanding up to 1000 ppm h of free chlorine exposure (Idrees & Tariq, 2021; Liu et al., 2021; Murphy et al., 2011). Polysulfone and substituted PVA Loeb–Sourirajan membranes are able to withstand high chlorine exposure and have found commercial application in water softening.

Early membrane development comprised many noncellulosic membranes that further advanced filtration processes relying entirely on size exclusion, such as microfiltration and ultrafiltration. The phase inversion technique gave rise to polysulfone (PSf), polyethersulfone (PES), PVDF, polyacrylonitrile (PAN), and polyethylene (PP) membranes for low-pressure filtration (M. K. Purkait et al., 2018). PVDF and PP remain commonly used materials for both MF and UF (Awwa, 2016), partly due to their solubility in common solvents and high mechanical strength (Delphos & American Water Works, 2016). PSf and PES are also commonly used in UF and as support substrates for NF and RO due to their superior permeability, high selectivity, and mechanical and chemical stability. The hydrophobicity of commonly used polymeric membranes is a drawback in pressure-driven processes as it increases their tendency to foul. We look at fouling and related issues more closely in subsequent chapters.

The challenge is to develop membranes with simultaneously high permeability, high flux, high stability, strong resistance to fouling, high chemical/thermal stability, mechanical strength, and long-term operational stability. Hydrophilic modification of materials through surface coating or plasma treatment is one method to improve hydrophilicity and increase fouling resistance. Other methods involve surface grafting, cross-linking, UV radiation, etc. Table 1.1 shows the characteristics of selected common membrane materials.

TABLE 1.1
Characteristics of Selected Membrane Materials.

Material	Process	Surface Tension (dynes cm⁻¹)	Melting Point (°C)	Hydrophobicity	pH Range	Chlorine Resistance
PVDF	MF, UF, MD	25–28.5	140–170	Hydrophobic	1–14	High
PSf	MF, UF	41		Hydrophilic	2–13	High
Polytetrafluoroethylene (PTFE)	MF, MD	22	310–385	Hydrophobic	1–14	High
Polypropylene (PP)	MF, MD	31	130–170	Hydrophobic	1–14	High
CA	MF, UF, NF, RO	30	230–300	Hydrophilic	4–8	High
Polyamide	NF, RO	47	223	Hydrophilic	2–10	Low
Polyethersulfone (PES)	MF, UF		340–380	Hydrophilic	2–12	High

Simultaneously, work has also been carried out in the area of inorganic membranes consisting of hydrous oxide formed on porous carbon or ceramic supports. While these membranes have demonstrated high flux, their low salt rejection compared to polymeric membranes prevented any commercially viable inorganic membranes for desalination at the time. Nevertheless, from 1980 to 1990, the limitations of polymeric membranes forced companies to participate in the manufacture of inorganic MF and UF membranes, primarily for uranium enrichment in nuclear applications (Gillot, 1991). Ceramic RO and UF membranes were prepared early on with dynamic zirconium hydroxide on a stainless steel support, which are currently marketed in the United States. Since then, a growing number of academic and industrial laboratories investigating inorganic membranes and applications have given rise to new companies entering the market (Gillot, 1991). Recently, new synthetic routes for inorganic materials and especially low-temperature synthesis have opened up pathways for growing ceramic layers onto porous polymer substrates without sintering (Fane et al., 2015). However, large-scale applications of inorganic membranes made of ceramics or zeolites are still limited due to high operation costs and challenges in special module housings and membrane geometries. Table 1.2 shows recent advances in polymer membrane materials.

TABLE 1.2
Recent Developments in Polymeric Membranes (Lee et al., 2016).

Membrane Material	Application	Performance	Key Features	Reference
Microfiltration				
Polyvinylidene fluoride (PVDF) with Ag nanoparticles and tannic acid	Oil/water separation	1431 L m^{-2} h^{-1} at 1 bar with 98% oil rejection	Hydrophilic modification of PVDF membranes	(Y. Zhang et al., 2022)
Silicon carbide	Oil/water separation	490 L m^{-2} h^{-1} at 0.5 bar	Dry-pressing and sintering	(Jiang et al., 2022)
PVDF/cellulose	Water flux	Pure water flux of >1000 L m^{-2} h^{-1} at 1 bar	Phase inversion	(Malucelli et al., 2022)
PVDF/polydopamine (PDA)/PVP-co-PMMA	Oil/water separation	Pure water flux quadruples with modification; 99% oil rejection	PDA coating combined with electrospinning to create hydrophilic surface on PVDF membrane	(Wanke et al., 2021)
Ultrafiltration				
Hollow fiber poly(biphenyl-trifluoroacetone) (PBT) membranes	BSA rejection	MWCO of 456.2 kDa; Water flux of 353.3 L m^{-2} h^{-1} bar^{-1} with 99.7% BSA rejection		(Dou et al., 2022)

TABLE 1.2 *(Continued)*

Recent Developments in Polymeric Membranes (Lee et al., 2016).

Membrane Material	Application	Performance	Key Features	Reference
Polyimide (PI)/ polyethersulfone (PES)		3565–1780	Controllable coating of PI on PES via ALD, enhanced thermal resistance and mechanical strength	
Sulfonated PVDF	BSA and humic acid rejection	$125\,L\,m^{-2}\,h^{-1}$ without pathogens; 96% and 97% rejection for bovine serum albumin (BSA) and humic acid (HA), respectively	Threefold increase in tensile strength compared to bare PVDF	(Ayyaru & Ahn, 2022)
PES/activated carbon	Concentration of proteins	Pure water flux of $136\,L\,m^{-2}\,h^{-1}$ at 1 bar with 8.4% BSA rejection	Phase inversion	(Zubia et al., 2022)
Polycarbonate (PC)/PSf/Chitin nanowhisker membrane	BSA rejection	Pure water flux of $262.6\,L\,m^{-2}\,h^{-1}$ with 94.4% BSA rejection	Chitin nanowhiskers improved tensile properties, permeability and hydrophilicity of composite membranes.	(Saijun et al., 2022)
CeO2/GO-modified PES	Dye retention	Pure water flux of $249\,L\,m^{-2}\,h^{-1}$ at 2 bars with 98% rejection for Yellow 105 organic dye	Pure water flux increases by 64% upon modification due to improved hydrophilicity.	(Safarpour et al., 2022)
Nanofiltration				
Polyester	Na_2SO_4 rejection	$14.7\,L\,m^{-2}\,h^{-1}\,bar^{-1}$ with 89% rejection	Permeability affected by alkali species used during polyester membrane formation due to varying diffusion rates.	(Hu et al., 2022)
Cellulose nanofiber/ graphene oxide (GO) cross-linked membrane	Removal of dyes	$13.9\,L\,m^{-2}\,h^{-1}\,bar^{-1}$ and >90% rejection for Rose Bengal and Brilliant blue dyes	Cross-linking improved selectivity	(Mohammed et al., 2022)
Thermally modified polyimide/SiO_2	Dye/salt separation	Water flux: 153 $L\,m^{-2}\,h^{-1}$; >90% rejection for various dyes	High dye rejection and low inorganic salt rejection	(Z. Liu et al., 2022)

(Continued)

TABLE 1.2 *(Continued)*

Recent Developments in Polymeric Membranes (Lee et al., 2016).

Membrane Material	Application	Performance	Key Features	Reference
Prussian blue (PB)/ GO/PEG/PSf	Removal of radioactive cesium and strontium from water	99.5% rejection of cesium (Cs^+, 99.5%) and 97.5% rejection of strontium (Sr^{2+}, 97.5%)	Cs^+ and Sr^{2+} removed by steric effect, Donnan effect, and two-step adsorption	(Ye et al., 2022)
L-menthol/10-camphorsulfonic acid (L-M/CSA)-modified PES	Pharmaceutical wastewater treatment	89.1% chemical oxygen demand (COD) removal; 100% turbidity removal; 51.6% total dissolved solids (TDS) removal from industrial pharmaceutical wastewater	Blending with L-M/ CSA increases pore size and improves antifouling properties.	(Moradi et al., 2022)
ZnO nanoparticles/ reduced graphene oxide (rGO)	Wastewater treatment	Water permeability of $225\,L\,m^{-2}\,h^{-1}\,bar^{-1}$; selectivity up to 98% in the size-exclusion separation of Methyl blue.	Metal oxide nanoparticles increase vertical interlayer spacing and lateral tortuous paths of the rGO membranes; control of membrane microstructure possible by varying nanoparticle size and loading.	(W. Zhang et al., 2022)
Reverse osmosis				
Nickel-based metal organic framework (MOF) nanosheet embedded in PA on PSf substrate	Brackish water desalination	Water permeance of $5\,L\,m^{-2}\,h^{-1}\,bar^{-1}$ w with NaCl rejection of 99.2%	Incorporation of 2D-MOF nanosheets can reduce transfer resistance and increase water diffusion because of the extra pores of 2D-MOF and H-bonds between free water molecules and the coordination water of 2D-MOF.	(Y. Liu et al., 2022)
Polyamide/ acetylated cellulose nanocrystals	Desalination	Water permeance of $2.6\,L\,m^{-2}\,h^{-1}\,bar^{-1}$ with 98–99% NaCl rejection	Acetylated cellulose nanocrystals simultaneously increase water permeability and salt rejection of PA membranes.	(Abedi et al., 2022)

1.6 CONCLUSION

Ease of use and versatility have rendered membrane technology an important field in separation science and engineering, especially in the areas of desalination and water treatment. This chapter presents historical context to the evolution of membrane separation processes and materials and discusses the fundamentals of membrane separation as relevant to desalination and water treatment. Processes differ in driving force and the type of matter that they can separate, enabling different uses for each. Modes of operation, materials, and research trends for common separation processes have been discussed.

BIBLIOGRAPHY

Abd Aziz, M. H., Othman, M. H. D., Hashim, N. A., Adam, M. R., & Mustafa, A. (2019). Fabrication and characterization of mullite ceramic hollow fiber membrane from natural occurring ball clay. *Applied Clay Science, 177*, 51–62.

Abedi, F., Emadzadeh, D., Dubé, M. A., & Kruczek, B. (2022). Modifying cellulose nanocrystal dispersibility to address the permeability/selectivity trade-off of thin-film nanocomposite reverse osmosis membranes. *Desalination, 538*, 115900. https://doi.org/10.1016/j.desal.2022.115900

Ahmad, A., Abdulkarim, A., Ooi, B., & Ismail, S. (2013). Recent development in additives modifications of polyethersulfone membrane for flux enhancement. *Chemical Engineering Journal, 223*, 246–267.

Ahmadi, R., Sedighian, R., Sanaeepur, H., Ebadi Amooghin, A., & Lak, S. (2022). Polyphenylsulfone/zinc ion-exchanged zeolite Y nanofiltration mixed matrix membrane for water desalination. *Journal of Applied Polymer Science, 139*(22), 52262. https://doi.org/10.1002/app.52262

Al Aani, S., Mustafa, T. N., & Hilal, N. (2020). Ultrafiltration membranes for wastewater and water process engineering: A comprehensive statistical review over the past decade. *Journal of Water Process Engineering, 35*, 101241. https://doi.org/10.1016/j.jwpe.2020.101241

Alshahrani, A. A., Alsohaimi, I. H., Alshehri, S., Alawady, A. R., El-Aassar, M. R., Nghiem, L. D., & Panhuis, M. I. H. (2020). Nanofiltration membranes prepared from pristine and functionalised multiwall carbon nanotubes/biopolymer composites for water treatment applications. *Journal of Materials Research and Technology, 9*(4), 9080–9092. https://doi.org/10.1016/j.jmrt.2020.06.055

Amjad, M., Gardy, J., Hassanpour, A., & Wen, D. (2018). Novel draw solution for forward osmosis based solar desalination. *Applied Energy, 230*, 220–231. https://doi.org/10.1016/j.apenergy.2018.08.021

Angelakis, A. N., Valipour, M., Choo, K.-H., Ahmed, A. T., Baba, A., Kumar, R., . . . Wang, Z. (2021). Desalination: From ancient to present and future. *Water, 13*(16), 2222.

Anis, S. F., Hashaikeh, R., & Hilal, N. (2019). Microfiltration membrane processes: A review of research trends over the past decade. *Journal of Water Process Engineering, 32*, 100941. https://doi.org/10.1016/j.jwpe.2019.100941

Ariono, D., & Wardani, A. K. (2017). *Modification and applications of hydrophilic polypropylene membrane.* Paper presented at the IOP Conference Series: Materials Science and Engineering.

Awwa. (2016). *M53 microfiltration and ultrafiltration membranes for drinking water.* American Water Works Association.

Ayyaru, S., & Ahn, Y.-H. (2022). Fabrication and application of novel high strength sulfonated PVDF ultrafiltration membrane for production of reclamation water. *Chemosphere, 305*, 135416. https://doi.org/10.1016/j.chemosphere.2022.135416

Baig, M. I., Durmaz, E. N., Willott, J. D., & de Vos, W. M. (2020). Sustainable membrane production through polyelectrolyte complexation induced aqueous phase separation. *Advanced Functional Materials, 30*(5), 1907344. https://doi.org/10.1002/adfm.201907344

Baker, R. W. (2012). *Membrane technology and applications*: John Wiley & Sons.

Balaji, K. R., Hardian, R., Kumar, V. G. D., Viswanatha, R., Kumar, S., Kumar, S., . . . Szekely, G. (2021). Composite nanofiltration membrane comprising one-dimensional erdite, two-dimensional reduced graphene oxide, and silkworm pupae binder. *Materials Today Chemistry, 22*, 100602. https://doi.org/10.1016/j.mtchem.2021.100602

Balster, J. (2016). Hollow fiber membrane module. In E. Drioli & L. Giorno (Eds.), *Encyclopedia of membranes* (pp. 955–957). Springer Berlin Heidelberg.

Batool, M., Shafeeq, A., Haider, B., & Ahmad, N. M. (2021). TiO2 nanoparticle filler-based mixed-matrix PES/CA Nanofiltration membranes for enhanced desalination. *Membranes, 11*(6), 433.

Baumgarten, C. M., & Feher, J. J. (2012). Chapter 16—osmosis and regulation of cell volume. In N. Sperelakis (Ed.), *Cell physiology source book* (4th ed., pp. 261–301). Academic Press.

Bechhold, H. (1907). Kolloidstudien mit der Filtrationsmethode. *Zeitschrift für Elektrochemie und angewandte physikalische Chemie, 13*(32), 527–533. https://doi.org/10.1002/bbpc.19070133207

Beda-Maluga, K., Pisarek, H., Romanowska, I., Komorowski, J., Świętosławski, J., & Winczyk, K. (2015). Ultrafiltration–an alternative method to polyethylene glycol precipitation for macroprolactin detection. *Archives of Medical Science, 11*(5), 1001–1007.

Bhave, R. R. (2014). Chapter 9—cross-flow filtration. In H. C. Vogel & C. M. Todaro (Eds.), *Fermentation and biochemical engineering handbook* (3rd ed., pp. 149–180). William Andrew Publishing.

Bowen, W. R., Mohammad, A. W., & Hilal, N. (1997). Characterisation of nanofiltration membranes for predictive purposes—use of salts, uncharged solutes and atomic force microscopy. *Journal of Membrane Science, 126*(1), 91–105.

Boyd, B. J. (2003). Characterisation of drug release from cubosomes using the pressure ultrafiltration method. *International Journal of Pharmaceutics, 260*(2), 239–247.

Cadotte, J. E. (1981). *Interfacially synthesized reverse osmosis membrane*. Google Patents.

Campos, J. C., Borges, R. M. H., Oliveira Filho, A. M., Nobrega, R., & Sant'Anna, G. L. (2002). Oilfield wastewater treatment by combined microfiltration and biological processes. *Water Research, 36*(1), 95–104. https://doi.org/10.1016/S0043-1354(01)00203-2

Cath, T. Y., Childress, A. E., & Elimelech, M. (2006). Forward osmosis: Principles, applications, and recent developments. *Journal of Membrane Science, 281*(1), 70–87. https://doi.org/10.1016/j.memsci.2006.05.048

Childress, A. E., & Elimelech, M. (2000). Relating nanofiltration membrane performance to membrane charge (electrokinetic) characteristics. *Environmental Science & Technology, 34*(17), 3710–3716. doi:10.1021/es0008620

Christensen, E. R., & Plaumann, K. W. (1981). Waste reuse: Ultrafiltration of industrial and municipal wastewaters. *Journal (Water Pollution Control Federation)*, 1206–1212.

Chung, T. S., & Feng, Y. (2021). *Hollow fiber membranes: Fabrication and applications*. Elsevier Science.

Cohen, M. E., Grable, M. A., & Riggleman, B. M. (1972). Hollow-fiber reverse osmosis membranes. In H. K. Lonsdale & H. E. Podall (Eds.), *Reverse osmosis membrane research: Based on the symposium on "polymers for desalination" held at the 162nd national meeting of the American chemical society in Washington, D.C., September 1971* (pp. 331–340). Springer US.

da Silva Biron, D., dos Santos, V., & Zeni, M. (2018). Ceramic membrane modules. In D. da Silva Biron, V. dos Santos, & M. Zeni (Eds.), *Ceramic membranes applied in separation processes* (pp. 81–91). Springer International Publishing.

Delphos, P. J., & American Water Works, A. (2016). *Microfiltration and ultrafiltration membranes for drinking water* (2nd ed.). American Water Works Association.

Demeuse, M. T. (2009). 15—Production and applications of hollow fibers. In S. J. Eichhorn, J. W. S. Hearle, M. Jaffe, & T. Kikutani (Eds.), *Handbook of textile fibre structure* (Vol. 2, pp. 485–499). Woodhead Publishing.

Diffenbaugh, N. S., Swain, D. L., & Touma, D. (2015). Anthropogenic warming has increased drought risk in California. *Proceedings of the National Academy of Sciences, 112*(13), 3931. doi:10.1073/pnas.1422385112

Dipak, R., Matsuura, T., Mohd Azraai, K., & Ismail, A. F. (2015). *Reverse osmosis membrane handbook of membrane separations.* CRC Press.

Dong, L.-X., Huang, X.-C., Wang, Z., Yang, Z., Wang, X.-M., & Tang, C. Y. (2016). A thin-film nanocomposite nanofiltration membrane prepared on a support with in situ embedded zeolite nanoparticles. *Separation and Purification Technology, 166*, 230–239. https://doi.org/10.1016/j.seppur.2016.04.043

Dou, Y., Dong, X., Ma, Y., Ge, P., Li, C., Zhu, A., . . . Zhang, Q. (2022). Hollow fiber ultrafiltration membranes of poly(biphenyl-trifluoroacetone). *Journal of Membrane Science, 659*, 120779. https://doi.org/10.1016/j.memsci.2022.120779

Eke, J., Yusuf, A., Giwa, A., & Sodiq, A. (2020). The global status of desalination: An assessment of current desalination technologies, plants and capacity. *Desalination, 495*, 114633. https://doi.org/10.1016/j.desal.2020.114633

Elford, W. J. (1930). Structure in very permeable collodion gel films and its significance in filtration problems. *Proceedings of the Royal Society of London. Series B, Containing Papers of a Biological Character, 106*(743), 216–228.

Eyvaz, M., Arslan, S., İmer, D., Yüksel, E., & Koyuncu, İ. (2018). *Forward osmosis membranes–a review: Part I osmotically driven membrane processes-approach, development and current status.* IntechOpen.

Fane, A. G., Wang, R., & Hu, M. X. (2015). Synthetic membranes for water purification: Status and Future. *Angewandte Chemie International Edition, 54*(11), 3368–3386. https://doi.org/10.1002/anie.201409783

Feher, J. (2017). 2.7—Osmosis and osmotic pressure. In J. Feher (Ed.), *Quantitative human physiology* (2nd ed., pp. 182–198). Academic Press.

Ficheux, A., Ronco, C., Brunet, P., & Argilés, À. (2013). The ultrafiltration coefficient: This old 'grand inconnu' in dialysis. *Nephrology Dialysis Transplantation, 30*(2), 204–208. doi:10.1093/ndt/gft493

Flörke, M., Schneider, C., & McDonald, R. I. (2018). Water competition between cities and agriculture driven by climate change and urban growth. *Nature Sustainability, 1*(1), 51–58. doi:10.1038/s41893-017-0006-8

Frost, K., & Hua, I. (2019). Quantifying spatiotemporal impacts of the interaction of water scarcity and water use by the global semiconductor manufacturing industry. *Water Resources and Industry, 22*, 100115. https://doi.org/10.1016/j.wri.2019.100115

Ganji, P., Nazari, S., Zinatizadeh, A. A., & Zinadini, S. (2022). Chitosan-wrapped multi-walled carbon nanotubes (CS/MWCNT) as nanofillers incorporated into nanofiltration (NF) membranes aiming at remarkable water purification. *Journal of Water Process Engineering, 48*, 102922. https://doi.org/10.1016/j.jwpe.2022.102922

Geens, J., Boussu, K., Vandecasteele, C., & Van der Bruggen, B. (2006). Modelling of solute transport in non-aqueous nanofiltration. *Journal of Membrane Science, 281*, 139–148. https://doi.org/10.1016/j.memsci.2006.03.028

Geise, G. M., Lee, H.-S., Miller, D. J., Freeman, B. D., McGrath, J. E., & Paul, D. R. (2010). Water purification by membranes: The role of polymer science. *Journal of Polymer Science Part B: Polymer Physics*, *48*(15), 1685–1718. https://doi.org/10.1002/polb.22037

Geise, G. M., Park, H. B., Sagle, A. C., Freeman, B. D., & McGrath, J. E. (2011). Water permeability and water/salt selectivity tradeoff in polymers for desalination. *Journal of Membrane Science*, *369*(1), 130–138. https://doi.org/10.1016/j.memsci.2010.11.054

Gholami, F., Zinatizadeh, A. A., Zinadini, S., Rittmann, B. E., & Torres, C. I. (2022). Enhanced antifouling and flux performances of a composite membrane via incorporating TiO2 functionalized with hydrophilic groups of L-cysteine for nanofiltration. *Polymers for Advanced Technologies*, *33*(5), 1544–1560. https://doi.org/10.1002/pat.5620

Gillot, J. (1991). *The developing use of inorganic membranes: A historical perspective inorganic membranes synthesis, characteristics and applications* (pp. 1–9). Springer.

Glater, J. (1998). The early history of reverse osmosis membrane development. *Desalination*, *117*(1), 297–309. https://doi.org/10.1016/S0011-9164(98)00122-2

Haddeland, I., Heinke, J., Biemans, H., Eisner, S., Flörke, M., Hanasaki, N., . . . Wisser, D. (2014). Global water resources affected by human interventions and climate change. *Proceedings of the National Academy of Sciences*, *111*(9), 3251. doi:10.1073/pnas.1222475110

Hamilton, C. E., Testing, A. S. F., & Materials. (1983). *Supplement to manual on water* (4th ed.). American Society for Testing and Materials.

Hammel, H. T., & Scholander, P. F. (1976). *Osmosis and tensile solvent*. Springer Berlin Heidelberg.

Hardenbergh, W. A. (1946). Army water supply in World War II. *Journal (American Water Works Association)*, *38*(8), 952–958.

Hendricks, D. (2010). *Fundamentals of water treatment unit processes: Physical, chemical, and biological*. Taylor & Francis Group.

Hermes, M. E. (1996). *Enough for one lifetime: Wallace Carothers, inventor of nylon*. Chemical Heritage Foundation.

Hilal, N., Al-Zoubi, H., Darwish, N. A., & Mohammad, A. W. (2005). Characterisation of nanofiltration membranes using atomic force microscopy. *Desalination*, *177*(1), 187–199. https://doi.org/10.1016/j.desal.2004.12.008

Hilal, N., Ismail, A. F., & Wright, C. (2015). *Membrane fabrication*. Taylor & Francis Group.

Hoek, E. M. V., Tarabara, V. V., & Kucera, J. (2013). Membrane materials and module development, historical perspective. In E. M. V. Hoek & V. V. Tarabara (Eds.), *Encyclopedia of membrane scienc and technology*. John Wiley & Sons, Inc. https://doi.org/10.1002/9781118522318.emst033

Howe, K. J., Hand, D. W., Crittenden, J. C., Trussell, R. R., Tchobanoglous, G., Howe, K. J., & Crittenden, J. C. (2012). *Principles of water treatment*. John Wiley & Sons, Incorporated.

Hu, Q., Li, D., Liu, S., Yan, G., Yang, J., & Zhang, G. (2022). Effects of alkali on the polyester membranes based on cyclic polyphenols for nanofiltration. *Desalination*, *533*, 115774. https://doi.org/10.1016/j.desal.2022.115774

Huang, G., Meng, F., Zheng, X., Wang, Y., Wang, Z., Liu, H., & Jekel, M. (2011). Biodegradation behavior of natural organic matter (NOM) in a biological aerated filter (BAF) as a pretreatment for ultrafiltration (UF) of river water. *Applied Microbiology and Biotechnology*, *90*(5), 1795–1803.

Idrees, M. F., & Tariq, U. (2021). Enhancing chlorine resistance in polyamide membranes with surface & structure modification strategies. *Water Supply*, *22*(2), 1199–1215. doi:10.2166/ws.2021.358

Ismail, A. F., & Goh, P. S. (2015). Microfiltration membrane. In S. Kobayashi & K. Müllen (Eds.), *Encyclopedia of polymeric nanomaterials* (pp. 1250–1255). Springer Berlin Heidelberg.

Jiang, Q., Wang, Y., Xie, Y., Zhou, M., Gu, Q., Zhong, Z., & Xing, W. (2022). Silicon carbide microfiltration membranes for oil-water separation: Pore structure-dependent wettability matters. *Water Research*, *216*, 118270. https://doi.org/10.1016/j.watres.2022.118270

Johnson, D. J., & Hilal, N. (2022). Nanocomposite nanofiltration membranes: State of play and recent advances. *Desalination*, *524*, 115480. https://doi.org/10.1016/j.desal.2021.115480

Johnson, D. J., Suwaileh, W. A., Mohammed, A. W., & Hilal, N. (2018). Osmotic's potential: An overview of draw solutes for forward osmosis. *Desalination*, *434*, 100–120. https://doi.org/10.1016/j.desal.2017.09.017

Juang, Y., Nurhayati, E., Huang, C., Pan, J. R., & Huang, S. (2013). A hybrid electrochemical advanced oxidation/microfiltration system using BDD/Ti anode for acid yellow 36 dye wastewater treatment. *Separation and Purification Technology*, *120*, 289–295. https://doi.org/10.1016/j.seppur.2013.09.042

Kang, G.-D., & Cao, Y.-M. (2014). Application and modification of poly (vinylidene fluoride) (PVDF) membranes–a review. *Journal of Membrane Science*, *463*, 145–165.

Katz, M. G., & Wydeven Jr, T. (1982). Selective permeability of PVA membranes. II. Heat-treated membranes. *Journal of Applied Polymer Science*, *27*(1), 79–87.

Kaur, S., Sundarrajan, S., Rana, D., Matsuura, T., & Ramakrishna, S. (2012). Influence of electrospun fiber size on the separation efficiency of thin film nanofiltration composite membrane. *Journal of Membrane Science*, *392*, 101–111.

Kesting, R. E. (1985). *Phase inversion membranes materials science of synthetic membranes* (Vol. 269, pp. 131–164). American Chemical Society.

Khan, A. A., & Boddu, S. (2021). Chapter 13—Hybrid membrane process: An emerging and promising technique toward industrial wastewater treatment. In M. P. Shah & S. Rodriguez-Couto (Eds.), *Membrane-based hybrid processes for wastewater treatment* (pp. 257–277). Elsevier.

Khan, A. A., & Chandra, A. (2021). Chapter 15—Hybrid membrane technique: A technological advancement of textile waste effluent treatment. In M. P. Shah & S. Rodriguez-Couto (Eds.), *Membrane-based hybrid processes for wastewater treatment* (pp. 313–340). Elsevier.

Konapala, G., Mishra, A. K., Wada, Y., & Mann, M. E. (2020). Climate change will affect global water availability through compounding changes in seasonal precipitation and evaporation. *Nature communications*, *11*(1), 3044. doi:10.1038/s41467-020-16757-w

Kong, G., Fan, L., Zhao, L., Feng, Y., Cui, X., Pang, J., . . . Mintova, S. (2021). Spray-dispersion of ultra-small EMT zeolite crystals in thin-film composite membrane for high-perme ability nanofiltration process. *Journal of Membrane Science*, *622*, 119045. https://doi.org/10.1016/j.memsci.2020.119045

Krishna, H. J. (2004). Introduction to desalination technologies. *Texas Water Development*, *2*, 1–7.

Kucera, J. (2010). Reverse osmosis principles. In J. Kucera (Ed.), *Reverse osmosis*. Scrivener Publishing LLC. https://doi.org/10.1002/9780470882634.ch2

Kumar, M., Culp, T., & Shen, Y. (2017). *Water desalination: History, advances, and challenges*. Paper presented at the Proc., Frontiers of Engineering: Reports on Leading-Edge Engineering from the 2016 Symp.

Lai, G. S., Lau, W. J., Goh, P. S., Ismail, A. F., Yusof, N., & Tan, Y. H. (2016). Graphene oxide incorporated thin film nanocomposite nanofiltration membrane for enhanced salt removal performance. *Desalination*, *387*, 14–24. https://doi.org/10.1016/j.desal.2016.03.007

Laîné, J. M., Vial, D., & Moulart, P. (2000). Status after 10 years of operation—overview of UF technology today. *Desalination*, *131*(1), 17–25. https://doi.org/10.1016/S0011-9164(00)90002-X

Lalia, B. S., Kochkodan, V., Hashaikeh, R., & Hilal, N. (2013). A review on membrane fabrication: Structure, properties and performance relationship. *Desalination*, *326*, 77–95. https://doi.org/10.1016/j.desal.2013.06.016

Lee, A., Elam, J. W., & Darling, S. B. (2016). Membrane materials for water purification: Design, development, and application. *Environmental Science: Water Research & Technology*, *2*(1), 17–42.

Li, K., Liang, H., Qu, F., Shao, S., Yu, H., Han, Z.-s., . . . Li, G. (2014). Control of natural organic matter fouling of ultrafiltration membrane by adsorption pretreatment: Comparison of mesoporous adsorbent resin and powdered activated carbon. *Journal of Membrane Science, 471*, 94–102.

Li, M., Han, J., Xue, Y., Dai, Y., Liu, J., Gan, L., . . . Long, M. (2019). Hydrogen peroxide pretreatment efficiently assisting enzymatic hydrolysis of chitosan at high concentration for chitooligosaccharides. *Polymer Degradation and Stability, 164*, 177–186.

Li, S., Du, S., Liu, S., Su, B., & Han, L. (2022). Ultra-smooth and ultra-thin polyamide thin film nanocomposite membranes incorporated with functionalized MoS2 nanosheets for high performance organic solvent nanofiltration. *Separation and Purification Technology, 291*, 120937. https://doi.org/10.1016/j.seppur.2022.120937

Li, X., & Li, J. (2016). Cross-flow filtration. In E. Drioli & L. Giorno (Eds.), *Encyclopedia of membranes* (pp. 477–478). Springer Berlin Heidelberg.

Li, X.-X., Guo, X.-F., Zhang, M., Zhang, H.-W., Wang, Y.-W., Chao, S.-L., . . . Han, X. (2022). Enhanced permeate flux by air micro-nano bubbles via reducing apparent viscosity during ultrafiltration process. *Chemosphere, 302*, 134782. https://doi.org/10.1016/j.chemosphere.2022.134782

Liangxiong, L., Whitworth, T. M., & Lee, R. (2003). Separation of inorganic solutes from oil-field produced water using a compacted bentonite membrane. *Journal of Membrane Science, 217*(1–2), 215–225.

Liu, C., Wang, W., Yang, B., Xiao, K., & Zhao, H. (2021). Separation, anti-fouling, and chlorine resistance of the polyamide reverse osmosis membrane: From mechanisms to mitigation strategies. *Water Research, 195*, 116976. https://doi.org/10.1016/j.watres.2021.116976

Liu, Y., Wang, X.-P., Zong, Z.-A., Lin, R., Zhang, X.-Y., Chen, F.-S., . . . Hou, J. (2022). Thin film nanocomposite membrane incorporated with 2D-MOF nanosheets for highly efficient reverse osmosis desalination. *Journal of Membrane Science, 653*, 120520. https://doi.org/10.1016/j.memsci.2022.120520

Liu, Z., Qiang, R., Lin, L., Deng, X., Yang, X., Zhao, K., . . . Xu, M. (2022). Thermally modified polyimide/SiO2 nanofiltration membrane with high permeance and selectivity for efficient dye/salt separation. *Journal of Membrane Science, 658*, 120747. https://doi.org/10.1016/j.memsci.2022.120747

Loeb, S. (1960). *Sea water demineralization by means of a semipermeable membrane*. UCLA Dept. of Engineering Report.

Loeb, S. (1981). *The Loeb-Sourirajan membrane: How it came about*. ACS Publications.

Lohokare, H. R., Muthu, M. R., Agarwal, G. P., & Kharul, U. K. (2008). Effective arsenic removal using polyacrylonitrile-based ultrafiltration (UF) membrane. *Journal of Membrane Science, 320*(1), 159–166. https://doi.org/10.1016/j.memsci.2008.03.068

Lonsdale, H. K. (1982). The growth of membrane technology. *Journal of Membrane Science, 10*(2), 81–181. https://doi.org/10.1016/S0376-7388(00)81408-8

Lonsdale, H. K., Merten, U., & Riley, R. (1965). Transport properties of cellulose acetate osmotic membranes. *Journal of Applied Polymer Science, 9*(4), 1341–1362.

Lowe, K. (2012). *Savage continent: Europe in the aftermath of World War II*. St. Martin's Press.

Lu, Q., Zhang, X., Hing Wong, N., Sunarso, J., & Li, N. (2022). Anti-biofouling polyvinylidene fluoride/quaternized polyvinyl alcohol ultrafiltration membrane selectively separates aromatic contaminants from wastewater by host–guest interactions. *Separation and Purification Technology, 296*, 121387. https://doi.org/10.1016/j.seppur.2022.121387

Luo, H., Wang, Q., Zhang, T. C., Tao, T., Zhou, A., Chen, L., & Bie, X. (2014). A review on the recovery methods of draw solutes in forward osmosis. *Journal of Water Process Engineering, 4*, 212–223. https://doi.org/10.1016/j.jwpe.2014.10.006

Luo, J., & Wan, Y. (2013). Effects of pH and salt on nanofiltration—a critical review. *Journal of Membrane Science, 438*, 18–28. https://doi.org/10.1016/j.memsci.2013.03.029

Ma, H., Burger, C., Hsiao, B. S., & Chu, B. (2014). Fabrication and characterization of cellulose nanofiber based thin-film nanofibrous composite membranes. *Journal of Membrane Science, 454,* 272–282. https://doi.org/10.1016/j.memsci.2013.11.055

Madaeni, S. S. (2015). Nanofiltration membranes. In E. Drioli & L. Giorno (Eds.), *Encyclopedia of membranes* (pp. 1–3). Springer Berlin Heidelberg.

Malik, N., Bulasara, V. K., & Basu, S. (2020). Preparation of novel porous ceramic microfiltration membranes from fly ash, kaolin and dolomite mixtures. *Ceramics International, 46*(5), 6889–6898.

Malucelli, L. C., Ozeri, I., Matos, M., Magalhães, W. L. E., Filho, M. A. S. C., & Eisen, M. S. (2022). High-flux, porous and homogeneous PVDF/cellulose microfiltration membranes. *Cellulose, 29*(3), 1943–1953. doi:10.1007/s10570-022-04422-y

Mamah, S. C., Goh, P. S., Ismail, A. F., Suzaimi, N. D., Yogarathinam, L. T., Raji, Y. O., & El-badawy, T. H. (2021). Recent development in modification of polysulfone membrane for water treatment application. *Journal of Water Process Engineering, 40,* 101835.

Manorma, Ferreira, I., Alves, P., Gil, M. H., & Gando-Ferreira, L. M. (2021). Lignin separation from black liquor by mixed matrix polysulfone nanofiltration membrane filled with multiwalled carbon nanotubes. *Separation and Purification Technology, 260,* 118231. https://doi.org/10.1016/j.seppur.2020.118231

Matchette, R. B. (1998). *Guide to federal records in the national archives of the United States* (Vol. 1). National Archives and Records Administration.

Matsuura, T. (2020). Tribute to S. Sourirajan: Great scientist, inventor and philosopher. *Chemical Engineering Research and Design, 160,* 351–355. https://doi.org/10.1016/j.cherd.2020.05.030

McCutcheon, J. R., McGinnis, R. L., & Elimelech, M. (2006). Desalination by ammonia–carbon dioxide forward osmosis: Influence of draw and feed solution concentrations on process performance. *Journal of Membrane Science, 278*(1), 114–123. https://doi.org/10.1016/j.memsci.2005.10.048

Mohamed Bazin, M., Ahmad, N., & Nakamura, Y. (2019). Preparation of porous ceramic membranes from Sayong ball clay. *Journal of Asian Ceramic Societies, 7*(4), 417–425.

Mohammed, S., M. Hegab, H., & Ou, R. (2022). Nanofiltration performance of glutaraldehyde crosslinked graphene oxide-cellulose nanofiber membrane. *Chemical Engineering Research and Design, 183,* 1–12. https://doi.org/10.1016/j.cherd.2022.04.039

Moradi, G., Rahimi, M., Zinadini, S., Shamsipur, M., & Babajani, N. (2022). Natural deep eutectic solvent modified nanofiltration membranes with superior antifouling properties for pharmaceutical wastewater treatment. *Chemical Engineering Journal, 448,* 137704. https://doi.org/10.1016/j.cej.2022.137704

Mulder, M. (1996). *Preparation of synthetic membranes basic principles of membrane technology* (pp. 71–156). Springer Netherlands.

Mulder, M., & Mulder, J. (1996). *Basic principles of membrane technology.* Springer Science & Business Media.

Murphy, A. P., Murugaverl, B., & Riley, R. L. (2011). *Chlorine resistant polyamides and membranes made from the same.* Google Patents.

Nataraj, S. K., Sridhar, S., Shaikha, I. N., Reddy, D. S., & Aminabhavi, T. M. (2007). Membrane-based microfiltration/electrodialysis hybrid process for the treatment of paper industry wastewater. *Separation and Purification Technology, 57*(1), 185–192. https://doi.org/10.1016/j.seppur.2007.03.014

Nieminen, J., Anugwom, I., Pihlajamäki, A., & Mänttäri, M. (2022). TEMPO-mediated oxidation as surface modification for cellulosic ultrafiltration membranes: Enhancement of ion rejection and permeability. *Journal of Membrane Science, 659,* 120786. https://doi.org/10.1016/j.memsci.2022.120786

Nollet, J.-A. (1752). Recherches sur les causes du bouillonnement des liquides. *Histoire de l'Académie Royale des Sciences, 1,* 57–104.

Nollet, J. A. (1764). *Lecons de physique experimentale. Par m. l'abbe Nollet.* H. L. Guérin et L. F. Delatour.

Organization, W. H. (2017). Progress on drinking water, sanitation and hygiene: 2017 update and SDG baselines. World Health Organization (WHO) and United Nations Children's Fund (UNICEF).

Otero, J. A., Mazarrasa, O., Villasante, J., Silva, V., Prádanos, P., Calvo, J. I., & Hernández, A. (2008). Three independent ways to obtain information on pore size distributions of nanofiltration membranes. *Journal of Membrane Science, 309*(1), 17–27. https://doi.org/10.1016/j.memsci.2007.09.065

Ozer, B. H., & Brazy, P. C. (2009). 2—body fluid compartments and their regulation. In A. V. Moorthy (Ed.), *Pathophysiology of kidney disease and hypertension* (pp. 17–27). W. B. Saunders.

Padrón, R. S., Gudmundsson, L., Decharme, B., Ducharne, A., Lawrence, D. M., Mao, J., . . . Seneviratne, S. I. (2020). Observed changes in dry-season water availability attributed to human-induced climate change. *Nature Geoscience, 13*(7), 477–481. doi:10.1038/s41561-020-0594-1

Pal, D. (2003). Membrane techniques | principles of ultrafiltration. In B. Caballero (Ed.), *Encyclopedia of food sciences and nutrition* (2nd ed., pp. 3837–3842). Academic Press.

Pal, P. (2020). Chapter 2—Introduction to membrane-based technology applications. In P. Pal (Ed.), *Membrane-based technologies for environmental pollution control* (pp. 71–100). Butterworth-Heinemann.

Park, H. B., Choo, K.-H., Park, H.-S., Choi, J., & Hoffmann, M. R. (2013). Electrochemical oxidation and microfiltration of municipal wastewater with simultaneous hydrogen production: Influence of organic and particulate matter. *Chemical Engineering Journal, 215–216*, 802–810. https://doi.org/10.1016/j.cej.2012.11.075

Park, H. B., Kamcev, J., Robeson, L. M., Elimelech, M., & Freeman, B. D. (2017). Maximizing the right stuff: The trade-off between membrane permeability and selectivity. *Science, 356*(6343), eaab0530. doi:10.1126/science.aab0530

Peeva, L. G., Sairam, M., & Livingston, A. G. (2010). 2.05—Nanofiltration operations in non-aqueous systems. In E. Drioli & L. Giorno (Eds.), *Comprehensive membrane science and engineering* (pp. 91–113). Elsevier.

Pillay, V. L., Townsend, B., & Buckley, C. A. (1994). Improving the performance of anaerobic digesters at wastewater treatment works: The coupled cross-flow microfiltration/digester process. *Water Science and Technology, 30*(12), 329–337.

Porter, M. C. (1986). Microfiltration. In P. M. Bungay, H. K. Lonsdale, & M. N. de Pinho (Eds.), *Synthetic membranes: Science, engineering and applications* (pp. 225–247). Springer Netherlands.

Priyadarshini, A., Tay, S. W., Ong, P. J., & Hong, L. (2018). Zeolite Y-carbonaceous composite membrane with a pseudo solid foam structure assessed by nanofiltration of aqueous dye solutions. *Journal of Membrane Science, 567*, 146–156. https://doi.org/10.1016/j.memsci.2018.09.025

Purkait, M. K., & Singh, R. (2020). *Membrane technology in separation science.* Taylor & Francis Limited.

Purkait, M. K., Sinha, M. K., Mondal, P., & Singh, R. (2018). *Stimuli responsive polymeric membranes: Smart polymeric membranes.* Elsevier Science.

Qasim, M., Badrelzaman, M., Darwish, N. N., Darwish, N. A., & Hilal, N. (2019). Reverse osmosis desalination: A state-of-the-art review. *Desalination, 459*, 59–104. https://doi.org/10.1016/j.desal.2019.02.008

Ray, P., Singh, P. S., & Polisetti, V. (2020). 2—Synthetic polymeric membranes for the removal of toxic pollutants and other harmful contaminants from water. In M. P. Shah (Ed.), *Removal of toxic pollutants through microbiological and tertiary treatment* (pp. 43–99). Elsevier.

Rekik, S. B., Bouaziz, J., Deratani, A., & Beklouti, S. (2017). Study of ceramic membrane from naturally occurring-kaolin clays for microfiltration applications. *Periodica Polytechnica Chemical Engineering*, *61*(3), 206.

Robeson, L. M. (1991). Correlation of separation factor versus permeability for polymeric membranes. *Journal of Membrane Science*, *62*(2), 165–185.

Robeson, L. M. (2008). The upper bound revisited. *Journal of Membrane Science*, *320*(1–2), 390–400.

Rozelle, L. T., Cadotte, J. E., McClure, D. J., Wong, C. M., Gillam, W. S., Podall, H. E., & Kindley, L. M. (1970). Development of new reverse osmosis membranes for desalination. *U S, Office of Saline Waters' Research and Development Progress Reports*, *531*.

Safarpour, M., Najjarizad-Peyvasti, S., Khataee, A., & Karimi, A. (2022). Polyethersulfone ultrafiltration membranes incorporated with CeO2/GO nanocomposite for enhanced fouling resistance and dye separation. *Journal of Environmental Chemical Engineering*, *10*(3), 107533. https://doi.org/10.1016/j.jece.2022.107533

Saijun, D., Boonsuk, P., & Chinpa, W. (2022). Conversion of polycarbonate from waste compact discs into antifouling ultrafiltration membrane via phase inversion. *Journal of Polymer Research*, *29*(6), 220. doi:10.1007/s10965-022-03073-8

Saja, S., Bouazizi, A., Achiou, B., Ouaddari, H., Karim, A., Ouammou, M., . . . Younssi, S. A. (2020). Fabrication of low-cost ceramic ultrafiltration membrane made from bentonite clay and its application for soluble dyes removal. *Journal of the European Ceramic Society*, *40*(6), 2453–2462.

Saleh, T. A., & Gupta, V. K. (2016). *Nanomaterial and polymer membranes: Synthesis, characterization, and applications*. Elsevier Science.

Sandhya Rani, S. L., & Kumar, R. V. (2021). Insights on applications of low-cost ceramic membranes in wastewater treatment: A mini-review. *Case Studies in Chemical and Environmental Engineering*, *4*, 100149. https://doi.org/10.1016/j.cscee.2021.100149

Schaep, J., Van der Bruggen, B., Vandecasteele, C., & Wilms, D. (1998). Influence of ion size and charge in nanofiltration. *Separation and Purification Technology*, *14*(1), 155–162. https://doi.org/10.1016/S1383-5866(98)00070-7

Schäfer, A., Fane, A. G., & Waite, T. D. (2005). *Nanofiltration: Principles and applications*. Elsevier Science.

Scholz, M., Wessling, M., & Balster, J. (2011). Design of membrane modules for gas separations. *Membrane Engineering for the Treatment of Gases*, *1*, 125–149.

Services, U. S. P. H. S. D. o. E., & Center, R. A. T. S. E. (1960). *Recent developments in water bacteriology: A course conducted by the training program of the Robert a. Taft sanitary engineering center*. U.S. Department of Health, Education, and Wealfare, Public Health Service, Bureau of State Services, Division of Sanitary Engineering Services.

Shah, M. P., & Rodriguez-Couto, S. (2021). *Membrane-based hybrid processes for wastewater treatment*. Elsevier Science.

Shao, L., Cheng, X., Wang, Z., Ma, J., & Guo, Z. (2014). Tuning the performance of polypyrrole-based solvent-resistant composite nanofiltration membranes by optimizing polymerization conditions and incorporating graphene oxide. *Journal of Membrane Science*, *452*, 82–89. https://doi.org/10.1016/j.memsci.2013.10.021

Shimizu, Y., Okuno, Y.-I., Uryu, K., Ohtsubo, S., & Watanabe, A. (1996). Filtration characteristics of hollow fiber microfiltration membranes used in membrane bioreactor for domestic wastewater treatment. *Water Research*, *30*(10), 2385–2392. https://doi.org/10.1016/0043-1354(96)00153-4

Siekierka, A., Smolińska-Kempisty, K., & Wolska, J. (2021). Enhanced specific mechanism of separation by polymeric membrane modification—a short review. *Membranes*, *11*(12), 942.

Singh, R. (2015). Chapter 1—introduction to membrane technology. In R. Singh (Ed.), *Membrane technology and engineering for water purification* (2nd ed., pp. 1–80). Butterworth-Heinemann.

Singh, R., & Hankins, N. P. (2016). Chapter 2—introduction to membrane processes for water treatment. In N. P. Hankins & R. Singh (Eds.), *Emerging membrane technology for sustainable water treatment* (pp. 15–52). Elsevier Science.

Singh, R., & Purkait, M. K. (2019). Chapter 4—microfiltration membranes. In A. F. Ismail, M. A. Rahman, M. H. D. Othman, & T. Matsuura (Eds.), *Membrane separation principles and applications* (pp. 111–146). Elsevier Science.

Snehal Mohite, P. M., & Prasad, E. (2022). *Process water treatment market by technology (activated carbon filters, chlorination, distillation, electrodeionization, ion exchange, microfiltration, nanofiltration, reverse osmosis, ultrafiltration, and others), manufacturing process (boiler make-up water, cooling tower make-up water, coating & plating, rinsing & spraying, washing, and others), and application (municipal, industrial and others): Global opportunity analysis and industry forecast 2021–2030.* www.allied marketresearch.com/process-water-treatment-market-A15544

Spivakov, B., & Shkinev, V. (2005). Membrane techniques | ultrafiltration. In P. Worsfold, A. Townshend, & C. Poole (Eds.), *Encyclopedia of analytical science* (2nd ed., pp. 524–530). Elsevier Science.

Tai, Z. S., Aziz, M. H. A., Othman, M. H. D., Ismail, A. F., Rahman, M. A., & Jaafar, J. (2019). Chapter 8—an overview of membrane distillation. In A. F. Ismail, M. A. Rahman, M. H. D. Othman, & T. Matsuura (Eds.), *Membrane separation principles and applications* (pp. 251–281). Elsevier.

Tan, X., Tan, S., Teo, W. K., & Li, K. (2006). Polyvinylidene fluoride (PVDF) hollow fibre membranes for ammonia removal from water. *Journal of Membrane Science, 271*(1–2), 59–68.

Van der Bruggen, B., Mänttäri, M., & Nyström, M. (2008). Drawbacks of applying nanofiltration and how to avoid them: A review. *Separation and Purification Technology, 63*(2), 251–263. https://doi.org/10.1016/j.seppur.2008.05.010

Van der Bruggen, B., Schaep, J., Wilms, D., & Vandecasteele, C. (1999). Influence of molecular size, polarity and charge on the retention of organic molecules by nanofiltration. *Journal of Membrane Science, 156*(1), 29–41. https://doi.org/10.1016/S0376-7388(98)00326-3

Van Lente, H., & Rip, A. (1998). The rise of membrane technology: From rhetorics to social reality. *Social Studies of Science, 28*(2), 221–254.

Vatanpour, V., Esmaeili, M., & Farahani, M. H. D. A. (2014). Fouling reduction and retention increment of polyethersulfone nanofiltration membranes embedded by amine-functionalized multi-walled carbon nanotubes. *Journal of Membrane Science, 466*, 70–81. https://doi.org/10.1016/j.memsci.2014.04.031

Vatanpour, V., Madaeni, S. S., Moradian, R., Zinadini, S., & Astinchap, B. (2011). Fabrication and characterization of novel antifouling nanofiltration membrane prepared from oxidized multiwalled carbon nanotube/polyethersulfone nanocomposite. *Journal of Membrane Science, 375*(1), 284–294. https://doi.org/10.1016/j.memsci.2011.03.055

Vrentas, J., & Vrentas, C. (2002). Transport in nonporous membranes. *Chemical Engineering Science, 57*(19), 4199–4208.

Vrouwenvelder, J., & Kruithof, J. (2011). *Biofouling of spiral wound membrane systems.* Iwa Publishing.

Wagner, J. (2001). *Membrane filtration handbook: Practical tips and hints* (Vol. 129). Osmonics.

Wang, K., Wang, S., Gu, K., Yan, W., Zhou, Y., & Gao, C. (2022). Ultra-low pressure PES ultrafiltration membrane with high-flux and enhanced anti-oil-fouling properties prepared via in-situ polycondensation of polyamic acid. *Science of the Total Environment, 842*, 156661. https://doi.org/10.1016/j.scitotenv.2022.156661

Wang, X., Xiao, Q., Wu, C., Li, P., & Xia, S. (2021). Fabrication of nanofiltration membrane on MoS2 modified PVDF substrate for excellent permeability, salt rejection, and structural stability. *Chemical Engineering Journal, 416*, 129154. https://doi.org/10.1016/j.cej.2021.129154

Wanke, D., da Silva, A., & Costa, C. (2021). Modification of PVDF hydrophobic microfiltration membrane with a layer of electrospun fibers of PVP-co-PMMA: Increased fouling resistance. *Chemical Engineering Research and Design, 171*, 268–276. https://doi.org/10.1016/j.cherd.2021.05.004

Water Desalination Equipment Market Size, share & trends analysis report by technology (RO, MSF, MED), by source (seawater, river water), by application (municipal, industrial), by region, and segment forecasts, 2020–2027. (2020). www.grandviewresearch.com/industry-analysis/water-desalination-equipment-market

Wei, S., Chen, Y., Hu, X., Wang, C., Huang, X., Liu, D., & Zhang, Y. (2020). Monovalent/Divalent salts separation via thin film nanocomposite nanofiltration membrane containing aminated TiO2 nanoparticles. *Journal of the Taiwan Institute of Chemical Engineers, 112*, 169–179. https://doi.org/10.1016/j.jtice.2020.06.014

Woodard, F., & Curran, I. (2006). 2—Fundamentals. In F. Woodard & I. Curran (Eds.), *Industrial waste treatment handbook* (2nd ed., pp. 29–49). Butterworth-Heinemann.

Wu, L., Wang, H., Xu, T. W., & Xu, Z. L. (2017). Chapter 12—polymeric membranes. In L. Y. Jiang & N. Li (Eds.), *Membrane-based separations in metallurgy* (pp. 297–334). Elsevier.

WWAP. (2018). *The united nations world water development report 2018: Nature-based solutions.* WWAP.

Xia, S., Nan, J., Liu, R., & Li, G. (2004). Study of drinking water treatment by ultrafiltration of surface water and its application to China. *Desalination, 170*(1), 41–47. https://doi.org/10.1016/j.desal.2004.03.014

Xie, F., Li, W.-X., Gong, X.-Y., Taymazov, D., Ding, H.-Z., Zhang, H., . . . Xu, Z.-L. (2022). MoS2 @PDA thin-film nanocomposite nanofiltration membrane for simultaneously improved permeability and selectivity. *Journal of Environmental Chemical Engineering, 10*(3), 107697. https://doi.org/10.1016/j.jece.2022.107697

Yang, G. C. C., Yang, T.-Y., & Tsai, S.-H. (2003). Crossflow electro-microfiltration of oxide-CMP wastewater. *Water Research, 37*(4), 785–792. https://doi.org/10.1016/S0043-1354(02)00388-3

Yang, Z., Guo, H., & Tang, C. Y. (2019). The upper bound of thin-film composite (TFC) polyamide membranes for desalination. *Journal of Membrane Science, 590*, 117297. https://doi.org/10.1016/j.memsci.2019.117297

Ye, Z., Zhang, Y., Hou, L.-A., Zhang, M., Zhu, Y., & Yang, Y. (2022). Preparation of a GO/PB-modified nanofiltration membrane for removal of radioactive cesium and strontium from water. *Chemical Engineering Journal, 446*, 137143. https://doi.org/10.1016/j.cej.2022.137143

Yoon, K., Kim, K., Wang, X., Fang, D., Hsiao, B. S., & Chu, B. (2006). High flux ultrafiltration membranes based on electrospun nanofibrous PAN scaffolds and chitosan coating. *Polymer, 47*(7), 2434–2441.

Yu, C., Gao, B., Wang, W., Xu, X., & Yue, Q. (2019). Alleviating membrane fouling of modified polysulfone membrane via coagulation pretreatment/ultrafiltration hybrid process. *Chemosphere, 235*, 58–69.

Yuan, X., Wang, L., Wu, P., Ji, P., Sheffield, J., & Zhang, M. (2019). Anthropogenic shift towards higher risk of flash drought over China. *Nature Communications, 10*(1), 4661–4661. doi:10.1038/s41467-019-12692-7

Zare, S., & Kargari, A. (2018). 4—Membrane properties in membrane distillation. In V. G. Gude (Ed.), *Emerging technologies for sustainable desalination handbook* (pp. 107–156). Butterworth-Heinemann.

Zhang, H., Liu, M., Sun, D., Li, B., & Li, P. (2016). Evaluation of commercial PTFE membranes for desalination of brine water through vacuum membrane distillation. *Chemical Engineering and Processing: Process Intensification, 110*, 52–63.

Zhang, W., Xu, H., Xie, F., Ma, X., Niu, B., Chen, M., . . . Long, D. (2022). General synthesis of ultrafine metal oxide/reduced graphene oxide nanocomposites for ultrahigh-flux nanofiltration membrane. *Nature Communications, 13*(1), 471. doi:10.1038/s41467-022-28180-4

Zhang, Y., Duan, X., Tan, B., Jiang, Y., Wang, Y., & Qi, T. (2022). PVDF microfiltration membranes modified with AgNPs/tannic acid for efficient separation of oil and water emulsions. *Colloids and Surfaces A: Physicochemical and Engineering Aspects, 644*, 128844. https://doi.org/10.1016/j.colsurfa.2022.128844

Zhao, C., Xue, J., Ran, F., & Sun, S. (2013). Modification of polyethersulfone membranes–A review of methods. *Progress in Materials Science, 58*(1), 76–150.

Zubia, A., Abbas, A., Mahmood, A., Nosheen Fatima, R., Khan, S. J., Duclaux, L., . . . Ahmad, N. M. (2022). Water treatment using high performance antifouling ultrafiltration polyether sulfone membranes incorporated with activated carbon. *Polymers, 14*(11), 2264. https://doi.org/10.3390/polym14112264

Zuo, K., Chen, M., Liu, F., Xiao, K., Zuo, J., Cao, X., . . . Huang, X. (2018). Coupling microfiltration membrane with biocathode microbial desalination cell enhances advanced purification and long-term stability for treatment of domestic wastewater. *Journal of Membrane Science, 547*, 34–42. https://doi.org/10.1016/j.memsci.2017.10.034

2 Fouling and Related Phenomena

2.1 INTRODUCTION

The gradual domination of membrane technology for water treatment and desalination applications from the 1960s to date has not been without challenges. Of these, one that has greatly troubled scientists and engineers is the eventual decline in membrane performance over time due to membrane fouling. Membrane fouling occurs when the membrane surface or pores are blocked due to sieving and adsorption of particulates and compounds, or foulants (Mallevialle et al., 1996). Membrane fouling limits productivity and efficiency (Du et al., 2019; Munla et al., 2012) and is universal to all membrane-based liquid separation processes. Fouling in pressure-driven systems causes a significant increase in hydraulic resistance and is observed either as permeate flux decline or as an increase in transmembrane pressure. In later chapters, we will touch upon the practicality of using these indicators to detect fouling. In order to improve processes for fouling detection and control, one must first familiarize with the types and mechanisms of fouling in membrane processes.

Flux decline behavior can typically be broken down into three stages, as shown in Figure 2.1. In the first stage, a rapid decline of flux is observed as some particles may be blocked in the membrane pore or on the surface, which reduces the effective filtration area (Sun et al., 2018). In the second stage, flux decline continues as fouling increases. In the third stage, flux only decreases slightly with time as steady-state is reached.

Membrane fouling is a complex interfacial phenomenon governed by interactions among membrane surface, foulants, and water, as well as foulant mass transport toward the membrane (Cirillo et al., 2021; H. Xu et al., 2020; Zainith et al., 2021). Fouling is the process through which unwanted feed constituents are deposited or adsorbed onto membrane pores or the membrane surface by physical, chemical, and/or biological interactions (Lei et al., 2020). Depending on the type and intensity, fouling can be dealt with by backwashing, chemical cleaning, and eventually replacement of the membrane element (Lingling Liu et al., 2019). For the remainder of the chapter, we consider fouling in high pressure filtration processes such as nanofiltration (NF), reverse osmosis (RO), low pressure filtration processes such as microfiltration (MF) and ultrafiltration (UF), and emerging technologies such as membrane distillation (MD).

Membrane fouling is a dynamic process that occurs in phases and consists of several complex surface and hydrodynamic interactions (C. Liu, 2014; Lohaus et al., 2018). Three mechanisms of mass transport are responsible for membrane fouling: permeation drag, shear-induced diffusion (orthokinetic diffusion) and Brownian diffusion (C. Liu, 2014; Wiesner & Aptel, 1996). Permeate drag force is the force

DOI: 10.1201/9781003144991-2

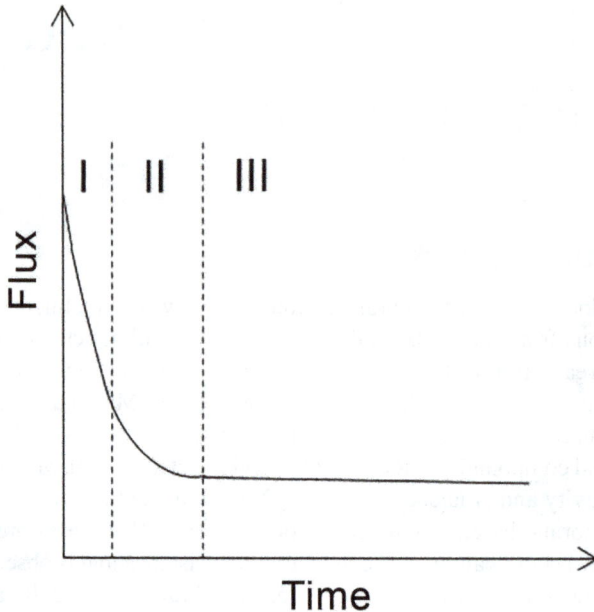

FIGURE 2.1   Three stages of flux decline: rapid initial drop, longer-term decline, and quasi-steady-state period.

exerted by the flow of water permeating through the membrane on the foulant and is responsible for transport of foulants toward the membrane (Ramon & Hoek, 2012). Brownian diffusion is the diffusion caused by the collision of particles with water molecules (C. Liu, 2014). Feed characteristics and particularly the size of the particles present can help determine the dominant mass transport mechanism and therefore the propensity to foul, which enables appropriate pretreatment processes to be chosen in order to minimize fouling (C. Liu, 2014).

Fouling is typically categorized as either colloidal, organic, inorganic, or biofouling. Inorganic fouling is also referred to as scaling. Figure 2.2 shows a schematic of various fouling types. Other fouling-related phenomena that are specific to the type of process include concentration polarization, which occurs predominantly in high pressure filtration, and wetting, which is a major drawback in membrane distillation. In this chapter, the reader will also be introduced to each of these phenomena, including causes, mechanism, and effects.

2.2 PRESSURE-DRIVEN MEMBRANE PROCESSES

2.2.1 High Pressure Membrane Processes (NF, RO)

Fouling is arguably the single largest operational challenge in RO systems and is directly linked to reduction in plant efficiency and increased cost of the product water (Melián-Martel et al., 2012). To iterate the significance of RO fouling, one

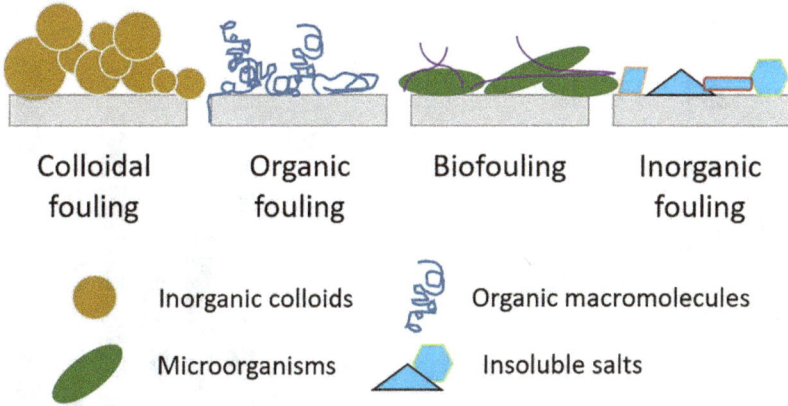

FIGURE 2.2 Types of fouling.

need not look beyond the ample and ever increasing literature found on this topic (Figure 2.3). Membrane fouling is recognized as a major challenge restraining the growth of the RO market today (Insights, 2021), especially under harsh feedwater conditions. In fact, the RO membrane chemicals market alone, which focuses on the prevention and control of membrane fouling and scaling, is set to reach USD 2751 million by 2025 with a 7.77 CAGR (Ocean, 2022). Simultaneously, the market for fouling-resistant compound RO membranes is also expected to grow (Reports, 2021). Jafari et al. estimated fouling costs as approximately 24% of OPEX for surface water RO (Jafari et al., 2021). This include costs of increases in feed channel pressure drop, reduction in water permeability, early membrane replacement, and extensive cleaning-in-place (CIP). According to some reports, membrane cleaning is considered inevitable and contributes to as much as 50% of RO operational costs (Beyer et al., 2017; Ridgway, 2003).

A relatively new definition of RO membrane fouling proposed by Hoek et al. categorizes fouling of RO membranes as external or internal (E. M. V. Hoek et al., 2008). External, or surface, fouling consists of (1) scale formation, (2) cake formation, and (3) biofilm formation, or a combination of these. Internal fouling in RO membranes shows itself as a change in solute and solvent transport caused by an unintended modification in membrane structure due to physical compaction or chemical degradation. As RO membranes are compact and non-porous, surface fouling on the active layer of the thin film composite membrane is the dominant class of fouling in RO systems (Greenlee et al., 2009).

2.2.1.1 Colloidal Fouling

Colloidal or particulate fouling is caused by the deposition of particulates on the membrane surface (Singh, 2015). Colloidal fouling of RO membranes reduces productivity and can also have an adverse effect on salt rejection (Ismail et al., 2019). Some examples of colloids found in aqueous streams include clay minerals, colloidal silica, iron, aluminum, and manganese oxides (Ismail et al., 2019). Colloidal fouling

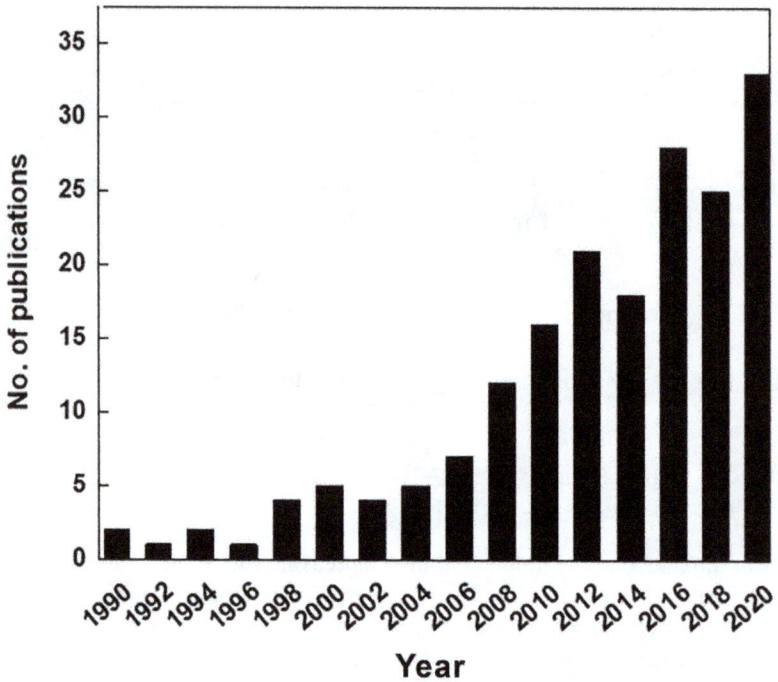

FIGURE 2.3   Increase in publications on reverse osmosis fouling between 1990 and 2020.

Source: Scopus.

consists of particulates on the membrane surface that gradually form a cake layer; this cake layer provides an additional hydraulic resistance to water flow (X. Zhu & Elimelech, 1995).

Buffle et al. point out that colloids found in natural waters fall under three main groups (Buffle & Leppard, 1995; Buffle et al., 1998): (1) inorganic colloids, which include iron oxides/hydroxides, elemental sulfur and metal sulfides, and various aluminum silicate minerals; (2) organic macromolecules, which include polysaccharides, large proteins, and NOM; (3) biocolloids such as bacteria, microorganisms, viruses, etc. Colloidal fouling may also be in the form of large organics as is the case in oily wastewaters.

Colloids in aqueous solutions range from 1 to 1000 nm in particle size (Stumm, 1993), which makes them disposed to cause fouling. Larger particles can be removed by lateral migration or shear-induced diffusion, whereas smaller particles can diffuse away from the membrane surface via molecular diffusion (Bacchin et al., 2006).

In RO membranes, the rate of colloidal fouling depends on the coupling of physical and chemical interactions (Ismail et al., 2019). The dynamics of cake layer formation are strongly dependent on permeation drag force as well as particulate size (Ismail et al., 2019). Apart from routine tests for conductivity, pH, turbidity, and dissolved oxygen, some parameters help assess the fouling potential of feed solutions.

Literature on colloidal fouling is plentiful for low pressure filtration, and recently a growing number of systematic studies have led to a deeper understanding of colloidal fouling in NF and RO as well (E. M. Hoek & Elimelech, 2003; C. Y. Tang et al., 2009). Many of these include membrane autopsy studies that reveal the nature and characteristics of foulants and the properties of foulant layers in NF and RO (Fortunato et al., 2020; F. Tang et al., 2014; P. Xu et al., 2010), as is elaborated in Chapter 3. Additionally, laboratory studies have been carried out to elucidate the relationship between membrane properties, hydrodynamic conditions, and solution chemistry with respect to colloidal fouling.

The interaction of colloidal matter and eventually the propensity of colloidal fouling is dependent on the size, shape, charge, and feedwater ions. Many major inorganic colloids including silica and aluminum silicates are negatively charged in water at neutral pH (Buffle et al., 1998). AFM studies have shown that in the initial stages of NF/RO fouling, rough surfaces are more prone to colloidal deposition as particles preferentially aggregate in the gaps, resulting in "valley clogging" and causing more severe flux decline than in membranes with smooth surfaces (Vrijenhoek et al., 2001). More recently however, Jiang and coworkers showed that the peak-valley morphology of rough TFC membranes does not affect flux decline due to fouling; instead, initial flux is the dominating factor (Jiang et al., 2020).

2.2.1.1.1 DLVO Theory of Colloidal Interactions

The attachment of a colloidal particle to a surface is widely described by the Derjaguin–Landau–Verwey–Overbeek (DLVO) theory of colloidal dispersion stability (Missana & Adell, 2000; Ortega-Vinuesa et al., 1996). The DLVO theory describes whether colloidal particles flocculate or coalesce. According to the DLVO theory, the total interaction energy is the sum of van der Waals and electric double layer interactions (Elimelech & O'Melia, 1990).

Assuming a constant surface potential boundary condition, the electric double layer interaction energy E_{EDL} is given by:

$$E_{EDL} = \pi \epsilon_0 \epsilon_r a_p \left[2\xi_p \xi_m \ln\left(\frac{1+e^{-\frac{d_s}{\lambda_d}}}{1-e^{-\frac{d_s}{\lambda_d}}}\right) + \left(\xi_p^2 + \xi_m^2\right) \ln\left(1-e^{-\frac{2d_s}{\lambda_d}}\right) \right] \quad \text{Equation 2.1}$$

where ϵ_0 is the permittivity of vacuum, and ϵ_r is the relative permittivity of water; ζ_p and ζ_m are the zeta potentials of the particle and membrane surface, respectively; λ_d is the Debye length, which is inversely proportional to the square root of the solution ionic strength I_S (λ_d is ~ 10, 3, and 1 nm, respectively for ionic strengths of 1, 10, and 100 mM) (Stumm, 1993; C. Y. Tang et al., 2011).

The van der Waals attractive energy between a spherical particle and an infinite planar surface is given by (Gregory, 2005; Hamaker, 1937):

$$E_{VDW} = \frac{A_H}{6}\left[\frac{a_p}{d_s} + \frac{a_p}{d_s+2a_p} + \ln\left(\frac{d_s}{d_s+2a_p}\right)\right] \quad \text{Equation 2.2}$$

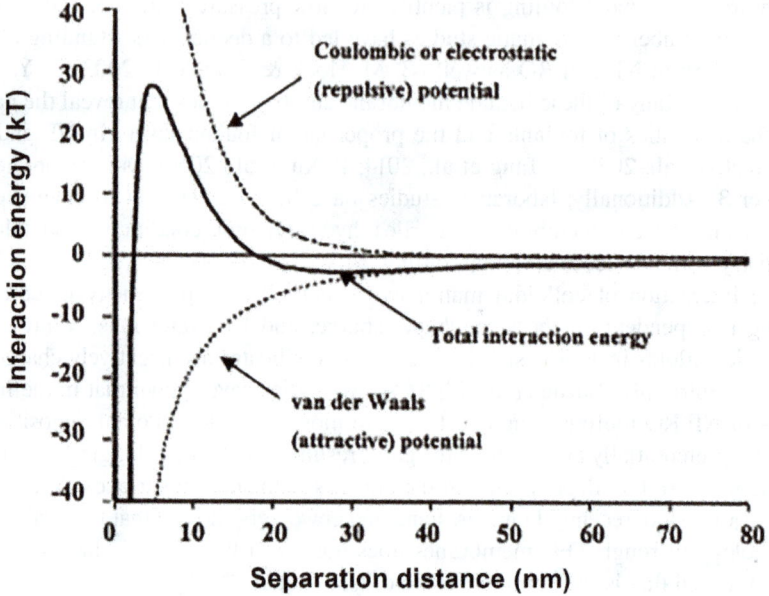

FIGURE 2.4   Schematic of interaction energy vs. distance profiles for DLVO interaction.

Source: Adair, J. H., Suvaci, E., & Sindel, J. (2001). Surface and colloid chemistry. In K. H. J. Buschow, R. W. Cahn, M. C. Flemings, B. Ilschner, E. J. Kramer, S. Mahajan, & P. Veyssière (Eds.), *Encyclopedia of materials: Science and technology* (pp. 1–10). Elsevier. Reprinted with permission.

where A_H is the Hamaker constant, and a_p is the particle radius. Van der waals forces are not sensitive to ionic strength (Adair et al., 2001). Figure 2.4 shows a schematic of the interaction energy vs. distance profiles of the DLVO interaction. The total interaction energy is a sum of the van der Waals energy and the electrostatic double layer potential energy.

The sum of the two energy components gives rise to an energy barrier that needs to be overcome before the particle coalesces to the surface. The frequency of particle collision and the attachment coefficient (i.e. ratio of successful attachment over total number of collision events) determine the rate of colloidal aggregation, according to classical colloidal stability and coagulation theories (Filella, 2007; C. Y. Tang et al., 2011). The attachment coefficient depends strongly on the energy barrier and the permeate flux, which is related to hydrodynamic drag force as discussed in Section 2.1.

Many factors affect the interaction energy, including particle size, surface charge, and surface roughness of both the colloids and the membrane. Solution chemistry, i.e. pH and ionic strength and the interaction of specific ions present in the water with colloids, also plays an important role. In fact, solution chemistry may affect surface charge. With respect to ionic strength, the double layer interaction energy is low at higher ionic strength solutions. When particles are under electrostatic repulsion, a higher ionic strength will reduce the size of the electrical double layer and therefore reduce the energy barrier required for attachment to the surface. Under high ionic strength environments, acid–base interactions may dominate colloidal interactions (Jin et al., 2009). When the distance between surface and colloidal particles is less than a few nanometers, the DLVO theory fails to explain interactions as solvation/hydration, and steric and hydrophobic forces take over (Ninham, 1999). For a more detailed description of the role of these forces in determining colloidal fouling, the reader is referred to this review on colloidal fouling (V. Kochkodan et al., 2014).

2.2.1.1.2 Silt Density Index (SDI)

The silt density index (SDI) is commonly used to predict the colloidal fouling potential of RO or NF feedwater (Alhadidi et al., 2011). SDI relies on measuring the time required to filter a fixed volume of feedwater through a 0.45 µm MF membrane at a constant pressure of 207 kPa in dead-end mode (Salinas Rodriguez et al., 2019). SDI is calculated as the percentage decrease in filtration rate per minute, from the difference between the initial measurement and a second measurement taken at 15 minutes. RO membrane manufacturers recommend feed SDI values under 3–5 without pretreatment and recommend pretreatment if SDI > 5 (Dennis E. Williams, 2018). Carefully designed intake and pretreatment systems ensure that the feed meets SDI requirements to prevent fouling (Dennis E Williams, 2011).

2.2.1.2 Organic Fouling

Organic fouling is caused mainly by natural organic matter (NOM) present in feedwater (Sangyoup Lee et al., 2005; Seidel & Elimelech, 2002). Although NOM is a heterogeneous mixture that contains a wide range of molecular weight and functional groups, it consists primarily of such components as humic and fulvic acids (Zularisam et al., 2006). Humic substances contain both aromatic and aliphatic components with mainly carboxylic and phenolic functional groups. Due to the presence of carboxylic functional groups, humic substances carry a negative charge above a pH of 5 (Newcombe, 1999; Sillanpää et al., 2015). At acidic pH levels, organic fouling can be enhanced due to the neutral charge of humic acid at low pH values (Al-Amoudi, 2010).

Physical, chemical, and electrostatic interactions between the organic compounds and the membrane induce organic fouling (Fane et al., 2006). Chemical adsorption is more difficult to remove compared to physical adsorption due to the strong attachment between foulant and membrane. Electrostatic interactions depend mainly on the charged functional groups and presence of other ionic species in the feed. Solution chemistry, specifically pH and the presence of divalent cations, play an important role on NOM adsorption (Ang et al., 2006; S. Hong & Elimelech, 1997; Sangyoup Lee et al., 2006; Q Li & M Elimelech, 2004a, 2004b).

Common non-humic NOM foulants include proteins, amino sugars, polysaccharides, and polyoxyaromatics (Wiesner et al., 1992). The complexity of organic fouling arises from different compounds causing different kinds of fouling (Abdelrasoul et al., 2013; Moonkhum et al., 2010). Factors that affect membrane fouling in the presence of NOM include membrane surface properties; chemistry of feed solution, i.e. ionic strength, pH, concentration of both monovalent and divalent ions, NOM composition; and operating conditions such as permeate flux and pressure (Al-Amoudi, 2010; Sangyoup Lee et al., 2006; Q. Li et al., 2007).

Measurement of foulant–foulant and foulant–membrane interactions using atomic force microscopy (AFM) can help predict the rate of organic fouling as studies indicate a strong correlation between measured values of interaction forces and rate of fouling with NOM (Sangyoup Lee & Elimelech, 2006; C. Y. Tang et al., 2009). An AFM investigation by Li and Elimelech revealed that the presence of divalent calcium ions enhances NOM fouling of NF membranes by complexation and formation of intermolecular bridges among organic foulant molecules (Qilin Li & Menachem Elimelech, 2004).

Researchers agree that identifying the composition of NOM present in the feed solution will lead to enhanced fouling mitigation measures in the form of appropriate pretreatment processes to reduce NOM levels, as well as optimization of membrane surface characteristics (Anjum et al., 2019). Still, the complexity of NOM components makes it challenging to devise a single fouling remediation strategy; instead, each type of foulant may need to be dealt with separately.

2.2.1.3 Inorganic Fouling

Inorganic fouling, also known as (mineral) scaling, is the deposition of sparingly soluble salts on the membrane surface (Matin et al., 2021), caused by the concentration of dissolved salts beyond their solubility limits. This occurs through two mechanisms: crystallization and particulate fouling. In the former, crystals form and grow at the membrane surface, while the latter consists of the formation of crystals in the bulk solution, which are then transported to the membrane surface by convection (Sangho Lee & Lee, 2000). The formation of salt crystals begins with nucleation of the salt ions that eventually come together to form salt crystals. Pervov identified crystallization within the feed solution followed by transport via convection as the dominant route of inorganic fouling with $CaCO_3$ and $CaSO_4$ on RO membranes (Pervov, 1991). Others have attributed scaling to crystallization at the membrane surface (Uchymiak et al., 2008).

The three steps in scale formation are (1) saturation and aggregation, (2) nucleation, and (3) crystal growth (Matin et al., 2019). The formation of calcium carbonate ($CaCO_3$) and gypsum ($CaSO_4 \cdot 2H_2O$) reduces membrane permeability and causes performance to deteriorate over time. Bhattacharjee et al. point out that the nature of flux decline observed during scaling differs from fouling curves obtained during colloidal fouling or flux decline due to concentration polarization (Bhattacharjee & Johnston, 2002). High recovery rates increase the prospect of precipitation and deposition of sparingly soluble salts present in the feedwater, which eventually affects process efficiency (Jawor & Hoek, 2009). Concentration polarization, i.e. the formation of a concentration gradient of salts near the membrane surface, increases the likelihood of scale formation on the membrane. High permeate flux and low cross-flow velocities enhance this concentration gradient, which is likely to induce membrane scaling (Tzotzi et al., 2007). Often, calcium and magnesium ions in the feed form sparingly soluble precipitates when carbonate and sulfate ions are present (Bhattacharjee & Johnston, 2002). Matin et al. explain that mineral salts such as $CaCO_3$, $CaSO_4$, and $BaSO_4$ are mostly present near their saturation levels in natural brackish water feed solutions (Matin et al., 2019). This makes inorganic fouling likely even at moderate recovery rates

(S. Zhao et al., 2012). When these salts exceed their solubility limit, the rate of deposition exceeds the removal rate, causing a foulant layer to form on the surface (Qureshi et al., 2013). The foulant layer increases in thickness as more scale formation occurs. This increases mass transfer resistance and reduces permeate flux and eventually shortens membrane life (Tu et al., 2011). For this reason, scaling potential is a considerable design obstacle in the way of operating at high water recovery. Salt crystals may deposit on the membrane itself as well as on the feed spacers, causing an increase in pressure drop (Amiri & Samiei, 2007). Spacers may enhance local scale formation. Removal of inorganic foulants requires system downtime and increased maintenance costs for membrane cleaning (Matin et al., 2019). In RO membranes, inorganic fouling can increase operation cost by 10–15% (Matin et al., 2019). The degree of scaling is also a function of precipitation kinetics (Au et al., 2007) and the abundance of nucleation sites (Bartman et al., 2011).

Despite the higher concentration of ions in seawater, brackish water RO is more prone to scale formation due to the higher recoveries used. In seawater RO, $CaCO_3$ is the prevalent source of scale formation (Matin et al., 2019). In brackish water RO, where recovery rate ranges from 70 to 90%, $CaCO_3$ and $CaSO_4$ are both prevalent inorganic foulants (Ochando-Pulido et al., 2015). Silica also contributes to scale formation as it is present near saturation levels in groundwater (Cob et al., 2012).

As with other types of fouling, inorganic fouling depends on several factors including membrane characteristics, module geometry, feed solution characteristics, and operating conditions (Shirazi et al., 2010). To better understand how fouling occurs in RO systems, one must consider the behavior of water and ion transport through the RO membrane, for which the reader is referred to Section 1.3.4.

2.2.1.4 Biofouling

Biofouling is the development of a colony of microorganisms on the membrane surface (Panchal & Knudsen, 1998). Biofilm formation is the primary step through which biofouling occurs (Patel et al., 2021). A biofilm is a community of microorganisms attached to a surface through a polymeric matrix known as an extracellular polymeric substance (EPS) (Carpentier & Cerf, 1999; Kannan et al., 2017).

Biofouling, or the growth of pathogenic or other microbes, is arguably the most difficult type of fouling. It has been described as the "Achilles heel" of membrane systems, due to the ability of microorganisms to multiply and thus the possibility of growth even after efficient removal (Hillis, 2000).

Initial attachment of bacteria on membrane surfaces is determined by both physicochemical factors and bacterial properties. Physicochemical factors include solution chemistry, surface properties of the membrane, and hydrodynamic conditions (Eshed et al., 2008). On the other hand, cellular adhesiveness resulting from production of specific surface proteins, bacterial motility, etc. are some bacterial factors that affect adhesion and subsequent biofilm growth on the surface of the membrane (Eshed et al., 2008). As with colloidal fouling, the permeation drag force supports adhesion, but in the case of biofouling, it also impacts biofilm development by (1) speeding up and enhancing cell attachment, (2) increasing convective migration of nutrients, and (3) improving the removal of metabolites (Eshed et al., 2008). Figure 2.5 shows the stages of biofilm development.

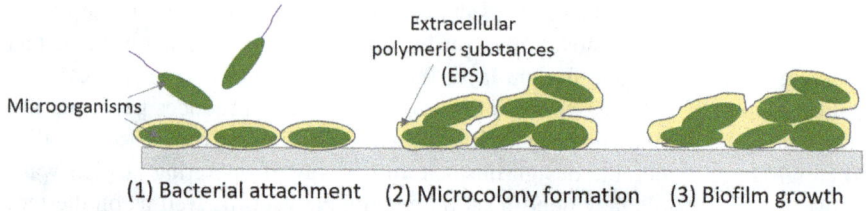

FIGURE 2.5   Stages of biofilm development on a surface.

Despite much progress, the mechanisms of biofouling, especially under different membrane configurations, are not yet fully understood due to the complexity of hydrodynamic conditions and biofilm growth (Vanysacker et al., 2014; Minglu Zhang et al., 2011). For example, the effect of microscale variabilities on full-scale biofilm systems is not clear (Bishop, 2007). While some researchers have linked biofilm development to membrane surface properties such as roughness and hydrophobicity (AlSawaftah et al., 2021), suggesting that hydrophobic membranes are more prone to biofouling, others have found that the surface properties of RO membrane are not linked to biofouling development (Baek et al., 2011). One must not be discouraged by the disparities that warrant the need for further research into the dynamics of biofilm development, especially interactions with the membrane and other foulants.

The mechanics of biofouling are further complicated by mutual interactions between different types of fouling, as discussed by Singh (Singh, 2015). An example of this is the effect of biofouling on mineral scaling and vice versa. Kumar et al. state that while measures can be taken to mitigate the effects of biofouling, there is no specific technique to eliminate it completely or to identify and measure its rate (P. Kumar et al., 2018). Kochkodan et al. add that while surface modification may reduce organic fouling, even the most resistant modification does not prevent biofouling once foulants have been deposited on the surface (V. Kochkodan & Hilal, 2015). This is largely due to foulant-deposited-foulant dominating the progression of fouling. They suggest the use of a sacrificial protective antifouling layer in the form of a polyelectrolyte coating that can be removed from the membrane surface and replaced once fouling occurs (V. Kochkodan & Hilal, 2015). Table 2.1 shows the typical composition of a biofilm.

Several surface modification strategies have been employed to prevent biofouling of thin film composite (TFC) membranes. One approach involves imparting PEGylated materials into the active layer as PEG is highly hydrophilic and is able to form hydrogen bonds with water to lower the interaction of the membrane surface with foulants (Gol et al., 2014). Surface coating with polymers containing hydrophilic end groups such as dendritic or hyperbranched polymers has also been applied to enhance the protein resistance of the membrane surface (Sarkar et al., 2010). Another class of materials that has attracted interest in fouling-resistant RO membranes is zwitterionic polymers (Azari & Zou, 2012). Zwitterionic polymers form a class of materials containing the same number of cations and anions along their polymer chains (L. Zheng et al., 2017). Their unique structure makes them highly hydrophilic and imparts antifouling property; they can resist protein adsorption, bacterial

TABLE 2.1

Typical Composition of a Biofilm (Baker & Dudley, 1998).

Parameter	Composition
Moisture content of dried deposit	>90%
Total organic matter (TOM)	>50%
Humic substances as % of total organic matter	≤40%
Microbiological counts	$>10^6$ cfu cm^{-2}

adhesion, and biofilm formation (Cheng et al., 2007), making them a promising material for antibiofouling modification. Another approach to prevent biofilm growth is through the immobilization of antimicrobial polymers as biocidal agents on the TFC membrane surface (Misdan et al., 2016). These polymers include polyamino acids (Nguyen et al., 2013) and tertiary and/or quaternary ammonium-group-containing polymers (T. Zhang et al., 2014; M.-M. Zhu et al., 2021). Antimicrobial activity can also be enhanced by incorporating biocidal nanoparticles that inactivate microorganisms and prevent biofilm growth (Rahaman et al., 2014). Common materials include Ag (Karkhanechi et al., 2013; Linhares et al., 2020; Madhavan et al., 2014) and Cu nanoparticles (Ben-Sasson et al., 2016; W. Ma et al., 2016). As we see in later chapters, carbon nanotubes (CNTs) and graphene-based materials are also used in the active layer of the membrane to prevent biofouling. Recent developments in the preceding strategies can be found in (Misdan et al., 2016).

Fouling can be minimized with pretreatment, and its effects are mitigated by cleaning and eventual replacement of the membrane element. In high pressure membrane processes, mitigation often involves process interruption and the use of harsh chemicals, both of which also adversely affect costs. It comes as no surprise, then, that academic and industrial researchers are focusing their efforts to develop solutions that prevent and/or mitigate RO membrane fouling. In parallel, extraordinary advances in automation and artificial intelligence (Bhargava & Sharma, 2021; Dash et al., 2016; Wirtz et al., 2019) dictate that even the well established process of RO desalination needs to accommodate newer, faster, and more accurate prediction and control tools to tackle fouling. Therefore, overcoming this grand challenge in RO systems today targets three main aspects:

1. An increased and comprehensive understanding of foulant–foulant, membrane–foulant, membrane–foulant–ion, foulant–ion interactions in realistic feed solutions
2. Development of low fouling and antifouling membrane materials and efficient pretreatment systems
3. Use of automation and artificial intelligence to replace outdated methods and introduce real-time prediction and fouling control tools

Electrically conductive membrane systems have already contributed to advances in the two latter areas, as we will see in Chapters 5, 6, and 7.

2.2.2 Low Pressure Membranes (MF, UF)

Membrane fouling remains a challenge in low-pressure processes such as microfiltration and ultrafiltration, both of which are commonly used for drinking water treatment, for wastewater treatment, and as pretreatment for RO feed. The water flux for ultrafiltration membranes can drop from over 500 LMH for pure water to 50 LMH for real seawater containing colloids, macromolecules, bacteria, and viruses (Gilabert-Oriol, 2021). This drastic decline in filtration performance results from the deposition/adsorption of foulants on the membrane or within its pores. The eventual development of a gel-like cake layer adds mass transfer resistance to the filtration. Contrary to NF and RO, most low pressure membranes suffer from internal fouling resulting from adsorption and clogging of the pores (AlSawaftah et al., 2021; Fan et al., 2008; Jim et al., 1992). In this section, we consider the mechanisms of fouling in low pressure separation processes.

Both MF and UF rely on a size exclusion mechanism for separation (Filippov et al., 1994), making the membrane pore size distribution an important factor in determining separation performance. Pore size not only plays a role in separation efficiency but can also help in understanding the evolution of fouling (Miyoshi et al., 2015). Findings by Meng et al. confirm that the effective dimension of foulants determines the fouling mechanism, when studying the fouling propensity and mechanism specifically of polysaccharides during MF (Meng et al., 2018).

Three sequential mechanisms for fouling exist (typically in this order): partial pore blockage, complete pore blockage, and cake formation (J. W. Chew et al., 2020; Duclos-Orsello et al., 2006; Hakami et al., 2020; L. Huang & Morrissey, 1998; Mora et al., 2019). First, partial pore blockage reduces the pore size. Next, pore blockage prevents further internal fouling. Finally, in the late stages of fouling, a cake layer is formed on the membrane surface (Duclos-Orsello et al., 2006; Enfrin et al., 2020; Gul et al., 2021). Partial blockage occurs when foulant molecules are similar in size to the membrane pores. More severe fouling can result in the form of pore blockage if the membrane pores are larger than the foulant molecules (Mora et al., 2019). Typically, the cake layer formation described in Section 2.2.1 is prevalent in MF and UF when foulant molecules are larger than the membrane pores, which can also take place due to pore blockage from the other two mechanisms. However, optimal pore size and operating conditions for fouling mitigation depend strongly on the membrane material, as shown by Miyoshi et al. (Miyoshi et al., 2015). Furthermore, for the same size colloidal foulants, flux decline behavior is also affected by the zeta potentials of foulants, as illustrated by Trinh et al. (Trinh et al., 2020).

Factors that affect membrane fouling include (Gul et al., 2021; Koo et al., 2012):

1. membrane properties such as pore size distribution, pore geometry, hydrophilicity, and charge density (R. Kumar & Ismail, 2015).
2. solution properties such as particle size and concentration (Tian et al., 2013), pH, ionic strength, and turbidity (Park et al., 2020).
3. operating conditions such as transmembrane pressure (Hube et al., 2021), temperature (France et al., 2021), and cross-flow velocity (S. Zhang et al., 2019).

2.2.2.1 Membrane Properties

2.2.2.1.1 Surface Morphology

In general, the greater the roughness of the surface, the higher the fouling rate. Many studies involving polyamide TFC membranes have revealed such a relationship for high pressure membranes. Few studies correlate roughness to fouling in MF/UF, of which many are focused exclusively on membrane bioreactors (MBR) for wastewater treatment. An assessment of interfacial interactions with a digitally modeled membrane surface revealed that an increase in fractal roughness enhances and lengthens the interfacial interactions between membranes and foulants by increasing the interaction surface area for adhesion (Feng et al., 2017). Li and coworkers applied the extended DLVO approach and density functional theory (DFT) to retrieve molecular insights into membrane fouling specifically with alginate. They used thermodynamic analyses to calculate the energy of a single typical alginate chain adhering to a rough membrane surface and found that a rough membrane leads to reduced alginate adhesion (R. Li et al., 2019). The role of surface roughness also depends on particle size of the foulants. Xing's group found that surface roughness affected the adhesion of nanosized particles but had no effect on microsized particles (Zhong et al., 2012).

Another factor affecting fouling propensity is the hydrophilicity of the membrane surface. Many studies have shown that hydrophilic membranes are less prone to fouling than hydrophobic membranes (V. M. Kochkodan et al., 2006). Shen et al. found high statistical correlations of membrane hydrophilicity with adhesive fouling in MBRs based on 36 membrane materials studied (L. Shen et al., 2017). Interestingly, these correlations were not dependent on the hydrophilicity/hydrophobicity of foulants. Another group found that hydrophilicity does not mitigate sludge adhesion on flat-sheet PVDF membranes (Meijia Zhang et al., 2015). The predominant mechanism affecting adsorptive fouling in PVDF MF membranes is hydrophobic interactions rather than electrostatic interactions (K. Xiao et al., 2011).

Much of the effort around achieving low fouling membrane materials focuses on developing smooth, hydrophilic surfaces (J. Hong & He, 2012).

Membrane fouling may also be affected by its surface charge. As an example, blending PAN membranes with hydrophilically modified PAN containing -SO_3H groups allows electrostatic repulsion of negatively charged BSA, which in turn reduces BSA fouling during UF (Jung, 2004). Similarly, a higher degree of carboxylation and sulfonation yields a negatively charged PSf membrane, which in turn reduces electrostatic adsorption of foulant molecules (Möckel et al., 1999). In another study, microporous PP was modified via UV-induced graft polymerization and post-treatment to generate a positively charged membrane surface with 100% antibacterial activity (Y.-F. Yang et al., 2011). Many studies have demonstrated the efficacy of introducing surface charges through modification for antifouling surfaces MF and UF membranes (Deng et al., 2010; M. Kumar & Ulbricht, 2013; Y.-H. Zhao et al., 2010).

2.2.2.1.2 Pore Size

Pore morphology impacts interactions between membrane and foulants, and in turn fouling evolution. Xiao and coworkers found that the rate of organic fouling decreased with pore size for PVDF MF membranes (K. Xiao et al., 2014). For PTFE/PVDF blend membranes with a fibrous network of interconnected pores, pore size did not have a significant impact on the organic fouling rate. Hwang et al. compared particulate fouling for two polycarbonate (PC) membranes with mean pore diameters of 0.2 and 0.4 μm to filter a suspension of 0.15 μm polymethyl methacrylate (PMMA) particles. At a larger pore size, the fouling mechanism changed from membrane blocking to cake filtration at low flux (Hwang et al., 2008). Larger pore sizes have been credited for low fouling in MF and UF membranes (Bowen et al., 2005; L.-Q. Shen et al., 2003). For activated sludge filtration, pore fouling is more severe in membranes with a sponge-like microstructure as compared to uniform cylindrical pores (Fang & Shi, 2005). Non-interconnected pores also show a more rapid flux decline due to fouling than membranes with an interconnected pore structure (Ho & Zydney, 2000a).

2.2.2.2 Operating Conditions

2.2.2.2.1 Cross-Flow Velocity

Cross-flow velocity affects the transport of particles away from the membrane surface and hence the rate of deposition of foulants (I.-S. Chang et al., 2002). Cross-flow velocity controls the shear and shear-induced diffusion. Other strategies that can assist in increasing shear stress involve vibrating membranes (F. Zhao et al., 2018), in which little research has been conducted to date. At high cross-flow velocities, flow perpendicular to the membrane surface "peels off" the foulant on the surface and makes it difficult for more foulant to deposit (L. Wang et al., 2008). Vyas et al. studied the fouling rate with cross-flow velocity for MF of particulate suspensions (Vyas et al., 2000). They found that internal fouling first decreased with increasing cross-flow velocity but increased above 1.5 m s^{-1} due to the particles size classification effect of cross-flow velocity.

2.2.2.2.2 Feed Pressure

Fouling severity increases with feed pressure (Loh et al., 2009). At higher operating pressures, the fouling in cross-flow MF is in agreement with the cake filtration model (H. Rezaei et al., 2011).

2.2.2.2.3 Temperature

Makardij et al. found that flux decline due to fouling decreases with increasing temperature, during MF and UF of milk (Makardij et al., 1999). Vincent-Vela et al. confirmed this but suggested that a noticeable reduction of the foulant layer resistance was not noticeable above 25 °C during UF.

The concept of critical flux (J_c), below which fouling does not occur (Field et al., 1995; Howell, 1995), is widely accepted. The value of the critical flux depends on particle size, hydrodynamics, and membrane–colloid interactions (Howell, 1995).

Operators are encouraged to stay below this value during operation to avoid fouling. Although filtration is stable over a longer period of time at subcritical flux values (Wu et al., 2018), there is evidence of low-level fouling even significantly below J_c (Pollice et al., 2005). One of the downsides of the critical flux concept is that values cannot be theoretically determined but must be measured for the system in question (X. Li & Li, 2015).

A study by Howe and Clark showed that particulate matter (> 0.45 μm) does not contribute to fouling as much as very small colloids (3–20 nm) and dissolved matter (Howe & Clark, 2002). Fouling during MF in the presence of natural surface water is often the result of adsorption of small molecules on the membrane as well as pore blockage by larger colloidal organics (Howe & Clark, 2002).

Electrostatic forces between membrane surface and foulants can worsen or mitigate fouling. Zhang et al. studied the interaction between humic substances and algal organic matter (AOM) on the fouling behavior of commercial ceramic membranes made of ZrO_2 and TiO_2 (X. Zhang et al., 2018). Negatively charged mixtures of humic acid (HA), fulvic acid (FA), and AOM caused an increase in resistance to irreversible fouling due to electrostatic repulsion between the mixture and the negatively charged membrane. Eshed et al. found that biofilm development during cross-flow ultrafiltration was enhanced in the presence of permeate flow due to the buildup of complex three-dimensional structures on the membrane (Eshed et al., 2008). They found bacterial transport by permeate drag to be the dominant factor in the buildup of the biofouling layer in cross-flow filtration. Lay et al. emphasized that an emphasis on the dynamic movement of protein cakes formed during microfiltration is needed to better understand protein fouling in low-pressure membrane processes (Lay et al., 2021).

2.3 MODELING OF FOULING

Although no single model for membrane fouling has been established, numerous empirical models have been used to explain blocking mechanisms during filtration and to predict fouling. Perhaps the most comprehensive of these is the Hermia model introduced in 1982, which illustrates laws for four idealized modes of blocking (Hermia, 1982):

- *Complete pore blocking*: Assumes that all the particles are larger than the membrane pore size and that every particle reaching the membrane surface will completely block the entrance to a pore.
- *Intermediate blocking*: Assumes that every particle reaching the membrane surface will contribute to fouling, with an expression for the probability that particles will deposit on an already blocked pore; some will block off the pore, while the rest will accumulate on top.
- *Standard blocking*: Assumes that particles are smaller than the membrane pore size and will accumulate inside the pore walls reducing the effective size of the pore; in this type of blocking, pore constriction causes membrane permeability to drop.

- *Cake layer filtration*: Particles accumulate on the membrane surface forming a cake layer that increases in thickness.

These mechanistic concepts were used to develop so-called blocking laws that we now know as Hermia's laws (Fouladitajar et al., 2013).

$$\frac{d^2t}{dV^2} = k\left(\frac{dt}{dV}\right)^N \ or \ \frac{dR}{dV} = kR^N \qquad\qquad \text{Equation 2.3}$$

where the characteristic exponent N indicates fouling mode: $N = 2$ for complete blocking, $N = 1$ for intermediate blocking, $N = 1.5$ for standard blocking, and $N = 0$ for cake filtration; t is filtration time, V is specific permeate volume, R is filtration resistance, and k is a model coefficient.

Filtration resistance is a combination of the intrinsic membrane resistance R_m, measured by filtration of pure water through a clean membrane, resistance due to internal membrane fouling R_{if}, and resistance due to formation of cake layer on the membrane surface R_c.

The sequence of fouling depends on membrane properties such as pore size distribution and surface porosity (F. Wang & Tarabara, 2008).

The preceding mathematical model can be used to explain experimental observations of fouling. This is of interest during the early stages as it determines cleaning efficiency and the tendency for irreversible fouling (Liderfelt & Royce, 2018). The characteristic exponent N can be determined by fitting this model to experimental data and thus determining the fouling mechanism. However, real systems are far more complicated as multiple modes may be present simultaneously. Ho and Zydney find little usefulness in adhering to a single law to explain fouling in real systems; they used a combination of pore blockage and cake filtration models to explain protein fouling during MF (Ho & Zydney, 2000a). Another drawback of this model is that it applies only to constant pressure filtration, whereas real systems often operate at constant flux. Hermia's model is based on constant transmembrane pressure (TMP) dead-end filtration. Citing these drawbacks, efforts to revise these classical filtration laws ensued. Arnot et al. modified Hermia's laws to derive equivalent equations for cross-flow filtration (Arnot et al., 2000). Huang et al. revised the Hermia model for application to both constant pressure and constant flux, from which they developed a unified membrane fouling index to quantify the fouling in low pressure membrane systems (H. Huang et al., 2008). Other numerical models for blocking mechanisms during membrane filtration have also been developed (Chellam & Xu, 2006; Kirschner et al., 2019; Miller et al., 2014).

Several studies have applied membrane blocking laws to interpret experimental data, often modifying the models for different experimental conditions (Khan et al., 2020). Gao et al. used Hermia's models to recognize fouling mechanisms of UF membranes exposed to sodium hypochlorite (Gao et al., 2016; Yang Zhang et al., 2017). Mohammadi and Esmaeelifar applied the Hermia model without any modification to cross-flow filtration of UF membranes (Mohammadi & Esmaeelifar, 2005).

When comparing different modes to evaluate ultrafiltration of HA, Ma et al. found cake filtration to be the best fit (B. Ma et al., 2018). Ng and coworkers used Hermia's model to evaluate membrane fouling during UF of skimmed coconut milk and discovered that fouling was dominated by standard, intermediate blocking and cake formation (Ng et al., 2014). Chang and coworkers assessed fouling during NF of natural water and suggested that intermediate blocking may be the dominant mechanism of fouling (E. E. Chang et al., 2011).

2.4 MEMBRANE DISTILLATION

2.4.1 WETTING

A closely related phenomenon that is unique to MD is the wetting of membrane pores by the feed solution. In membrane distillation, wetting is widely regarded as the principal obstacle to the industrial use of MD (M. Rezaei et al., 2018). Wetting occurs when the liquid feed solution penetrates into the membrane pores. Depending on the type, wetting causes a reduction in rejection, loss of vapor flux, and concurrent loss of energy efficiency (Lu & Chung, 2019).

Until 2016, very few publications addressed wetting; since then, interest has grown exponentially, with particular curiosity in understanding wetting and associated phenomena (J. Liu et al., 2022; Lou et al., 2022) and in developing special wettability membranes (Hou et al., 2021; Z. Wang & Lin, 2017). Here, we analyze literature present on understanding the complex phenomenon of wetting in MD systems, with an emphasis on types and mechanisms of wetting, as well as the intricate relationship between wetting, fouling, and scaling in MD systems.

The four states of pore wetting, namely non-wetted, surface wetted, partially wetted, and fully wetted, were first introduced by Gryta et al. in 2007 (Gryta, 2007). Figure 2.6 shows a schematic of these four wetting states in MD. These states reflect the extent of membrane wetting by the liquid (S. Yang et al., 2022):

1. *Non-wetted state*: Highest vapor transport within pores, i.e. maximum flux and salt rejection
2. *Surface wetted state:* Reduction of vapor transport, with permeate quality still high as there is no liquid permeation through pores
3. *Partially wetted state:* Some pores allow water molecules to pass, and permeate quality decreases; other pores exhibit reduced vapor transport.

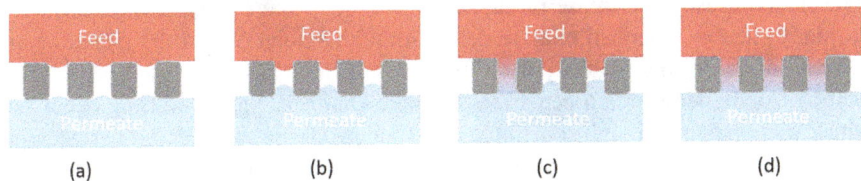

(a)   (b)   (c)   (d)

FIGURE 2.6   States of membrane wetting: (a) non-wetted, (b) surface wetted, (c) partial wetted, and (d) wetted.

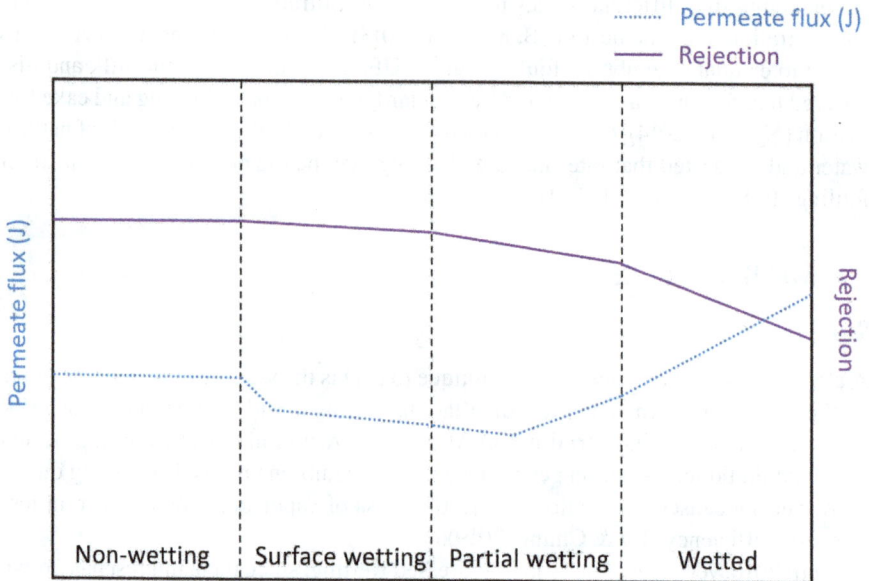

FIGURE 2.7 Degree of membrane pore wetting in the MD process: (a) no wetting state, (b) surface wetting state, (c) partial wetting state, and (d) full wetting state.

Source: Adapted from Sinha Ray, S., Lee, H.-K., & Kwon, Y.-N. (2020). Review on blueprint of designing antiwetting polymeric membrane surfaces for enhanced membrane distillation performance. *Polymers*, *12*(1), 23. Reprinted with permission.

4. *Fully wetted state:* All pores wetted by liquid feed; operation failure; decline of permeate quality observed

While it is generally agreed that permeate quality begins to drop at the onset of partial wetting, there seems to be a slight disagreement concerning the effect of wetting state on permeate flux. Figure 2.7 shows the dependence of permeate flux and salt rejection during the four states of pore wetting (Sinha Ray et al., 2020). For example, Sinha Ray et al. (Sinha Ray et al., 2020) and Gryta et al. both suggest that permeate flux gradually decreases upon surface wetting due to a shift of the liquid–vapor interface resulting from the interaction between the feed and membrane materials. Yet Chamani et al. argue that a shorter distance of the vapor layer in this state should cause an increase in membrane flux (Chamani et al., 2021). However, the study cited by Chamani et al. to demonstrate this point (Gryta, 2005) shows an increase in flux in the *partial* pore wetting state, which is not the same as surface wetting. As the feed stream penetrates the pore channels, partial wetting occurs in which water passes through some pores, and the gap between feed and permeate is narrowed in others (Gryta, 2007). In the partially wetted state, a simultaneous reduction of permeate quality and increase in flux take place. Eventually, full wetting takes place as the liquid is able to pass through the membrane pores, severely deteriorating permeate quality and causing operation failure.

2.4.1.1 Mechanisms of Wetting

Several mechanisms for wetting have been proposed. These include capillary effects, diffusion, fouling and scaling, chemical oxidative degradation, adsorption, hydrophobic–hydrophobic interactions, electrostatic attractions, microbial growth, and secretion of hydrophilic EPS (E. Guillen-Burrieza et al., 2016). All of these induce wetting through localized losses in the hydrophobicity or, in some cases, enlargement of pores.

By far, the most commonly considered mechanism of wetting is related to the premise of MD operation. When a high surface tension liquid is in contact with a non-wetting surface, capillary action prevents the liquid from entering the membrane pore by forming a convex meniscus (Brodskaya & Piotrovskaya, 1994; G. Yang et al., 2021). For non-wetting, the pressure on either side of the liquid must be smaller than the capillary pressure of the membrane. When the capillary pressure, caused by surface tension (Dimri et al., 2012), exceeds the pressure drop due to the vapor pressure difference across the membrane, the liquid beings to penetrate the pores (Kammerhofer et al., 2018; M. Rezaei et al., 2018; Sarti et al., 1985).

Another commonly cited mechanism, especially in the case of surfactant-induced wetting, is adsorption–desorption–adsorption, which allows passage of non-volatile organic compounds through membranes without exceeding the liquid entry pressure (LEP) (Yao et al., 2020). In a detailed review on the mechanisms of failure in MD (Horseman et al., 2021), Horseman et al. iterate that surfactants:

1. readily adsorb onto hydrophobic surfaces immersed in water.
2. are very effective in reducing liquid surface tension, and a low concentration of surfactants can lower LEP at the interface to under ΔP.

Surfactant-induced wetting could depend on the competitive adsorption and enrichment near the liquid–vapor interface (Horseman et al., 2021).

Another mechanism for wetting, in particular scaling-induced wetting, is through pore enlargement (J. Liu et al., 2022). Crystallization and growth of scaling agents such $CaSO_4$ can cause pores to expand, which decreases effective LEP and may induce wetting. Crystal growth of scaling agents including $CaSO_4$ and $CaCO_3$ can also cause localized loss of hydrophobicity (David M. Warsinger et al., 2015). Liu et al. investigated wetting induced by $CaSO_4$ scaling and correlated scaling-induced wetting with the rate of concentration in order to extend understanding of crystal–membrane interactions in wetting (J. Liu et al., 2022). They found that increasing the concentration of $CaSO_4$ correlates with the incidence of wetting as a higher concentration promotes supersaturation and crystallization. Oxidative degradation of the membrane and the resulting formation of hydrophilic groups such as hydroxyl (OH) and carbonyl (C=O) also increase membrane hydrophilicity and accelerate membrane wetting (Gryta, 2007; Luo et al., 2014).

Localized loss of hydrophobicity, through adsorption of hydrophilic compounds, or formation of hydrophilic groups, etc. is a primary factor in wetting. The other, pore enlargement due to crystal growth, is less commonly observed. All aforementioned mechanisms of wetting are in some way related to ΔP exceeding the LEP liquid–solid–air interface, but incorrect determination of the dynamic LEP makes it challenging to predict wetting accurately. It is evident that existing methods to determine LEP and attempts to link this static value to wetting behavior are insufficient when dealing with real-time complex feed solutions. The dynamic progression of the LEP at the water–air interface necessitates the need for in situ dynamic LEP determination that would accurately predict wetting behavior at the interface.

2.4.1.2 Effect of Temperature on Wetting

Operational parameters such as flow rate and temperature can affect membrane wetting (E. Guillen-Burrieza et al., 2016). Since the temperature affects the physical properties of the solution including viscosity and surface tension, it indirectly controls wetting. At higher feed temperatures, liquid surface tension and viscosity drop, making it easier for the liquid to penetrate the pores (Mosadegh-Sedghi et al., 2014). However, He et al. found no effect of temperature on the pore wetting of PTFE membranes (K. He et al., 2011). Saffarini et al. considered the microstructure evolution of PTFE membranes at elevated temperatures and its subsequent effect on wetting (Saffarini et al., 2013). They found that experimentally measured LEP decreases with increasing temperature in the range of 25–75 °C due to microstructural changes within the PTFE layer. A drop in LEP with increasing feed temperature can also be attributed to the surface tension of the liquid. As temperature rises, molecules in the liquid gain kinetic energy and cohesive forces between molecules drop, which translates into a reduction of the liquid's surface tension (Pallas & Pethica, 1983; Palmer, 1976; Shahbaz et al., 2012).

2.4.1.3 Effect of Flow Rate on Wetting

Adjusting feed and permeate flow rates directly impacts the transmembrane pressure applied. Lowering the feed flow rate lowers the pressure difference, which reduces the risk of wetting (M. Rezaei et al., 2018). Literature shows that a slightly higher pressure on the permeate side of the membrane as compared to the feed side can lower the tendency of the membrane to wet (Zakrzewska-Trznadel et al., 1999).

2.4.1.4 Surfactant-Induced Wetting

Using a single-frequency impedance measurement, Wang et al. compared the wetting behavior of PVDF membranes during DCMD using either surfactant (Triton X-100) or alcohol (ethanol) as the wetting agent (Z. Wang et al., 2018). Wetting dynamics differ drastically between the two agents: wetting induced by ethanol was instantaneous, whereas wetting induced by surfactant was gradual, as shown by flux and salt rejection data in Figure 2.8. The kinetic rate of surfactant-induced wetting is related to the kinetics of surfactant adsorption by the pore surface and is affected by the bulk concentration of surfactants as well as MD water vapor flux. Interestingly, surfactant adsorption does not promote wetting, but instead it deters pore wetting by reducing the surface tension of the feed solution at the liquid–air interface (Z. Wang et al., 2018). The authors warn that more complex feed solutions that involve mineral scaling or fouling by NOM may involve different mechanisms and determining factors.

Surfactant physical properties and concentration also affect wetting behavior. Chew et al. found that surfactants with a lower hydrophilic–lipophilic balance (HLB) were more readily adsorbed on the membrane surface, causing an earlier

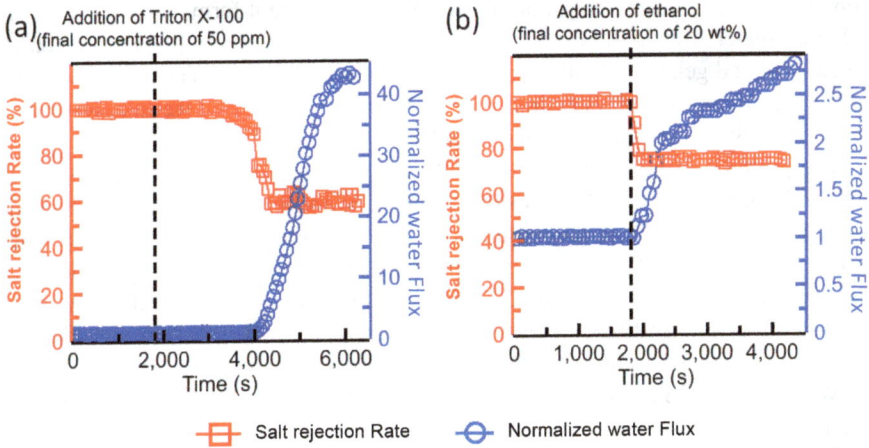

FIGURE 2.8 Different wetting behavior for (a) surfactant-induced and (b) alcohol-induced wetting given by normalized water flux and salt rejection rate for PVDF membranes during DCMD.

Source: Wang, Z., Chen, Y., Sun, X., Duddu, R., & Lin, S. (2018). Mechanism of pore wetting in membrane distillation with alcohol vs. surfactant. *Journal of Membrane Science, 559,* 183–195. Reprinted with permission.

onset of wetting (N. G. P. Chew et al., 2017). HLB is an index related to the size and strength of the hydrophilic and lipophilic moieties of a surfactant molecule (Y. Zheng et al., 2015). A higher HLB corresponds to greater hydrophilicity (Premlal Ranjith & Wijewardene, 2006), i.e. a surfactant with a lower HLB will be more lipophilic and will be easily adsorbed on a hydrophobic surface due to the hydrophobic effect (Vaziri Hassas et al., 2014; Yue Zhang et al., 2006). The onset of wetting was delayed by the presence of oils. Eykens et al. used surface tension measurements to show that surfactant-induced wetting behavior does not depend just on the concentration of surfactant, in this case sodium dodecyl sulfate (SDS), but also on the NaCl solution, which affects the critical micelle concentration (CMC). The greater the membrane hydrophicity, the more susceptible the membrane is to wetting in the presence of SDS.

Beyond the realms of academia lie real-life complex feed solutions, whose behavior may drastically differ from homogeneous solutions commonly investigated for fouling. As membrane distillation inches toward commercial use, one important aspect is gaining a better insight into the interactions between feed constituents and their effect on wetting, fouling, and scaling dynamics. Scaling-induced wetting depends strongly on crystallization kinetics. The coupled effect of scaling and membrane wetting, especially when the feed contains $CaCO_3$ or $CaSO_4$ is well-known (Christie et al., 2020; El-Bourawi et al., 2006; Gryta, 2005; Kiefer et al., 2019; David M. Warsinger et al., 2015). The mechanical stress exerted by the rapid, oriented growth of gypsum crystals (Desarnaud et al., 2016; Steiger, 2005) into the pores causes pore deformation and induces wetting (Xie et al., 2022). Under the same operating conditions, silica scaling does not induce wetting because it forms as a thin layer of submicrometer silica particles only on the membrane surface via polymerization of silicic acid and gelation of silica particles (Christie et al., 2020). The hydrophobicity of the membrane also plays an important role in scaling-induced wetting, as well as in the reversibility of scaling (Yin et al., 2020).

The presence of humic acid (HA) inhibits gypsum scaling and prevents gypsum-induced wetting during MD desalination (P. Wang et al., 2022). Wang et al. attribute the role of HA in mitigating wetting to the formation of a compact protective layer of gypsum/HA scale, which prevents the intrusion of scale particles into membrane pores (P. Wang et al., 2022). Deprotonated HA delays scaling by interrupting the crystallization and growth of gypsum (Yan et al., 2021). A considerable implication of these findings is that organic removal pretreatment may not be suited to MD of high-salinity feed solutions; in fact it may worsen the effects of gypsum scaling and wetting (Yan et al., 2021).

Wetting will not take place as long as the transmembrane pressure is less than a critical entry pressure, i.e. the liquid entry pressure (LEP). LEP is the minimum hydrostatic pressure applied to the feed solution before it penetrates through the pores (García-Fernández et al., 2015). A high LEP reflects greater resistance to wetting. The Young–Laplace model governs this behavior such that (Rahimpour & Esmaeilbeig, 2019):

$$\text{LEP} = -\frac{(\beta \gamma_l \cos\theta)}{r_{max}}$$

Equation 2.4

where β is the pore geometry coefficient (= 1 for cylindrical pores and $0 < \beta < 1$ for non-cylindrical), γ_l is the liquid surface tension in N m^{-1}, θ is the contact angle between the liquid and the membrane surface in °, and r_{max} is the maximum pore size in m.

Although long believed to be an acceptable indicator of wetting prediction, LEP as it is typically determined does not take into account wetting induced by surface tension, scaling, or fouling (M. Rezaei et al., 2018; M. Rezaei et al., 2017; Tijing et al., 2015), as has been demonstrated (Drioli, 2018). Other researchers have also highlighted the ineffectiveness of LEP in determining membrane wetting (Bilad et al., 2015). Horseman et al. argue that the criterion for wetting based on LEP applies well as long as LEP is accurately determined as a function of surfactant concentration at the wetting frontier (Horseman et al., 2021), which is rarely the case.

Effective wetting agents that are commonly found in feed solutions include low-surface-tension and water-miscible liquids, such as alcohols and amphiphilic molecules (surfactants) (Z. Wang, Chen, Sun et al., 2018). Wetting mechanisms remain poorly understood with painstakingly limited studies elucidating which constituents of the feed solutions cause pore wetting (N. G. P. Chew et al., 2017; Jacob et al., 2019; McGaughey & Childress, 2022; Velioğlu et al., 2018; Z. Wang et al., 2018). One reason for this is the uniqueness of pore wetting to membrane distillation, as compared to the more investigated phenomena of fouling and scaling, which are common to all membrane separation processes.

Wetting occurs as a loss of membrane hydrophobicity and is especially challenging in the presence of highly wetting compounds such as oils and surfactants (Villalobos García et al., 2018; Y. Wang et al., 2020)

2.4.1.5 Foulant-Induced Wetting

Membrane fouling is considered an influencing factor in pore wetting as the wetting behavior changes depending on feed solution constituents and the presence of oils, surfactants, and scaling agents (Han et al., 2019; Yu et al., 2021).

Gilron et al. investigated the effect of silica solutions on fouling hollow fiber and flat-sheet DCMD modules. They found that fouling via deposition of colloidal silica in the pore mouths caused partial wetting, followed by increased temperature and concentration polarization (Gilron et al., 2013). Initial foulants and scalants often lead to secondary fouling due to nucleation or growth, which causes progressive wetting of the membrane (Z. Xiao et al., 2020).

Jacob et al. developed an optical tool for in situ visualization of membrane wetting (Jacob et al., 2020). Using controlled wetting through the addition of a surfactant, they applied their tool in parallel to a previously developed the detection of dissolved tracer intrusion (DDTI) method to detect wetting in a vacuum MD system.

Considering the repercussions of wetting in MD systems, literature on the subject is surprisingly scarce; this explains why wetting still presents a major roadblock in MD commercialization. In particular, the role of wetting using realistic feed solutions, especially in large-scale long-term operations is currently not at all understood (G. Yang et al., 2021). Membrane wettability is getting most of the attention with respect to wetting resistance in MD systems; both theoretical and experimental studies have added to our understanding of special wettability materials to prevent wetting.

2.4.2 FOULING

Membrane distillation also suffers from fouling, although to a much less extent than pressure-driven processes such as MF, UF, NF, and RO. Accumulation of species can block membrane pores and reduce vapor flux, eventually leading to wetting of the membrane by the liquid feed. As a result, although fouling leads to a flux decline, a more serious form of failure ensues in the form of wetting. The types of fouling in MD are the same as those discussed in Section 5.2. Due to the coupling of mass and thermal transfer, the foulant layer in MD may add thermal resistance in addition to hydraulic resistance. The thermal and hydraulic resistance caused by fouling depends on the physicochemical properties of the fouling layer such as porosity (Gryta, 2008).

Fouling results from adhesion of particulate materials to the membrane surface. The DLVO theory is often used to describe the net force of interaction between foulants and surface (Oliveira, 1997). According to this theory, the interactions between foulants and membrane surface are the combined effect of van der Waals attraction and electrostatic double layer forces. Accumulation of pharmaceutical foulants on hydrophobic PVDF and PTFE membranes is caused by electrical attraction of the negatively charged membrane with the positively charged antibiotic (Guo et al.,

2020). In MD systems, the likelihood of fouling is reduced when the surface and foulant are similarly charged or when the force of interaction between them decreases.

Organic foulants in MD include humic acids, proteins and emulsified oil droplets (Gryta, 2008). Oil fouling has been the topic of many recent papers, accompanying increasing interest in the use of MD for oil and gas produced water. Organic fouling in MD is believed to be dominated by long-range hydrophobic interaction (Elcik et al., 2021; Horseman et al., 2021). Hydrophobic interactions are caused by the unfavorable interactions of non-polar substances with water, termed the "hydrophobic effect" (Franks, 1975; Lazaridis, 2013). The hydrophobic interaction is much stronger than the van der Waal forces (J. Israelachvili & Pashley, 1982; J. N. Israelachvili & Pashley, 1984; Van Oss et al., 1986), especially in the case of hydrophobic membrane surfaces and when air is intrinsically present in the non-wetted membrane in a Cassie-Baxter state (Horseman et al., 2021). One of the methods of organic fouling limitation is based on the hydrophilization of the membrane surface. Coating the PTFE membranes with sodium alginate hydrogel allows the inhibiting of the adsorption of citrus oil to a significant degree (Xu et al., 2005).

Lack of standardization in lab-scale MD studies makes it difficult to extract fouling patterns. Furthermore, fouling evolution in real systems is, as discussed, an extremely complex phenomenon depending on various factors. When three model organic foulants, namely HA, BSA, and AA, were compared in a DCMD system, minimal fouling was observed with the more hydrophilic compound (AA) (Naidu et al., 2014) due to a weakened hydrophobic effect as well as negative electrostatic repulsion. As with pressure-driven filtration, both feed characteristics and operating parameters (such as permeate flux, feed temperature, and flow rate) affect fouling behavior in MD. In order to tackle fouling, closer inspection of the foulant–membrane interactions must be carried out.

2.4.3 SCALING

Inasmuch as mineral scaling and inorganic fouling are often used interchangeably, inorganic foulants can exist in the form of inorganic particles that already exist in the feedwater and that do not always include mineral scaling (Horseman et al., 2021). In MD, however, mineral scaling is the dominant form of inorganic fouling. Mineral scaling involves complex reactions that take place in solution as well as at the membrane–water interface (Z. Xiao et al., 2019). The evolution of mineral scaling, i.e. nucleation and growth kinetics, depends on factors such as solution chemical composition, temperature, pH, and the presence of NOM (Tong et al., 2019). Scaling causes reduction in permeate water quality, reduction in permeate flux, greater temperature and concentration polarization, and chemical degradation and membrane damage (David M. Warsinger et al., 2015).

2.4.3.1 Temperature

In the temperature range used for MD, the solubility of individual salts may vary with temperature. As an example, the solubility of $CaCO_3$, $MgOH$, and $Ca_3(PO_4)_2$ decreases with rising temperature, making scaling more probably with heated feed solutions (Stamatakis et al., 2005).

2.4.3.2 Feedwater Composition

As is the case with pressure-driven processes, some natural waters are more prone to cause scaling due to the concentration of salts found. Seawater is more prone to $CaSO_4$ and $CaCO_3$ scaling compared to other surface waters (Morel & Hering, 1993). Groundwater composition varies with time and geographical location and may contain high amounts of Ca^{2+}, HCO_3^-, Mg^{2+}, and SO_4^{2-}. Variation can make it difficult to predict scalants and take appropriate cleaning measures. Comparing water samples from a river, lake, and groundwater sources for MD, Gryta et al. found that scaling was the most significant when groundwater was used as feed (Gryta, 2010) and that $CaCO_3$ was the dominant scalant.

Classical nucleation theory is widely used to explain the kinetics of nucleation for inorganic scaling in MD systems (Prieto, 2018; Yin et al., 2021). The probability of inorganic fouling in MD relates to membrane hydrophobicity and surface porosity according to the classical nucleation theory (David M Warsinger et al., 2017; Z. Xiao et al., 2020). Mineral scaling is often based on heterogeneous nucleation, i.e. the solution contains foreign substances with active sites for nucleation (Li Liu et al., 2021). According to the classical nucleation theory, the first stage of scaling, heterogeneous nucleation, starts when a minimum concentration is reached; when concentrations exceed solubility, crystal growth takes place (Kim et al., 2022). Surface crystallization depends on membrane materials, interaction between particles and hydrodynamic conditions. Surface crystallization is more dominant at lower flow velocities, whereas high flow velocities point to bulk solution crystallization (Naidu et al., 2014).

With $CaSO_4$ scaling, fluorosilicone-coated polypropylene membranes inhibited surface nucleation and particle attachment (He et al., 2009). Therefore, the increase in scaling kinetics from decreasing membrane hydrophobicity suggests that a less hydrophobic surface is more favorable for heterogeneous nucleation. Decreasing membrane hydrophobicity enhances scaling kinetics (Gryta, 2008; F. He et al., 2008; Horseman et al., 2019). Curcio and coworkers studied the kinetics of $CaCO_3$ scaling during seawater desalination with DCMD. They predicted the Gibbs free energy barrier to the formation of critical nuclei for a membrane with 70% porosity and a contact angle of 115°. When the temperature difference between feed and permeate is 20 °C, permeate flux was reduced by 45% after 35 hours of operation, although cleaning with citric acid and NaOH restored flux completely (Curcio et al., 2010).

Scaling in MD systems also causes membrane wetting (Li Liu et al., 2021). Theoretical and experimental studies on understanding fouling, both organic and inorganic, are key to developing mitigation strategies through membrane fabrication and process optimization (Ye et al., 2019). The addition of antiscalants at dosage levels as low as 0.6 mg L^{-1} can prolong the induction period of $CaSO_4$ and $CaCO_3$ (F. He et al., 2009). Nghiem and coworkers use a membrane flushing method to reset the induction period and prevent $CaSO_4$ scaling, while operating at low recovery (Nghiem & Cath, 2011). Membranes with special wettability that demonstrate antiscaling properties have also been developed for MD (Chen et al., 2020; Liao et al., 2020).

2.5 IMPLICATIONS OF FOULING: THE CASE OF THE TAMPA BAY SEAWATER REVERSE OSMOSIS FACILITY

To understand the severe implications of fouling on construction timeline, plant operation, and financial obligations, we consider the case of the Tampa Bay seawater reverse osmosis (SWRO) facility. This was the first large-capacity seawater desalination plant in the United States. Contracted in 1998, plant design and construction were given to Covanta Tampa Construction in 2000 when the original contractors claimed bankruptcy. Covanta, however, was unable to meet performance testing deadlines and also declared bankruptcy in 2003, after which Tampa Bay Water resumed ownership of the facility when it was only 50% completed. Figure 2.9 shows the major obstacles and milestones during the project history.

During initial testing, the plant experienced problems and a 14-day performance test revealed 31 deficiencies. Due to excessive membrane fouling, elements had to be cleaned every 2 weeks as opposed to every 2–6 months. Cartridge filters in the pretreatment stage of the plant were clogged due to an infestation of Asian green mussels, causing severe fouling of the filters at 3–4 days of operation as opposed to the specified 90 days. To combat these issues, the strength of the cleaning solution was increased, but this resulted in large amounts of spent chemicals that could not be disposed of without violating sewage permits. Severe membrane fouling due to insufficient pretreatment was the main cause of difficulties faced during early operation.

Eventually, the pretreatment system was redesigned to account for these problems with the aim of extending the lifetime of the RO membranes. A more detailed description of the redesigned pretreatment system can be found elsewhere (Harvey et al., 2020). Separation efficiency of the sand filters to remove fine particles was enhanced using precoat filtration. Next, the staging of the sand filtration was modified to allow for lower feed rates and greater efficiency. Piping and controls were allowed such that incoming water could be uniformly dispersed. Measures were taken to maintain constant water temperature and stable water chemistry to mitigate the effects of biofouling. Biological growth was also impeded through improved screening. Within the RO process itself, pipes, pumps, and electrical systems were added to facilitate the cleaning process and render it more reliable. Even the post-treatment was modified by adding lime to protect distribution pipelines. It is clear that the repercussions of not taking appropriate measures to mitigate membrane fouling early on can be severe and cost millions of dollars as well as earn contractors a poor reputation. Several companies went bankrupt, as the delays and failures caused massive disruptions in deadlines to meet financial obligations.

Today, the plant still faces some operational challenges as the feedwater source is extremely biologically productive and also contains high concentrations of algae, marine bacteria, total organic carbon (TOC), transparent exopolymer particles, the biopolymer fraction of NOM, and phosphate (Harvey et al., 2020). Passage of transparent exopolymer particles (TEP) into the membranes enhances biofouling. Harvey and coworkers propose switching the feed to groundwater as a permanent solution and suggest considering advanced pretreatment design to include dissolved air flotation systems and UF membranes (Harvey et al., 2020).

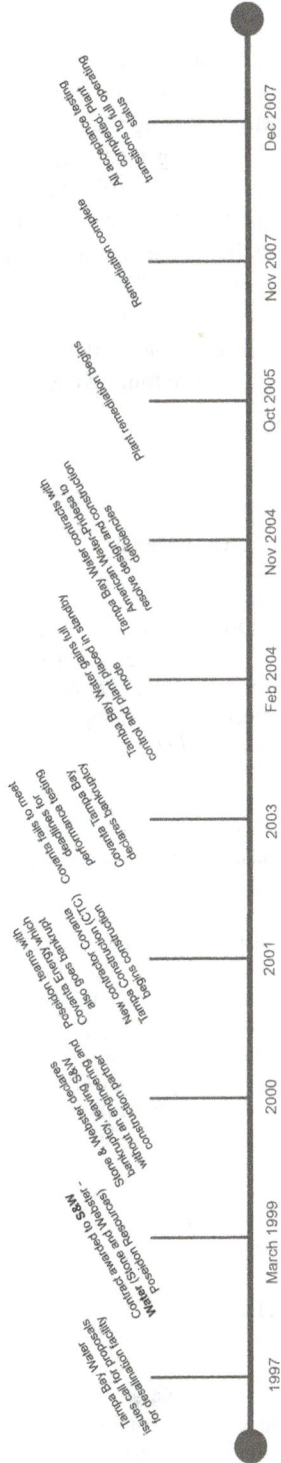

FIGURE 2.9 Tampa Bay seawater reverse osmosis project history.

2.6 CONCLUSION

Fouling is arguably the biggest challenge in membrane separation processes. Over time, the unwanted deposition of feed constituents blocks membrane surface area and causes flux to decline, resulting in a significant increase in operational costs. Fouling can be of many types, depending on the mechanism and type of foulant present: In fact, in real systems, fouling is often an interplay of the interactions between different types of foulants with the membrane surface. We explored fouling evolution, types, and mechanisms in various membrane processes as a first step to devising appropriate mitigation strategies. The chapter also familiarized the reader with models that describe common fouling mechanisms. The catastrophic effects of fouling were explained using the example of the Tampa Bay seawater RO plant, where insufficient pretreatment and severe fouling caused major setbacks in its commissioning and operation.

BIBLIOGRAPHY

Abdelrasoul, A., Doan, H., & Lohi, A. (2013). Fouling in membrane filtration and remediation methods. *Mass Transfer-Advances in Sustainable Energy and Environment Oriented Numerical Modeling, 195.*

Adair, J. H., Suvaci, E., & Sindel, J. (2001). Surface and colloid chemistry. In K. H. J. Buschow, R. W. Cahn, M. C. Flemings, B. Ilschner, E. J. Kramer, S. Mahajan, & P. Veyssière (Eds.), *Encyclopedia of materials: Science and technology* (pp. 1–10). Elsevier.

Al-Amoudi, A. S. (2010). Factors affecting natural organic matter (NOM) and scaling fouling in NF membranes: A review. *Desalination, 259*(1), 1–10. https://doi.org/10.1016/j.desal.2010.04.003

Alhadidi, A., Kemperman, A. J. B., Blankert, B., Schippers, J. C., Wessling, M., & van der Meer, W. G. J. (2011). Silt density index and modified fouling index relation, and effect of pressure, temperature and membrane resistance. *Desalination, 273*(1), 48–56. https://doi.org/10.1016/j.desal.2010.11.031

AlSawaftah, N., Abuwatfa, W., Darwish, N., & Husseini, G. (2021). A comprehensive review on membrane fouling: Mathematical modelling, prediction, diagnosis, and mitigation. *Water, 13*(9), 1327.

Amiri, M., & Samiei, M. (2007). Enhancing permeate flux in a RO plant by controlling membrane fouling. *Desalination, 207*(1–3), 361–369.

Ang, W. S., Lee, S., & Elimelech, M. (2006). Chemical and physical aspects of cleaning of organic-fouled reverse osmosis membranes. *Journal of Membrane Science, 272*(1), 198–210. https://doi.org/10.1016/j.memsci.2005.07.035

Anjum, M., Miandad, R., Waqas, M., Gehany, F., & Barakat, M. A. (2019). Remediation of wastewater using various nano-materials. *Arabian Journal of Chemistry, 12*(8), 4897–4919. https://doi.org/10.1016/j.arabjc.2016.10.004

Arnot, T., Field, R., & Koltuniewicz, A. (2000). Cross-flow and dead-end microfiltration of oily-water emulsions: Part II. Mechanisms and modelling of flux decline. *Journal of Membrane Science, 169*(1), 1–15.

Au, J., Kim, M.-M., Rahardianto, A., Lyster, E., & Cohen, Y. (2007). *Xinghua Sun, Raja Ghosh PM 305d kinetics of Ro Membrane scaling in the presence of antiscalants.* Paper presented at the 2007 Annual Meeting.

Azari, S., & Zou, L. (2012). Using zwitterionic amino acid l-DOPA to modify the surface of thin film composite polyamide reverse osmosis membranes to increase their fouling resistance. *Journal of Membrane Science, 401*, 68–75.

Bacchin, P., Aimar, P., & Field, R. W. (2006). Critical and sustainable fluxes: Theory, experiments and applications. *Journal of Membrane Science, 281*(1–2), 42–69.

Baek, Y., Yu, J., Kim, S.-H., Lee, S., & Yoon, J. (2011). Effect of surface properties of reverse osmosis membranes on biofouling occurrence under filtration conditions. *Journal of Membrane Science, 382*(1), 91–99. https://doi.org/10.1016/j.memsci.2011.07.049

Baker, J., & Dudley, L. (1998). Biofouling in membrane systems—a review. *Desalination, 118*(1–3), 81–89.

Bartman, A. R., Lyster, E., Rallo, R., Christofides, P. D., & Cohen, Y. (2011). Mineral scale monitoring for reverse osmosis desalination via real-time membrane surface image analysis. *Desalination, 273*(1), 64–71.

Ben-Sasson, M., Lu, X., Nejati, S., Jaramillo, H., & Elimelech, M. (2016). In situ surface functionalization of reverse osmosis membranes with biocidal copper nanoparticles. *Desalination, 388*, 1–8.

Beyer, F., Laurinonyte, J., Zwijnenburg, A., Stams, A. J., & Plugge, C. M. (2017). Membrane fouling and chemical cleaning in three full-scale reverse osmosis plants producing demineralized water. *Journal of Engineering, 2017*, 635–675. https://doi.org/10.1155/2017/6356751

Bhargava, C., & Sharma, P. K. (2021). *Artificial intelligence: Fundamentals and applications* (1st ed.). CRC Press.

Bhattacharjee, S., & Johnston, G. M. (2002). A model of membrane fouling by salt precipitation from multicomponent ionic mixtures in crossflow nanofiltration. *Environmental Engineering Science, 19*(6), 399–412.

Bilad, M. R., Guillen-Burrieza, E., Mavukkandy, M. O., Al Marzooqi, F. A., & Arafat, H. A. (2015). Shrinkage, defect and membrane distillation performance of composite PVDF membranes. *Desalination, 376*, 62–72. https://doi.org/10.1016/j.desal.2015.08.015

Bishop, P. (2007). The role of biofilms in water reclamation and reuse. *Water Science and Technology, 55*(1–2), 19–26.

Bowen, W. R., Cheng, S. Y., Doneva, T. A., & Oatley, D. L. (2005). Manufacture and characterisation of polyetherimide/sulfonated poly(ether ether ketone) blend membranes. *Journal of Membrane Science, 250*(1), 1–10. https://doi.org/10.1016/j.memsci.2004.07.004

Brodskaya, E. N., & Piotrovskaya, E. M. (1994). Monte Carlo simulations of the Laplace pressure dependence on the curvature of the convex meniscus in thin unwetted capillaries. *Langmuir, 10*(6), 1837–1840.

Buffle, J., & Leppard, G. G. (1995). Characterization of aquatic colloids and macromolecules. 2. Key role of physical structures on analytical results. *Environmental Science & Technology, 29*(9), 2176–2184. doi:10.1021/es00009a005

Buffle, J., Wilkinson, K. J., Stoll, S., Filella, M., & Zhang, J. (1998). A generalized description of aquatic colloidal interactions: The three-culloidal component approach. *Environmental Science and Technology, 32*(19), 2887–2899. doi:10.1021/es980217h

Carpentier, B., & Cerf, O. (1999). Biofilms. In R. K. Robinson (Ed.), *Encyclopedia of food microbiology* (pp. 252–259). Elsevier.

Chamani, H., Woloszyn, J., Matsuura, T., Rana, D., & Lan, C. Q. (2021). Pore wetting in membrane distillation: A comprehensive review. *Progress in Materials Science, 122*, 100843. https://doi.org/10.1016/j.pmatsci.2021.100843

Chang, E. E., Yang, S.-Y., Huang, C.-P., Liang, C.-H., & Chiang, P.-C. (2011). Assessing the fouling mechanisms of high-pressure nanofiltration membrane using the modified Hermia model and the resistance-in-series model. *Separation and Purification Technology, 79*(3), 329–336. https://doi.org/10.1016/j.seppur.2011.03.017

Chang, I.-S., Le Clech, P., Jefferson, B., & Judd, S. (2002). Membrane fouling in membrane bioreactors for wastewater treatment. *Journal of Environmental Engineering, 128*(11), 1018–1029.

Chellam, S., & Xu, W. (2006). Blocking laws analysis of dead-end constant flux microfiltration of compressible cakes. *Journal of Colloid and Interface Science, 301*(1), 248–257.

Chen, Y., Lu, K. J., & Chung, T.-S. (2020). An omniphobic slippery membrane with simultaneous anti-wetting and anti-scaling properties for robust membrane distillation. *Journal of Membrane Science, 595*, 117572. https://doi.org/10.1016/j.memsci.2019.117572

Cheng, G., Zhang, Z., Chen, S., Bryers, J. D., & Jiang, S. (2007). Inhibition of bacterial adhesion and biofilm formation on zwitterionic surfaces. *Biomaterials, 28*(29), 4192–4199. https://doi.org/10.1016/j.biomaterials.2007.05.041

Chew, J. W., Kilduff, J., & Belfort, G. (2020). The behavior of suspensions and macromolecular solutions in crossflow microfiltration: An update. *Journal of Membrane Science, 601*, 117865. https://doi.org/10.1016/j.memsci.2020.117865

Chew, N. G. P., Zhao, S., Loh, C. H., Permogorov, N., & Wang, R. (2017). Surfactant effects on water recovery from produced water via direct-contact membrane distillation. *Journal of Membrane Science, 528*, 126–134. https://doi.org/10.1016/j.memsci.2017.01.024

Christie, K. S. S., Yin, Y., Lin, S., & Tong, T. (2020). Distinct behaviors between gypsum and silica scaling in membrane distillation. *Environmental Science & Technology, 54*(1), 568–576. doi:10.1021/acs.est.9b06023

Cirillo, A. I., Tomaiuolo, G., & Guido, S. (2021). Membrane fouling phenomena in microfluidic systems: From technical challenges to scientific opportunities. *Micromachines, 12*(7), 820. doi:10.3390/mi12070820

Cob, S. S., Beaupin, C., Hofs, B., Nederlof, M., Harmsen, D., Cornelissen, E., . . . Witkamp, G. (2012). Silica and silicate precipitation as limiting factors in high-recovery reverse osmosis operations. *Journal of Membrane Science, 423*, 1–10.

Curcio, E., Ji, X., Di Profio, G., Sulaiman, A. O., Fontananova, E., & Drioli, E. (2010). Membrane distillation operated at high seawater concentration factors: Role of the membrane on CaCO3 scaling in presence of humic acid. *Journal of Membrane Science, 346*(2), 263–269. https://doi.org/10.1016/j.memsci.2009.09.044

Dash, S. S., Bhaskar, M. A., Panigrahi, B. K., & Das, S. (2016). *Artificial intelligence and evolutionary computations in engineering systems: Proceedings of ICAIECES 2015.* Springer.

Deng, B., Yu, M., Yang, X., Zhang, B., Li, L., Xie, L., . . . Lu, X. (2010). Antifouling microfiltration membranes prepared from acrylic acid or methacrylic acid grafted poly (vinylidene fluoride) powder synthesized via pre-irradiation induced graft polymerization. *Journal of Membrane Science, 350*(1), 252–258. https://doi.org/10.1016/j.memsci.2009.12.035

Desarnaud, J., Bonn, D., & Shahidzadeh, N. (2016). The Pressure induced by salt crystallization in confinement. *Scientific Reports, 6*(1), 30856. doi:10.1038/srep30856

Dimri, V. P., Srivastava, R. P., & Vedanti, N. (2012). Chapter 5—reservoir geophysics: Some basic concepts. In V. P. Dimri, R. P. Srivastava, & N. Vedanti (Eds.), *Handbook of geophysical exploration: Seismic exploration* (Vol. 41, pp. 89–118). Pergamon.

Drioli, E. (2018). *Membrane distillation.* MDPI AG.

Du, X., Zhang, K., Yang, H., Li, K., Liu, X., Wang, Z., . . . Liang, H. (2019). The relationship between size-segregated particles migration phenomenon and combined membrane fouling in ultrafiltration processes: The significance of shear stress. *Journal of the Taiwan Institute of Chemical Engineers, 96*, 45–52. https://doi.org/10.1016/j.jtice.2018.11.016

Duclos-Orsello, C., Li, W., & Ho, C.-C. (2006). A three mechanism model to describe fouling of microfiltration membranes. *Journal of Membrane Science, 280*(1), 856–866. https://doi.org/10.1016/j.memsci.2006.03.005

El-Bourawi, M. S., Ding, Z., Ma, R., & Khayet, M. (2006). A framework for better understanding membrane distillation separation process. *Journal of Membrane Science, 285*(1), 4–29. https://doi.org/10.1016/j.memsci.2006.08.002

Elcik, H., Fortunato, L., Vrouwenvelder, J. S., & Ghaffour, N. (2021). Real-time membrane fouling analysis for the assessment of reclamation potential of textile wastewater processed by membrane distillation. *Journal of Water Process Engineering, 43*, 102296. https://doi.org/10.1016/j.jwpe.2021.102296

Elimelech, M., & O'Melia, C. R. (1990). Kinetics of deposition of colloidal particles in porous media. *Environmental Science & Technology*, *24*(10), 1528–1536. doi:10.1021/es00080a012

Enfrin, M., Lee, J., Le-Clech, P., & Dumée, L. F. (2020). Kinetic and mechanistic aspects of ultrafiltration membrane fouling by nano- and microplastics. *Journal of Membrane Science*, *601*, 117890. https://doi.org/10.1016/j.memsci.2020.117890

Eshed, L., Yaron, S., & Dosoretz, C. (2008). Effect of permeate drag force on the development of a biofouling layer in a pressure-driven membrane separation system. *Applied and Environmental Microbiology*, *74*(23), 7338–7347.

Fan, L., Nguyen, T., Roddick, F. A., & Harris, J. L. (2008). Low-pressure membrane filtration of secondary effluent in water reuse: Pre-treatment for fouling reduction. *Journal of Membrane Science*, *320*(1), 135–142. https://doi.org/10.1016/j.memsci.2008.03.058

Fane, A. G., Xi, W., & Rong, W. (2006). Chapter 7—Membrane filtration processes and fouling. In G. Newcombe & D. Dixon (Eds.), *Interface science and technology* (Vol. 10, pp. 109–132). Elsevier.

Fang, H. H. P., & Shi, X. (2005). Pore fouling of microfiltration membranes by activated sludge. *Journal of Membrane Science*, *264*(1), 161–166. https://doi.org/10.1016/j.memsci.2005.04.029

Feng, S., Yu, G., Cai, X., Eulade, M., Lin, H., Chen, J., . . . Liao, B.-Q. (2017). Effects of fractal roughness of membrane surfaces on interfacial interactions associated with membrane fouling in a membrane bioreactor. *Bioresource Technology*, *244*, 560–568. https://doi.org/10.1016/j.biortech.2017.07.160

Field, R. W., Wu, D., Howell, J. A., & Gupta, B. B. (1995). Critical flux concept for microfiltration fouling. *Journal of Membrane Science*, *100*(3), 259–272. https://doi.org/10.1016/0376-7388(94)00265-Z

Filella, M. (2007). Colloidal properties of submicron particles in natural waters. *IUPAC Series on Analytical and Physical Chemistry of Environmental Systems*, *10*, 17.

Filippov, A., Starov, V. M., Llyod, D. R., Chakravarti, S., & Glaser, S. (1994). Sieve mechanism of microfiltration. *Journal of Membrane Science*, *89*(3), 199–213. https://doi.org/10.1016/0376-7388(94)80102-9

Fortunato, L., Alshahri, A. H., Farinha, A. S., Zakzouk, I., Jeong, S., & Leiknes, T. (2020). Fouling investigation of a full-scale seawater reverse osmosis desalination (SWRO) plant on the red sea: Membrane autopsy and pretreatment efficiency. *Desalination*, *496*, 114536.

Fouladitajar, A., Zokaee Ashtiani, F., Okhovat, A., & Dabir, B. (2013). Membrane fouling in microfiltration of oil-in-water emulsions: A comparison between constant pressure blocking laws and genetic programming (GP) model. *Desalination*, *329*, 41–49. https://doi.org/10.1016/j.desal.2013.09.003

France, T. C., Bot, F., Kelly, A. L., Crowley, S. V., & O'Mahony, J. A. (2021). The influence of temperature on filtration performance and fouling during cold microfiltration of skim milk. *Separation and Purification Technology*, *262*, 118256. https://doi.org/10.1016/j.seppur.2020.118256

Franks, F. (1975). The hydrophobic interaction. In F. Franks (Ed.), *Water A comprehensive treatise: Aqueous solutions of amphiphiles and macromolecules* (pp. 1–94). Springer US.

Gao, F., Wang, J., Zhang, H., Zhang, Y., & Hang, M. A. (2016). Effects of sodium hypochlorite on structural/surface characteristics, filtration performance and fouling behaviors of PVDF membranes. *Journal of Membrane Science*, *519*, 22–31. https://doi.org/10.1016/j.memsci.2016.07.024

García-Fernández, L., Khayet, M., & García-Payo, M. C. (2015). 11—Membranes used in membrane distillation: Preparation and characterization. In A. Basile, A. Figoli, & M. Khayet (Eds.), *Pervaporation, vapour permeation and membrane distillation* (pp. 317–359). Woodhead Publishing.

Gilabert-Oriol, G. (2021). *Ultrafiltration membrane cleaning processes: Optimization in seawater desalination plants*. De Gruyter. https://doi.org/10.1515/9783110715149

Gilron, J., Ladizansky, Y., & Korin, E. (2013). Silica fouling in direct contact membrane distillation. *Industrial & Engineering Chemistry Research, 52*(31), 10521–10529. doi:10.1021/ie400265b

Gol, R. M., Bera, A., Banjo, S., Ganguly, B., & Jewrajka, S. K. (2014). Effect of amine spacer of PEG on the properties, performance and antifouling behavior of poly (piperazine-amide) thin film composite nanofiltration membranes prepared by in situ PEGylation approach. *Journal of Membrane Science, 472*, 154–166.

Greenlee, L. F., Lawler, D. F., Freeman, B. D., Marrot, B., & Moulin, P. (2009). Reverse osmosis desalination: Water sources, technology, and today's challenges. *Water Research, 43*(9), 2317–2348.

Gregory, J. (2005). *Particles in water: Properties and processes*. CRC Press.

Gryta, M. (2005). Long-term performance of membrane distillation process. *Journal of Membrane Science, 265*(1), 153–159. https://doi.org/10.1016/j.memsci.2005.04.049

Gryta, M. (2007). Influence of polypropylene membrane surface porosity on the performance of membrane distillation process. *Journal of Membrane Science, 287*(1), 67–78. https://doi.org/10.1016/j.memsci.2006.10.011

Gryta, M. (2008). Fouling in direct contact membrane distillation process. *Journal of Membrane Science, 325*(1), 383–394. https://doi.org/10.1016/j.memsci.2008.08.001

Gryta, M. (2010). Desalination of thermally softened water by membrane distillation process. *Desalination, 257*(1–3), 30–35.

Guillen-Burrieza, E., Mavukkandy, M. O., Bilad, M. R., & Arafat, H. A. (2016). Understanding wetting phenomena in membrane distillation and how operational parameters can affect it. *Journal of Membrane Science, 515*, 163–174. https://doi.org/10.1016/j.memsci.2016.05.051

Gul, A., Hruza, J., & Yalcinkaya, F. (2021). Fouling and chemical cleaning of microfiltration membranes: A mini-review. *Polymers, 13*(6), 846.

Guo, J., Fortunato, L., Deka, B. J., Jeong, S., & An, A. K. (2020). Elucidating the fouling mechanism in pharmaceutical wastewater treatment by membrane distillation. *Desalination, 475*, 114148. https://doi.org/10.1016/j.desal.2019.114148

Hakami, M. W., Alkhudhiri, A., Al-Batty, S., Zacharof, M.-P., Maddy, J., & Hilal, N. (2020). Ceramic microfiltration membranes in wastewater treatment: Filtration behavior, fouling and prevention. *Membranes, 10*(9), 248.

Hamaker, H. C. (1937). The London—van der Waals attraction between spherical particles. *physica, 4*(10), 1058–1072.

Han, L., Tan, Y. Z., Xu, C., Xiao, T., Trinh, T. A., & Chew, J. W. (2019). Zwitterionic grafting of sulfobetaine methacrylate (SBMA) on hydrophobic PVDF membranes for enhanced anti-fouling and anti-wetting in the membrane distillation of oil emulsions. *Journal of Membrane Science, 588*, 117196. https://doi.org/10.1016/j.memsci.2019.117196

Harvey, N. J., ur Rehman, Z., Leiknes, T., Ghaffour, N., Urakawa, H., & Missimer, T. M. (2020). Organic compounds and microbial assessment of a seawater reverse osmosis facility at Tampa Bay Water, USA. *Desalination, 496*, 114735. https://doi.org/10.1016/j.desal.2020.114735

He, F., Gilron, J., Lee, H., Song, L., & Sirkar, K. K. (2008). Potential for scaling by sparingly soluble salts in crossflow DCMD. *Journal of Membrane Science, 311*(1–2), 68–80.

He, F., Sirkar, K. K., & Gilron, J. (2009). Effects of antiscalants to mitigate membrane scaling by direct contact membrane distillation. *Journal of Membrane Science, 345*(1), 53–58. https://doi.org/10.1016/j.memsci.2009.08.021

He, K., Hwang, H. J., & Moon, I. S. (2011). Air gap membrane distillation on the different types of membrane. *Korean Journal of Chemical Engineering, 28*(3), 770–777. doi:10.1007/s11814-010-0415-0

Hermia, J. (1982). Constant pressure blocking filtration laws: Application to power-law non-Newtonian fluids. *Transactions of the Institution of Chemical Engineers, 60*, 183–187.

Hillis, P. (2000). *Membrane technology in water and wastewater treatment*. Royal Society of Chemistry.

Ho, C.-C., & Zydney, A. L. (2000a). A combined pore blockage and cake filtration model for protein fouling during microfiltration. *Journal of Colloid and Interface Science, 232*(2), 389–399.

Ho, C.-C., & Zydney, A. L. (2000b). Measurement of membrane pore interconnectivity. *Journal of Membrane Science, 170*(1), 101–112. https://doi.org/10.1016/S0376-7388(99)00360-9

Hoek, E. M. V., Allred, J., Knoell, T., & Jeong, B.-H. (2008). Modeling the effects of fouling on full-scale reverse osmosis processes. *Journal of Membrane Science, 314*(1), 33–49. https://doi.org/10.1016/j.memsci.2008.01.025

Hoek, E. M., & Elimelech, M. (2003). Cake-enhanced concentration polarization: A new fouling mechanism for salt-rejecting membranes. *Environmental Science & Technology, 37*(24), 5581–5588.

Hong, J., & He, Y. (2012). Effects of nano sized zinc oxide on the performance of PVDF microfiltration membranes. *Desalination, 302*, 71–79. https://doi.org/10.1016/j.desal.2012.07.001

Hong, S., & Elimelech, M. (1997). Chemical and physical aspects of natural organic matter (NOM) fouling of nanofiltration membranes. *Journal of Membrane Science, 132*(2), 159–181. https://doi.org/10.1016/S0376-7388(97)00060-4

Horseman, T., Su, C., Christie, K. S. S., & Lin, S. (2019). Highly effective scaling mitigation in membrane distillation using a superhydrophobic membrane with gas purging. *Environmental Science & Technology Letters, 6*(7), 423–429. doi:10.1021/acs.estlett.9b00354

Horseman, T., Yin, Y., Christie, K. S. S., Wang, Z., Tong, T., & Lin, S. (2021). Wetting, scaling, and fouling in membrane distillation: State-of-the-art insights on fundamental mechanisms and mitigation strategies. *ACS ES&T Engineering, 1*(1), 117–140. doi:10.1021/acsestengg.0c00025

Hou, L., Liu, J., Li, D., Gao, Y., Wang, Y., Hu, R., . . . Wang, N. (2021). Electrospinning Janus nanofibrous membrane for unidirectional liquid penetration and its applications. *Chemical Research in Chinese Universities, 37*(3), 337–354. doi:10.1007/s40242-021-0010-4

Howe, K. J., & Clark, M. M. (2002). Fouling of microfiltration and ultrafiltration membranes by natural waters. *Environmental Science & Technology, 36*(16), 3571–3576. doi:10.1021/es025587r

Howell, J. A. (1995). Sub-critical flux operation of microfiltration. *Journal of Membrane Science, 107*(1), 165–171. https://doi.org/10.1016/0376-7388(95)00114-R

Huang, H., Young, T. A., & Jacangelo, J. G. (2008). Unified membrane fouling index for low pressure membrane filtration of natural waters: Principles and methodology. *Environmental Science & Technology, 42*(3), 714–720. doi:10.1021/es071043j

Huang, L., & Morrissey, M. T. (1998). Fouling of membranes during microfiltration of surimi wash water: Roles of pore blocking and surface cake formation. *Journal of Membrane Science, 144*(1), 113–123. https://doi.org/10.1016/S0376-7388(98)00038-6

Hube, S., Wang, J., Sim, L. N., Ólafsdóttir, D., Chong, T. H., & Wu, B. (2021). Fouling and mitigation mechanisms during direct microfiltration and ultrafiltration of primary wastewater. *Journal of Water Process Engineering, 44*, 102331. https://doi.org/10.1016/j.jwpe.2021.102331

Hwang, K.-J., Liao, C.-Y., & Tung, K.-L. (2008). Effect of membrane pore size on the particle fouling in membrane filtration. *Desalination, 234*(1), 16–23. https://doi.org/10.1016/j.desal.2007.09.065

Insights, Q. M. (2021). *Reverse osmosis (RO) membrane market, by filter module (plate and frame (PF), tubular, spiral wound, and hollow fiber), by type (thin-film composite membranes, cellulose based membranes), by application (wastewater treatment and reuse, desalination, utility water treatment, and process water), end-use industry (residential and commercial, municipal, energy and power, food and beverage, healthcare, chemical, others), filter module, application, and region (North America, Europe, Asia Pacific, Middle East & Africa, South America)—market size & forecasting to 2030.* www. quincemarketinsights.com/industry-analysis/reverse-osmosis-ro-membrane-market

Ismail, A. F., Khulbe, K. C., & Matsuura, T. (2019). Chapter 8—RO membrane fouling. In A. F. Ismail, K. C. Khulbe, & T. Matsuura (Eds.), *Reverse osmosis* (pp. 189–220). Elsevier.

Israelachvili, J. N., & Pashley, R. M. (1982). The hydrophobic interaction is long range, decaying exponentially with distance. *Nature, 300*(5890), 341–342. doi:10.1038/300341a0

Israelachvili, J. N., & Pashley, R. M. (1984). Measurement of the hydrophobic interaction between two hydrophobic surfaces in aqueous electrolyte solutions. *Journal of Colloid and Interface Science, 98*(2), 500–514. https://doi.org/10.1016/0021-9797(84)90177-2

Jacob, P., Dejean, B., Laborie, S., & Cabassud, C. (2020). An optical in-situ tool for visualizing and understanding wetting dynamics in membrane distillation. *Journal of Membrane Science, 595*, 117587. https://doi.org/10.1016/j.memsci.2019.117587

Jacob, P., Zhang, T., Laborie, S., & Cabassud, C. (2019). Influence of operating conditions on wetting and wettability in membrane distillation using detection of dissolved tracer intrusion (DDTI). *Desalination, 468*, 114086. https://doi.org/10.1016/j.desal.2019.114086

Jafari, M., Vanoppen, M., van Agtmaal, J. M. C., Cornelissen, E. R., Vrouwenvelder, J. S., Verliefde, A., . . . Picioreanu, C. (2021). Cost of fouling in full-scale reverse osmosis and nanofiltration installations in the Netherlands. *Desalination, 500*, 114865. https://doi.org/10.1016/j.desal.2020.114865

Jawor, A., & Hoek, E. M. (2009). Effects of feed water temperature on inorganic fouling of brackish water RO membranes. *Desalination, 235*(1–3), 44–57.

Jiang, Z., Karan, S., & Livingston, A. G. (2020). Membrane fouling: Does microscale roughness matter? *Industrial & Engineering Chemistry Research, 59*(12), 5424–5431. doi:10.1021/acs.iecr.9b04798

Jim, K. J., Fane, A. G., Fell, C. J. D., & Joy, D. C. (1992). Fouling mechanisms of membranes during protein ultrafiltration. *Journal of Membrane Science, 68*(1), 79–91. https://doi.org/10.1016/0376-7388(92)80151-9

Jin, X., Huang, X., & Hoek, E. M. (2009). Role of specific ion interactions in seawater RO membrane fouling by alginic acid. *Environmental Science & Technology, 43*(10), 3580–3587.

Jung, B. (2004). Preparation of hydrophilic polyacrylonitrile blend membranes for ultrafiltration. *Journal of Membrane Science, 229*(1), 129–136. https://doi.org/10.1016/j.memsci.2003.10.020

Kammerhofer, J., Fries, L., Dupas, J., Forny, L., Heinrich, S., & Palzer, S. (2018). Impact of hydrophobic surfaces on capillary wetting. *Powder Technology, 328*, 367–374. https://doi.org/10.1016/j.powtec.2018.01.033

Kannan, M., Rajarathinam, K., Venkatesan, S., Dheeba, B., & Maniraj, A. (2017). Chapter 19—silver iodide nanoparticles as an antibiofilm agent—a case study on gram-negative biofilm-forming bacteria. In A. Ficai & A. M. Grumezescu (Eds.), *Nanostructures for antimicrobial therapy* (pp. 435–456). Elsevier.

Karkhanechi, H., Razi, F., Sawada, I., Takagi, R., Ohmukai, Y., & Matsuyama, H. (2013). Improvement of antibiofouling performance of a reverse osmosis membrane through biocide release and adhesion resistance. *Separation and Purification Technology, 105*, 106–113.

Khan, I. A., Lee, Y.-S., & Kim, J.-O. (2020). A comparison of variations in blocking mechanisms of membrane-fouling models for estimating flux during water treatment. *Chemosphere, 259*, 127328. https://doi.org/10.1016/j.chemosphere.2020.127328

Kiefer, F., Präbst, A., Rodewald, K. S., & Sattelmayer, T. (2019). Membrane scaling in vacuum membrane distillation—part 1: In-situ observation of crystal growth and membrane wetting. *Journal of Membrane Science, 590*, 117294. https://doi.org/10.1016/j.memsci.2019.117294

Kim, J., Kim, H.-W., Tijing, L. D., Shon, H. K., & Hong, S. (2022). Elucidation of physicochemical scaling mechanisms in membrane distillation (MD): Implication to the control of inorganic fouling. *Desalination, 527*, 115573. https://doi.org/10.1016/j. desal.2022.115573

Kirschner, A. Y., Cheng, Y.-H., Paul, D. R., Field, R. W., & Freeman, B. D. (2019). Fouling mechanisms in constant flux crossflow ultrafiltration. *Journal of Membrane Science, 574*, 65–75.

Kochkodan, V. M., & Hilal, N. (2015). A comprehensive review on surface modified polymer membranes for biofouling mitigation. *Desalination, 356*, 187–207. https://doi. org/10.1016/j.desal.2014.09.015

Kochkodan, V. M., Hilal, N., Goncharuk, V. V., Al-Khatib, L., & Levadna, T. I. (2006). Effect of the surface modification of polymer membranes on their microbiological fouling. *Colloid Journal, 68*(3), 267–273. doi:10.1134/S1061933X06030021

Kochkodan, V. M., Johnson, D. J., & Hilal, N. (2014). Polymeric membranes: Surface modification for minimizing (bio)colloidal fouling. *Advances in Colloid and Interface Science, 206*, 116–140. https://doi.org/10.1016/j.cis.2013.05.005

Koo, C. H., Mohammad, A. W., Suja', F., & Meor Talib, M. Z. (2012). Review of the effect of selected physicochemical factors on membrane fouling propensity based on fouling indices. *Desalination, 287*, 167–177. https://doi.org/10.1016/j.desal.2011.11.003

Kumar, M., & Ulbricht, M. (2013). Novel antifouling positively charged hybrid ultrafiltration membranes for protein separation based on blends of carboxylated carbon nanotubes and aminated poly (arylene ether sulfone). *Journal of Membrane Science, 448*, 62–73. https://doi.org/10.1016/j.memsci.2013.07.055

Kumar, P., Bharti, R. P., Kumar, V., & Kundu, P. P. (2018). Chapter 4—polymer electrolyte membranes for microbial fuel cells: Part A. Nafion-Based membranes. In P. P. Kundu & K. Dutta (Eds.), *Progress and recent trends in microbial fuel cells* (pp. 47–72). Elsevier.

Kumar, R., & Ismail, A. F. (2015). Fouling control on microfiltration/ultrafiltration membranes: Effects of morphology, hydrophilicity, and charge. *Journal of Applied Polymer Science, 132*(21). https://doi.org/10.1002/app.42042

Lay, H. T., Yeow, R. J. E., Ma, Y., Zydney, A. L., Wang, R., & Chew, J. W. (2021). Internal membrane fouling by proteins during microfiltration. *Journal of Membrane Science, 637*, 119589. https://doi.org/10.1016/j.memsci.2021.119589

Lazaridis, T. (2013). Hydrophobic effect. *eLS*.

Lee, S., Ang, W. S., & Elimelech, M. (2006). Fouling of reverse osmosis membranes by hydrophilic organic matter: Implications for water reuse. *Desalination, 187*(1), 313–321. https://doi.org/10.1016/j.desal.2005.04.090

Lee, S., Cho, J., & Elimelech, M. (2005). Combined influence of natural organic matter (NOM) and colloidal particles on nanofiltration membrane fouling. *Journal of Membrane Science, 262*(1), 27–41. https://doi.org/10.1016/j.memsci.2005.03.043

Lee, S., & Elimelech, M. (2006). Relating organic fouling of reverse osmosis membranes to intermolecular adhesion forces. *Environmental Science & Technology, 40*(3), 980–987. doi:10.1021/es051825h

Lee, S., & Lee, C.-H. (2000). Effect of operating conditions on CaSO4 scale formation mechanism in nanofiltration for water softening. *Water Research, 34*(15), 3854–3866.

Lei, Z., Dzakpasu, M., Li, Q., & Chen, R. (2020). 6—Anaerobic membrane bioreactors for domestic wastewater treatment. In H. H. Ngo, W. Guo, H. Y. Ng, G. Mannina, & A. Pandey (Eds.), *Current developments in biotechnology and bioengineering* (pp. 143–165). Elsevier.

Li, Q., & Elimelech, M. (2004a). Natural organic matter fouling and chemical cleaning of nanofiltration membranes. *Water Science and Technology: Water Supply, 4*(5–6), 245–251.

Li, Q., & Elimelech, M. (2004b). Organic fouling and chemical cleaning of nanofiltration membranes: Measurements and mechanisms. *Environmental Science & Technology, 38*(17), 4683–4693. doi:10.1021/es0354162

Li, Q., Xu, Z., & Pinnau, I. (2007). Fouling of reverse osmosis membranes by biopolymers in wastewater secondary effluent: Role of membrane surface properties and initial permeate flux. *Journal of Membrane Science, 290*(1–2), 173–181.

Li, R., Lou, Y., Xu, Y., Ma, G., Liao, B.-Q., Shen, L., & Lin, H. (2019). Effects of surface morphology on alginate adhesion: Molecular insights into membrane fouling based on XDLVO and DFT analysis. *Chemosphere, 233*, 373–380. https://doi.org/10.1016/j.chemosphere.2019.05.262

Li, X., & Li, J. (2015). Critical flux. In E. Drioli & L. Giorno (Eds.), *Encyclopedia of membranes* (pp. 1–3). Springer Berlin Heidelberg.

Liao, Y., Zheng, G., Huang, J. J., Tian, M., & Wang, R. (2020). Development of robust and superhydrophobic membranes to mitigate membrane scaling and fouling in membrane distillation. *Journal of Membrane Science, 601*, 117962. https://doi.org/10.1016/j.memsci.2020.117962

Liderfelt, J., & Royce, J. (2018). Chapter 14—filtration principles. In G. Jagschies, E. Lindskog, K. Łącki, & P. Galliher (Eds.), *Biopharmaceutical processing* (pp. 279–293). Elsevier.

Linhares, A. M., Borges, C. P., & Fonseca, F. V. (2020). Investigation of biocidal effect of microfiltration membranes impregnated with silver nanoparticles by sputtering technique. *Polymers, 12*(8), 1686.

Liu, C. (2014). 2.5—advances in membrane technologies for drinking water purification. In S. Ahuja (Ed.), *Comprehensive water quality and purification* (pp. 75–97). Elsevier.

Liu, J., Wang, Y., Li, S., Li, Z., Liu, X., & Li, W. (2022). Insights into the wetting phenomenon induced by scaling of calcium sulfate in membrane distillation. *Water Research*, 118282. https://doi.org/10.1016/j.watres.2022.118282

Liu, L., Luo, X.-B., Ding, L., & Luo, S.-L. (2019). 4—application of nanotechnology in the removal of heavy metal from water. In X. Luo & F. Deng (Eds.), *Nanomaterials for the removal of pollutants and resource reutilization* (pp. 83–147). Elsevier.

Liu, L., Xiao, Z., Liu, Y., Li, X., Yin, H., Volkov, A., & He, T. (2021). Understanding the fouling/scaling resistance of superhydrophobic/omniphobic membranes in membrane distillation. *Desalination, 499*, 114864. https://doi.org/10.1016/j.desal.2020.114864

Loh, S., Beuscher, U., Poddar, T. K., Porter, A. G., Wingard, J. M., Husson, S. M., & Wickramasinghe, S. R. (2009). Interplay among membrane properties, protein properties and operating conditions on protein fouling during normal-flow microfiltration. *Journal of Membrane Science, 332*(1), 93–103. https://doi.org/10.1016/j.memsci.2009.01.031

Lohaus, J., Perez, Y. M., & Wessling, M. (2018). What are the microscopic events of colloidal membrane fouling? *Journal of Membrane Science, 553*, 90–98. https://doi.org/10.1016/j.memsci.2018.02.023

Lou, M., Fang, X., Huang, S., Li, J., Liu, Y., Chen, G., & Li, F. (2022). Effect of cations on surfactant induced membrane wetting during membrane distillation. *Desalination, 532*, 115739. https://doi.org/10.1016/j.desal.2022.115739

Lu, K.-J., & Chung, T.-S. (2019). *Membrane distillation: Membranes, hybrid systems and pilot studies*. Taylor & Francis Group.

Luo, Z., Wang, Y., Yang, Q., Luo, Y., Tan, S., Chen, T., & Xie, Z. (2014). Influence of engineering environment on wetting properties and long-term stability of a superhydrophobic polymer coating. *Journal of Polymer Research, 21*(5), 447. doi:10.1007/s10965-014-0447-y

Ma, B., Ding, Y., Li, W., Hu, C., Yang, M., Liu, H., & Qu, J. (2018). Ultrafiltration membrane fouling induced by humic acid with typical inorganic salts. *Chemosphere, 197*, 793–802. https://doi.org/10.1016/j.chemosphere.2018.01.037

Ma, W., Soroush, A., Luong, T. V. A., Brennan, G., Rahaman, M. S., Asadishad, B., & Tufenkji, N. (2016). Spray-and spin-assisted layer-by-layer assembly of copper nanoparticles on thin-film composite reverse osmosis membrane for biofouling mitigation. *Water Research, 99*, 188–199.

Madhavan, P., Hong, P.-Y., Sougrat, R., & Nunes, S. P. (2014). Silver-enhanced block copolymer membranes with biocidal activity. *ACS Applied Materials & Interfaces*, *6*(21), 18497–18501.

Makardij, A., Chen, X. D., & Farid, M. M. (1999). Microfiltration and ultrafiltration of milk: Some aspects of fouling and cleaning. *Food and Bioproducts Processing*, *77*(2), 107–113. https://doi.org/10.1205/096030899532394

Mallevialle, J., Odendaal, P. E., & Wiesner, M. R. (1996). The emergence of membranes in water and wastewater treatment. *Water Treatment Membrane Processes*, 1.1–1.10.

Matin, A., Laoui, T., Falath, W., & Farooque, M. (2021). Fouling control in reverse osmosis for water desalination & reuse: Current practices & emerging environment-friendly technologies. *Science of the Total Environment*, *765*, 142721.

Matin, A., Rahman, F., Shafi, H. Z., & Zubair, S. M. (2019). Scaling of reverse osmosis membranes used in water desalination: Phenomena, impact, and control; future directions. *Desalination*, *455*, 135–157. https://doi.org/10.1016/j.desal.2018.12.009

McGaughey, A. L., & Childress, A. E. (2022). Wetting indicators, modes, and trade-offs in membrane distillation. *Journal of Membrane Science*, *642*, 119947. https://doi.org/10.1016/j.memsci.2021.119947

Melián-Martel, N., Sadhwani, J. J., Malamis, S., & Ochsenkühn-Petropoulou, M. (2012). Structural and chemical characterization of long-term reverse osmosis membrane fouling in a full scale desalination plant. *Desalination*, *305*, 44–53. https://doi.org/10.1016/j.desal.2012.08.011

Meng, S., Fan, W., Li, X., Liu, Y., Liang, D., & Liu, X. (2018). Intermolecular interactions of polysaccharides in membrane fouling during microfiltration. *Water Research*, *143*, 38–46. https://doi.org/10.1016/j.watres.2018.06.027

Miller, D. J., Kasemset, S., Paul, D. R., & Freeman, B. D. (2014). Comparison of membrane fouling at constant flux and constant transmembrane pressure conditions. *Journal of Membrane Science*, *454*, 505–515.

Misdan, N., Ismail, A. F., & Hilal, N. (2016). Recent advances in the development of (bio) fouling resistant thin film composite membranes for desalination. *Desalination*, *380*, 105–111. https://doi.org/10.1016/j.desal.2015.06.001

Missana, T., & Adell, A. (2000). On the applicability of DLVO theory to the prediction of clay colloids stability. *Journal of Colloid and Interface Science*, *230*(1), 150–156.

Miyoshi, T., Yuasa, K., Ishigami, T., Rajabzadeh, S., Kamio, E., Ohmukai, Y., . . . Matsuyama, H. (2015). Effect of membrane polymeric materials on relationship between surface pore size and membrane fouling in membrane bioreactors. *Applied Surface Science*, *330*, 351–357. https://doi.org/10.1016/j.apsusc.2015.01.018

Möckel, D., Staude, E., & Guiver, M. D. (1999). Static protein adsorption, ultrafiltration behavior and cleanability of hydrophilized polysulfone membranes. *Journal of Membrane Science*, *158*(1–2), 63–75.

Mohammadi, T., & Esmaeelifar, A. (2005). Wastewater treatment of a vegetable oil factory by a hybrid ultrafiltration-activated carbon process. *Journal of Membrane Science*, *254*(1), 129–137. https://doi.org/10.1016/j.memsci.2004.12.037

Moonkhum, M., Lee, Y. G., Lee, Y. S., & Kim, J. H. (2010). Review of seawater natural organic matter fouling and reverse osmosis transport modeling for seawater reverse osmosis desalination. *Desalination and Water Treatment*, *15*(1–3), 92–107.

Mora, F., Pérez, K., Quezada, C., Herrera, C., Cassano, A., & Ruby-Figueroa, R. (2019). Impact of membrane pore size on the clarification performance of grape marc extract by microfiltration. *Membranes*, *9*(11), 146. doi:10.3390/membranes9110146

Morel, F. M., & Hering, J. G. (1993). *Principles and applications of aquatic chemistry*. John Wiley & Sons.

Mosadegh-Sedghi, S., Rodrigue, D., Brisson, J., & Iliuta, M. C. (2014). Wetting phenomenon in membrane contactors—causes and prevention. *Journal of Membrane Science*, *452*, 332–353. https://doi.org/10.1016/j.memsci.2013.09.055

Munla, L., Peldszus, S., & Huck, P. M. (2012). Reversible and irreversible fouling of ultrafiltration ceramic membranes by model solutions. *Journal AWWA, 104*(10), E540–E554. https://doi.org/10.5942/jawwa.2012.104.0137

Naidu, G., Jeong, S., Kim, S.-J., Kim, I. S., & Vigneswaran, S. (2014). Organic fouling behavior in direct contact membrane distillation. *Desalination, 347*, 230–239. https://doi.org/10.1016/j.desal.2014.05.045

Naidu, G., Jeong, S., & Vigneswaran, S. (2014). Influence of feed/permeate velocity on scaling development in a direct contact membrane distillation. *Separation and Purification Technology, 125*, 291–300. https://doi.org/10.1016/j.seppur.2014.01.049

Newcombe, G. (1999). Charge vs. porosity—Some influences on the adsorption of natural organic matter (NOM) by activated carbon. *Water Science and Technology, 40*(9), 191–198. https://doi.org/10.1016/S0273-1223(99)00656-3

Ng, C. Y., Mohammad, A. W., Ng, L. Y., & Jahim, J. M. (2014). Membrane fouling mechanisms during ultrafiltration of skimmed coconut milk. *Journal of Food Engineering, 142*, 190–200. https://doi.org/10.1016/j.jfoodeng.2014.06.005

Nghiem, L. D., & Cath, T. (2011). A scaling mitigation approach during direct contact membrane distillation. *Separation and Purification Technology, 80*(2), 315–322. https://doi.org/10.1016/j.seppur.2011.05.013

Nguyen, A., Azari, S., & Zou, L. (2013). Coating zwitterionic amino acid l-DOPA to increase fouling resistance of forward osmosis membrane. *Desalination, 312*, 82–87. https://doi.org/10.1016/j.desal.2012.11.038

Ninham, B. W. (1999). On progress in forces since the DLVO theory. *Advances in Colloid and Interface Science, 83*(1), 1–17. doi:10.1016/S0001-8686(99)00008-1

Ocean, R. (2022). *Global RO membrane chemicals market research report forecast to 2025.* https://reportocean.com/industry-verticals/sample-request?report_id=32848

Ochando-Pulido, J., Victor-Ortega, M., & Martinez-Ferez, A. (2015). On the cleaning procedure of a hydrophilic reverse osmosis membrane fouled by secondary-treated olive mill wastewater. *Chemical Engineering Journal, 260*, 142–151.

Oliveira, R. (1997). Understanding adhesion: A means for preventing fouling. *Experimental Thermal and Fluid Science, 14*(4), 316–322.

Ortega-Vinuesa, J. L., Martín-Rodríguez, A., & Hidalgo-Álvarez, R. (1996). Colloidal stability of polymer colloids with different interfacial properties: Mechanisms. *Journal of Colloid and Interface Science, 184*(1), 259–267. https://doi.org/10.1006/jcis.1996.0619

Pallas, N. R., & Pethica, B. A. (1983). The surface tension of water. *Colloids and Surfaces, 6*(3), 221–227. https://doi.org/10.1016/0166-6622(83)80014-6

Palmer, S. (1976). The effect of temperature on surface tension. *Physics Education, 11*(2), 119.

Panchal, C. B., & Knudsen, J. G. (1998). Mitigation of water fouling: Technology status and challenges**Work supported by the U.S. Department of energy, assistant secretary of energy efficiency and renewable energy, and the office of industrial technologies, under contract W-31–109-Eng-38. Accordingly, the U.S. Government retains a nonexclusive, royalty-free license to publish or reproduce the published form of this contribution, or allow others to do so, for U.S. Government purposes. In J. P. Hartnett, T. F. Irvine, Y. I. Cho, & G. A. Greene (Eds.), *Advances in heat transfer* (Vol. 31, pp. 431–474). Elsevier.

Park, W.-I., Jeong, S., Im, S.-J., & Jang, A. (2020). High turbidity water treatment by ceramic microfiltration membrane: Fouling identification and process optimization. *Environmental Technology & Innovation, 17*, 100578. https://doi.org/10.1016/j.eti.2019.100578

Patel, D. T., Solanki, J. D., Patel, K. C., & Nataraj, M. (2021). Chapter 13—application of biosurfactants as antifouling agent. In Inamuddin & C. O. Adetunji (Eds.), *Green sustainable process for chemical and environmental engineering and science* (pp. 275–289). Elsevier.

Pervov, A. G. (1991). Scale formation prognosis and cleaning procedure schedules in reverse osmosis systems operation. *Desalination, 83*(1–3), 77–118.

Pollice, A., Brookes, A., Jefferson, B., & Judd, S. (2005). Sub-critical flux fouling in membrane bioreactors—a review of recent literature. *Desalination, 174*(3), 221–230. https://doi.org/10.1016/j.desal.2004.09.012

Premlal Ranjith, H. M., & Wijewardene, U. (2006). 16—Lipid emulsifiers and surfactants in dairy and bakery products. In F. D. Gunstone (Ed.), *Modifying lipids for use in food* (pp. 393–428). Woodhead Publishing.

Prieto, M. (2018). Nucleation and supersaturation in porous media (revisited). *Mineralogical Magazine, 78*(6), 1437–1447. doi:10.1180/minmag.2014.078.6.11

Qureshi, B. A., Zubair, S. M., Sheikh, A. K., Bhujle, A., & Dubowsky, S. (2013). Design and performance evaluation of reverse osmosis desalination systems: An emphasis on fouling modeling. *Applied Thermal Engineering, 60*(1–2), 208–217.

Rahaman, M. S., Thérien-Aubin, H., Ben-Sasson, M., Ober, C. K., Nielsen, M., & Elimelech, M. (2014). Control of biofouling on reverse osmosis polyamide membranes modified with biocidal nanoparticles and antifouling polymer brushes. *Journal of Materials Chemistry B, 2*(12), 1724–1732.

Rahimpour, M. R., & Esmaeilbeig, M. A. (2019). Chapter 6—membrane wetting in membrane distillation. In A. Basile, E. Curcio, & Inamuddin (Eds.), *Current trends and future developments on (Bio-) membranes* (pp. 143–174). Elsevier.

Ramon, G. Z., & Hoek, E. M. V. (2012). On the enhanced drag force induced by permeation through a filtration membrane. *Journal of Membrane Science, 392–393*, 1–8. https://doi.org/10.1016/j.memsci.2011.10.056

Reports, P. (2021). *Fouling resistant compound RO membrane market 2022–2028.* https://www.marketsandresearch.biz/

Rezaei, H., Ashtiani, F. Z., & Fouladitajar, A. (2011). Effects of operating parameters on fouling mechanism and membrane flux in cross-flow microfiltration of whey. *Desalination, 274*(1), 262–271. https://doi.org/10.1016/j.desal.2011.02.015

Rezaei, M., Warsinger, D. M., Lienhard, V. J. H., Duke, M. C., Matsuura, T., & Samhaber, W. M. (2018). Wetting phenomena in membrane distillation: Mechanisms, reversal, and prevention. *Water Research, 139*, 329–352. https://doi.org/10.1016/j.watres.2018.03.058

Rezaei, M., Warsinger, D. M., Lienhard, V. J. H., & Samhaber, W. M. (2017). Wetting prevention in membrane distillation through superhydrophobicity and recharging an air layer on the membrane surface. *Journal of Membrane Science, 530*, 42–52. https://doi.org/10.1016/j.memsci.2017.02.013

Ridgway, H. F. (2003). *Biological fouling of separation membranes used in water treatment applications*: AWWA research Foundation.

Saffarini, R. B., Mansoor, B., Thomas, R., & Arafat, H. A. (2013). Effect of temperature-dependent microstructure evolution on pore wetting in PTFE membranes under membrane distillation conditions. *Journal of Membrane Science, 429*, 282–294. https://doi.org/10.1016/j.memsci.2012.11.049

Salinas Rodriguez, S. G., Sithole, N., Dhakal, N., Olive, M., Schippers, J. C., & Kennedy, M. D. (2019). Monitoring particulate fouling of North Sea water with SDI and new ASTM MFI0.45 test. *Desalination, 454*, 10–19. https://doi.org/10.1016/j.desal.2018.12.006

Sarkar, A., Carver, P. I., Zhang, T., Merrington, A., Bruza, K. J., Rousseau, J. L., . . . Dvornic, P. R. (2010). Dendrimer-based coatings for surface modification of polyamide reverse osmosis membranes. *Journal of Membrane Science, 349*(1–2), 421–428.

Sarti, G. C., Gostoli, C., & Matulli, S. (1985). Low energy cost desalination processes using hydrophobic membranes. *Desalination, 56*, 277–286. https://doi.org/10.1016/0011-9164(85)85031-1

Seidel, A., & Elimelech, M. (2002). Coupling between chemical and physical interactions in natural organic matter (NOM) fouling of nanofiltration membranes: Implications for fouling control. *Journal of Membrane Science, 203*(1), 245–255. https://doi.org/10.1016/S0376-7388(02)00013-3

Shahbaz, K., Mjalli, F. S., Hashim, M. A., & AlNashef, I. M. (2012). Prediction of the surface tension of deep eutectic solvents. *Fluid Phase Equilibria*, *319*, 48–54. https://doi.
org/10.1016/j.fluid.2012.01.025

Shen, L.-Q., Wang, X., Li, R., Yu, H., Hong, H., Lin, H., . . . Liao, B.-Q. (2017). Physicochemical correlations between membrane surface hydrophilicity and adhesive fouling in
membrane bioreactors. *Journal of Colloid and Interface Science*, *505*, 900–909. https://
doi.org/10.1016/j.jcis.2017.06.090

Shen, L.-Q., Xu, Z.-K., Liu, Z.-M., & Xu, Y.-Y. (2003). Ultrafiltration hollow fiber membranes
of sulfonated polyetherimide/polyetherimide blends: Preparation, morphologies and
anti-fouling properties. *Journal of Membrane Science*, *218*(1), 279–293. https://doi.
org/10.1016/S0376-7388(03)00186-8

Shirazi, S., Lin, C.-J., & Chen, D. (2010). Inorganic fouling of pressure-driven membrane processes—a
critical review. *Desalination*, *250*(1), 236–248. https://doi.org/10.1016/j.desal.2009.02.056

Sillanpää, M., Matilainen, A., & Lahtinen, T. (2015). Chapter 2—characterization of NOM. In
M. Sillanpää (Ed.), *Natural organic matter in water* (pp. 17–53). Butterworth-Heinemann.

Singh, R. (2015). Chapter 2—water and membrane treatment. In R. Singh (Ed.), *Membrane
technology and engineering for water purification* (2nd ed., pp. 81–178). Butterworth-
Heinemann.

Sinha Ray, S., Lee, H.-K., & Kwon, Y.-N. (2020). Review on blueprint of designing anti-wetting
polymeric membrane surfaces for enhanced membrane distillation performance. *Polymers*,
12(1), 23.

Stamatakis, E., Stubos, A., Palyvos, J., Chatzichristos, C., & Muller, J. (2005). An improved
predictive correlation for the induction time of CaCO3 scale formation during flow in
porous media. *Journal of Colloid and Interface Science*, *286*(1), 7–13.

Steiger, M. (2005). Crystal growth in porous materials—I: The crystallization pressure of
large crystals. *Journal of Crystal Growth*, *282*(3), 455–469. https://doi.org/10.1016/j.
jcrysgro.2005.05.007

Stumm, W. (1993). Aquatic colloids as chemical reactants: Surface structure and reactivity.
Colloids and Surfaces A: Physicochemical and Engineering Aspects, *73*, 1–18. https://
doi.org/10.1016/0927-7757(93)80003-W

Sun, Y., Qin, Z., Zhao, L., Chen, Q., Hou, Q., Lin, H., . . . Du, Z. (2018). Membrane fouling
mechanisms and permeate flux decline model in soy sauce microfiltration. *Journal of
Food Process Engineering*, *41*(1), e12599. https://doi.org/10.1111/jfpe.12599

Tang, C. Y., Chong, T. H., & Fane, A. G. (2011). Colloidal interactions and fouling of NF and
RO membranes: A review. *Advances in Colloid and Interface Science*, *164*(1), 126–143.
https://doi.org/10.1016/j.cis.2010.10.007

Tang, C. Y., Kwon, Y.-N., & Leckie, J. O. (2009). The role of foulant–foulant electrostatic
interaction on limiting flux for RO and NF membranes during humic acid fouling—The
oretical basis, experimental evidence, and AFM interaction force measurement. *Journal
of Membrane Science*, *326*(2), 526–532. https://doi.org/10.1016/j.memsci.2008.10.043

Tang, F., Hu, H.-Y., Sun, L.-J., Wu, Q.-Y., Jiang, Y.-M., Guan, Y.-T., & Huang, J.-J. (2014).
Fouling of reverse osmosis membrane for municipal wastewater reclamation: Autopsy
results from a full-scale plant. *Desalination*, *349*, 73–79.

Tian, J.-Y., Ernst, M., Cui, F., & Jekel, M. (2013). Effect of particle size and concentration on
the synergistic UF membrane fouling by particles and NOM fractions. *Journal of Membrane Science*, *446*, 1–9. https://doi.org/10.1016/j.memsci.2013.06.016

Tijing, L. D., Woo, Y. C., Choi, J.-S., Lee, S., Kim, S.-H., & Shon, H. K. (2015). Fouling
and its control in membrane distillation—a review. *Journal of Membrane Science*, *475*,
215–244. https://doi.org/10.1016/j.memsci.2014.09.042

Tong, T., Wallace, A. F., Zhao, S., & Wang, Z. (2019). Mineral scaling in membrane desalination:
Mechanisms, mitigation strategies, and feasibility of scaling-resistant membranes. *Journal of Membrane Science*, *579*, 52–69. https://doi.org/10.1016/j.memsci.2019.02.049

Trinh, T. A., Li, W., & Chew, J. W. (2020). Internal fouling during microfiltration with foulants of different surface charges. *Journal of Membrane Science, 602*, 117983. https://doi. org/10.1016/j.memsci.2020.117983

Tu, K. L., Chivas, A. R., & Nghiem, L. D. (2011). Effects of membrane fouling and scaling on boron rejection by nanofiltration and reverse osmosis membranes. *Desalination, 279*(1–3), 269–277.

Tzotzi, C., Pahiadaki, T., Yiantsios, S., Karabelas, A., & Andritsos, N. (2007). A study of CaCO3 scale formation and inhibition in RO and NF membrane processes. *Journal of Membrane Science, 296*(1–2), 171–184.

Uchymiak, M., Lyster, E., Glater, J., & Cohen, Y. (2008). Kinetics of gypsum crystal growth on a reverse osmosis membrane. *Journal of Membrane Science, 314*(1–2), 163–172.

Van Oss, C. J., Good, R. J., & Chaudhury, M. K. (1986). The role of van der Waals forces and hydrogen bonds in "hydrophobic interactions" between biopolymers and low energy surfaces. *Journal of Colloid and Interface Science, 111*(2), 378–390. https://doi. org/10.1016/0021-9797(86)90041-X

Vanysacker, L., Boerjan, B., Declerck, P., & Vankelecom, I. F. J. (2014). Biofouling ecology as a means to better understand membrane biofouling. *Applied Microbiology and Biotechnology, 98*(19), 8047–8072. doi:10.1007/s00253-014-5921-2

Vaziri Hassas, B., Karakaş, F., & Çelik, M. S. (2014). Ultrafine coal dewatering: Relationship between hydrophilic lipophilic balance (HLB) of surfactants and coal rank. *International Journal of Mineral Processing, 133*, 97–104. https://doi.org/10.1016/j. minpro.2014.10.010

Velioğlu, S., Han, L., & Chew, J. W. (2018). Understanding membrane pore-wetting in the membrane distillation of oil emulsions via molecular dynamics simulations. *Journal of Membrane Science, 551*, 76–84. https://doi.org/10.1016/j.memsci.2018.01.027

Villalobos García, J., Dow, N., Milne, N., Zhang, J., Naidoo, L., Gray, S., & Duke, M. (2018). Membrane distillation trial on textile wastewater containing surfactants using hydrophobic and hydrophilic-coated polytetrafluoroethylene (PTFE) membranes. *Membranes, 8*(2), 31.

Vrijenhoek, E. M., Hong, S., & Elimelech, M. (2001). Influence of membrane surface properties on initial rate of colloidal fouling of reverse osmosis and nanofiltration membranes. *Journal of Membrane Science, 188*(1), 115–128. https://doi.org/10.1016/ S0376-7388(01)00376-3

Vyas, H. K., Bennett, R. J., & Marshall, A. D. (2000). Influence of operating conditions on membrane fouling in crossflow microfiltration of particulate suspensions. *International Dairy Journal, 10*(7), 477–487. https://doi.org/10.1016/S0958-6946(00)00058-3

Wang, F., & Tarabara, V. V. (2008). Pore blocking mechanisms during early stages of membrane fouling by colloids. *Journal of Colloid and Interface Science, 328*(2), 464–469. https://doi.org/10.1016/j.jcis.2008.09.028

Wang, L., Wang, X., & Fukushi, K.-I. (2008). Effects of operational conditions on ultrafiltration membrane fouling. *Desalination, 229*(1), 181–191. https://doi.org/10.1016/j. desal.2007.08.018

Wang, P., Cheng, W., Zhang, X., Liu, Q., Li, J., Ma, J., & Zhang, T. (2022). Membrane scaling and wetting in membrane distillation: Mitigation roles played by humic substances. *Environmental Science & Technology, 56*(5), 3258–3266. doi:10.1021/acs.est.1c07294

Wang, Y., Han, M., Liu, L., Yao, J., & Han, L. (2020). Beneficial CNT intermediate layer for membrane fluorination toward robust superhydrophobicity and wetting resistance in membrane distillation. *ACS Applied Materials & Interfaces, 12*(18), 20942–20954. doi:10.1021/acsami.0c03577

Wang, Z., Chen, Y., & Lin, S. (2018). Kinetic model for surfactant-induced pore wetting in membrane distillation. *Journal of Membrane Science, 564*, 275–288. https://doi. org/10.1016/j.memsci.2018.07.010

Wang, Z., Chen, Y., Sun, X., Duddu, R., & Lin, S. (2018). Mechanism of pore wetting in membrane distillation with alcohol vs. surfactant. *Journal of Membrane Science, 559*, 183–195. https://doi.org/10.1016/j.memsci.2018.04.045

Wang, Z., & Lin, S. (2017). Membrane fouling and wetting in membrane distillation and their mitigation by novel membranes with special wettability. *Water Research, 112*, 38–47. https://doi.org/10.1016/j.watres.2017.01.022

Warsinger, D. M., Swaminathan, J., Guillen-Burrieza, E., Arafat, H. A., & Lienhard V, J. H. (2015). Scaling and fouling in membrane distillation for desalination applications: A review. *Desalination, 356*, 294–313. https://doi.org/10.1016/j.desal.2014.06.031

Warsinger, D. M., Tow, E. W., & Swaminathan, J. (2017). Theoretical framework for predicting inorganic fouling in membrane distillation and experimental validation with calcium sulfate. *Journal of Membrane Science, 528*, 381–390.

Wiesner, M. R., & Aptel, P. (1996). Mass transport and permeate flux and fouling in pressure-driven processes. *Water Treatment Membrane Processes*, 4.1–4.30.

Wiesner, M. R., Clark, M. M., Jacangelo, J. G., Lykins, B. W., Marinas, B. J., O'Melia, C. R., . . . Fiessinger, F. (1992). Committee report. Membrane processes in potable water treatment. *Journal/American Water Works Association, 84*(1), 59–67.

Williams, D. E. (2011). *Design and construction of slant and vertical wells for desalination intake*. Paper presented at the Proceeding, International Desalination Association World Congress on Desalination and Water Reuse.

Williams, D. E. (2018). Chapter 6—design and construction of subsurface intakes. In V. G. Gude (Ed.), *Sustainable desalination handbook* (pp. 227–258). Butterworth-Heinemann.

Wirtz, B. W., Weyerer, J. C., & Geyer, C. (2019). Artificial intelligence and the public sector—applications and challenges. *International Journal of Public Administration, 42*(7), 596–615. doi:10.1080/01900692.2018.1498103

Wu, X., Zhou, C., Li, K., Zhang, W., & Tao, Y. (2018). Probing the fouling process and mechanisms of submerged ceramic membrane ultrafiltration during algal harvesting under sub- and super-critical fluxes. *Separation and Purification Technology, 195*, 199–207. https://doi.org/10.1016/j.seppur.2017.12.001

Xiao, K., Sun, J., Mo, Y., Fang, Z., Liang, P., Huang, X., . . . Ma, B. (2014). Effect of membrane pore morphology on microfiltration organic fouling: PTFE/PVDF blend membranes compared with PVDF membranes. *Desalination, 343*, 217–225. https://doi.org/10.1016/j.desal.2013.09.026

Xiao, K., Wang, X., Huang, X., Waite, T. D., & Wen, X. (2011). Combined effect of membrane and foulant hydrophobicity and surface charge on adsorptive fouling during microfiltration. *Journal of Membrane Science, 373*(1), 140–151. https://doi.org/10.1016/j.memsci.2011.02.041

Xiao, Z., Guo, H., He, H., Liu, Y., Li, X., Zhang, Y., . . . He, T. (2020). Unprecedented scaling/fouling resistance of omniphobic polyvinylidene fluoride membrane with silica nanoparticle coated micropillars in direct contact membrane distillation. *Journal of Membrane Science, 599*, 117819. https://doi.org/10.1016/j.memsci.2020.117819

Xiao, Z., Li, Z., Guo, H., Liu, Y., Wang, Y., Yin, H., . . . He, T. (2019). Scaling mitigation in membrane distillation: From superhydrophobic to slippery. *Desalination, 466*, 36–43. https://doi.org/10.1016/j.desal.2019.05.006

Xie, S., Li, Z., Wong, N. H., Sunarso, J., Jin, D., Yin, L., & Peng, Y. (2022). Gypsum scaling mechanisms on hydrophobic membranes and its mitigation strategies in membrane distillation. *Journal of Membrane Science, 648*, 120297. https://doi.org/10.1016/j.memsci.2022.120297

Xu, H., Xiao, K., Wang, X., Liang, S., Wei, C., Wen, X., & Huang, X. (2020). Outlining the roles of membrane-foulant and foulant-foulant interactions in organic fouling during microfiltration and ultrafiltration: A mini-review. *Frontiers in Chemistry, 8*, 417.

Xu, J. B., Bartley, J. P., & Johnson, R. A. (2005). Application of sodium alginate-carrageenan coatings to PTFE membranes for protection against wet-out by surface-active agents. *Seperation Science and Technology*, *40*(5), 1067–1081.

Xu, P., Bellona, C., & Drewes, J. E. (2010). Fouling of nanofiltration and reverse osmosis membranes during municipal wastewater reclamation: Membrane autopsy results from pilot-scale investigations. *Journal of Membrane Science*, *353*(1), 111–121. https://doi.org/10.1016/j.memsci.2010.02.037

Yan, Z., Qu, F., Liang, H., Yu, H., Pang, H., Rong, H., . . . Van der Bruggen, B. (2021). Effect of biopolymers and humic substances on gypsum scaling and membrane wetting during membrane distillation. *Journal of Membrane Science*, *617*, 118638. https://doi.org/10.1016/j.memsci.2020.118638

Yang, G., Zhang, J., Peng, M., Du, E., Wang, Y., Shan, G., . . . Xie, Z. (2021). A mini review on anti-wetting studies in membrane distillation for textile wastewater treatment. *Processes*, *9*(2), 243.

Yang, S., Abdalkareem Jasim, S., Bokov, D., Chupradit, S., Nakhjiri, A. T., & El-Shafay, A. S. (2022). Membrane distillation technology for molecular separation: A review on the fouling, wetting and transport phenomena. *Journal of Molecular Liquids*, *349*, 118115. https://doi.org/10.1016/j.molliq.2021.118115

Yang, Y.-F., Hu, H.-Q., Li, Y., Wan, L.-S., & Xu, Z.-K. (2011). Membrane surface with antibacterial property by grafting polycation. *Journal of Membrane Science*, *376*(1–2), 132–141.

Yao, M., Tijing, L. D., Naidu, G., Kim, S.-H., Matsuyama, H., Fane, A. G., & Shon, H. K. (2020). A review of membrane wettability for the treatment of saline water deploying membrane distillation. *Desalination*, *479*, 114312. https://doi.org/10.1016/j.desal.2020.114312

Ye, Y., Yu, S., Liu, B., Xia, Q., Liu, G., & Li, P. (2019). Microbubble aeration enhances performance of vacuum membrane distillation desalination by alleviating membrane scaling. *Water Research*, *149*, 588–595.

Yin, Y., Jeong, N., Minjarez, R., Robbins, C. A., Carlson, K. H., & Tong, T. (2021). Contrasting behaviors between gypsum and silica scaling in the presence of antiscalants during membrane distillation. *Environmental Science & Technology*, *55*(8), 5335–5346.

Yin, Y., Jeong, N., & Tong, T. (2020). The effects of membrane surface wettability on pore wetting and scaling reversibility associated with mineral scaling in membrane distillation. *Journal of Membrane Science*, *614*, 118503. https://doi.org/10.1016/j.memsci.2020.118503

Yu, S., Kang, G., Zhu, Z., Zhou, M., Yu, H., & Cao, Y. (2021). Nafion-PTFE hollow fiber composite membranes for improvement of anti-fouling and anti-wetting properties in vacuum membrane distillation. *Journal of Membrane Science*, *620*, 118915. https://doi.org/10.1016/j.memsci.2020.118915

Zainith, S., Ferreira, L. F. R., Saratale, G. D., Mulla, S. I., & Bharagava, R. N. (2021). Chapter 11—membrane-based hybrid processes in industrial waste effluent treatment. In M. P. Shah & S. Rodriguez-Couto (Eds.), *Membrane-based hybrid processes for wastewater treatment* (pp. 205–226). Elsevier.

Zakrzewska-Trznadel, G., Harasimowicz, M., & Chmielewski, A. G. (1999). Concentration of radioactive components in liquid low-level radioactive waste by membrane distillation. *Journal of Membrane Science*, *163*(2), 257–264. https://doi.org/10.1016/S0376-7388(99)00171-4

Zhang, M., Jiang, S., Tanuwidjaja, D., Voutchkov, N., Hoek, E. M., & Cai, B. (2011). Composition and variability of biofouling organisms in seawater reverse osmosis desalination plants. *Applied and Environmental Microbiology*, *77*(13), 4390–4398.

Zhang, M., Liao, B.-Q., Zhou, X., He, Y., Hong, H., Lin, H., & Chen, J. (2015). Effects of hydrophilicity/hydrophobicity of membrane on membrane fouling in a submerged membrane bioreactor. *Bioresource Technology*, *175*, 59–67. https://doi.org/10.1016/j.biortech.2014.10.058

Zhang, S., Gao, Y., Liu, Q., Ye, J., Hu, Q., & Zhang, X. (2019). Harvesting of Isochrysis
    zhanjiangensis using ultrafiltration: Changes in the contribution ratios of cells and algo-
    genic organic matter to membrane fouling under different cross-flow velocities. *Algal
    Research, 41,* 101567. https://doi.org/10.1016/j.algal.2019.101567
Zhang, T., Zhu, C., Ma, H., Li, R., Dong, B., Liu, Y., & Li, S. (2014). Surface modification of
    APA-TFC membrane with quaternary ammonium cation and salicylaldehyde to improve
    performance. *Journal of Membrane Science, 457,* 88–94.
Zhang, X., Fan, L., & Roddick, F. A. (2018). Impact of the interaction between aquatic humic
    substances and algal organic matter on the fouling of a ceramic microfiltration mem-
    brane. *Membranes, 8*(1), 7. doi:10.3390/membranes8010007
Zhang, Y., Wang, J., Gao, F., Tao, H., Chen, Y., & Zhang, H. (2017). Impact of sodium
    hypochlorite (NaClO) on polysulfone (PSF) ultrafiltration membranes: The evolution of
    membrane performance and fouling behavior. *Separation and Purification Technology,
    175,* 238–247. https://doi.org/10.1016/j.seppur.2016.11.037
Zhang, Y., Zhang, G., & Han, F. (2006). The spreading and superspeading behavior of new
    glucosamide-based trisiloxane surfactants on hydrophobic foliage. *Colloids and Surfaces A:
    Physicochemical and Engineering Aspects, 276*(1), 100–106. https://doi.org/10.1016/j.
    colsurfa.2005.10.024
Zhao, F., Zhang, Y., Chu, H., Jiang, S., Yu, Z., Wang, M., . . . Zhao, J. (2018). A uniform shear-
    ing vibration membrane system reducing membrane fouling in algae harvesting. *Journal
    of Cleaner Production, 196,* 1026–1033.
Zhao, S., Zou, L., & Mulcahy, D. (2012). Brackish water desalination by a hybrid forward
    osmosis–nanofiltration system using divalent draw solute. *Desalination, 284,* 175–181.
    https://doi.org/10.1016/j.desal.2011.08.053
Zhao, Y.-H., Wee, K.-H., & Bai, R. (2010). Highly hydrophilic and low-protein-fouling poly-
    propylene membrane prepared by surface modification with sulfobetaine-based zwitteri-
    onic polymer through a combined surface polymerization method. *Journal of Membrane
    Science, 362*(1), 326–333. https://doi.org/10.1016/j.memsci.2010.06.037
Zheng, L., Sundaram, H. S., Wei, Z., Li, C., & Yuan, Z. (2017). Applications of zwitterionic
    polymers. *Reactive and Functional Polymers, 118,* 51–61. https://doi.org/10.1016/j.
    reactfunctpolym.2017.07.006
Zheng, Y., Zheng, M., Ma, Z., Xin, B., Guo, R., & Xu, X. (2015). 8—sugar fatty acid esters. In
    M. U. Ahmad & X. Xu (Eds.), *Polar lipids* (pp. 215–243). Elsevier.
Zhong, Z., Li, D., Zhang, B., & Xing, W. (2012). Membrane surface roughness characteriza-
    tion and its influence on ultrafine particle adhesion. *Separation and Purification Tech-
    nology, 90,* 140–146. https://doi.org/10.1016/j.seppur.2011.09.016
Zhu, M.-M., Fang, Y., Chen, Y.-C., Lei, Y.-Q., Fang, L.-F., Zhu, B.-K., & Matsuyama, H.
    (2021). Antifouling and antibacterial behavior of membranes containing quaternary
    ammonium and zwitterionic polymers. *Journal of Colloid and Interface Science, 584,*
    225–235.
Zhu, X., & Elimelech, M. (1995). Fouling of reverse osmosis membranes by aluminum oxide
    colloids. *Journal of Environmental Engineering, 121*(12), 884–892.
Zularisam, A. W., Ismail, A. F., & Salim, R. (2006). Behaviours of natural organic matter in
    membrane filtration for surface water treatment—a review. *Desalination, 194*(1), 211–
    231. https://doi.org/10.1016/j.desal.2005.10.030

3 Monitoring, Prevention, and Control of Fouling and Related Phenomena

3.1 INTRODUCTION

Although water-scarce regions such as the Middle East and North Africa have traditionally been key players in seawater desalination, climate change and unprecedented water stress have driven coastal communities worldwide to employ desalination as a means of obtaining potable water and diversifying their water supply (Eke et al., 2020). Several countries have prioritized desalination and water purification projects within their development frameworks (Adeel, 2017; Gadzalo et al., 2018; Muller et al., 2009; Xenarios et al., 2018). Membrane separation is replacing traditional technologies in many industries due to reliability, low operating and maintenance costs, and wide applicability (Jiaping Paul Chen et al., 2011; Jevons & Awe, 2010).

Simultaneously, or perhaps a little delayed, there has been considerable interest in monitoring and maintaining continuous membrane performance with optimized yield and throughput (Joe Qin, 1998). Now that we have familiarized ourselves with common operational challenges and mechanisms for the same in membrane systems, more questions arise with respect to performance monitoring and control: How are faults detected during operation? How effective and reliable are these methods? What information do they give us on fouling dynamics? Do they interrupt operation? Are reliable non-invasive real-time monitoring techniques available?

3.2 PROCESS MONITORING

In industrial processes, process monitoring and control strategies are aimed at identifying undesired states and taking timely and appropriate actions to manipulate the process to maintain a desired state (Hejnaes & Ransohoff, 2018) in order to avoid damage or loss of productivity (Isermann, 2011). Process monitoring involves many aspects: One is the detection of an anomaly that may or may not lead to a fault, second is the diagnosis of a fault and subsequent response, and the third is an emerging malfunction or performance deterioration that has not yet caused an alarm (Mannan, 2014). Remote efficiency monitoring allows for timely corrective recommendations for performance degradation recovery in industrial plants (X. Jiang & Foster, 2014; Lorenzo et al., 2020). Fault monitoring in industrial processes concerns approaches with detecting faults, that can be either in the form of a slowly evolving effect or

DOI: 10.1201/9781003144991-3

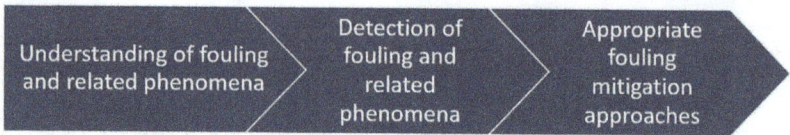

FIGURE 3.1   Road map to developing appropriate fouling mitigation measures.

a complete breakdown (Tzafestas et al., 2010). In membrane separation processes, continuously reliable operation and intervention efforts depend on the accuracy of monitoring systems (Virtanen et al., 2018). Monitoring systems should consider that faults are dynamic and should be discovered at early onset in order to devise optimized control methods (Ochando-Pulido, 2016; Virtanen et al., 2018). In this book, we will concern ourselves only with membrane integrity monitoring and monitoring of membrane-based faults, i.e. those described in Chapter 2. These include various types of membrane fouling in MF, UF, NF, RO, and MD as well as the associated phenomena of wetting/scaling in MD systems.

The lack of non-invasive real-time monitoring techniques for fouling detection in membrane processes is partially why we have such limited understanding of fouling and related phenomena. The reverse also holds true, i.e. our limited understanding of fouling has made it difficult to develop and employ suitable monitoring methods. An understanding of the evolution of fouling and related phenomena, as well as the availability of low-cost non-invasive real-time techniques to detect fouling are both prerequisites to effective and timely corrective cleaning measures (Figure 3.1). Local information on the process of fouling and related phenomena will not only help analyze the mechanisms involved but may help predict process operation and plan preventive and corrective actions (Mendret et al., 2009). Other aspects also contribute to the lack of substantial development and use of early detection techniques: On the research front, little consideration was given to monitoring techniques in desalination and water treatment prior to 2009; on the industry front, unwillingness to adopt new methods has stalled the transition to more effective monitoring systems.

In this section, we explore standard practices, recent developments, and challenges in (1) monitoring membrane integrity and (2) detection of fouling and related phenomena. We also discuss the potential of electrochemical impedance spectroscopy in deconstructing the structure of membrane and foulant layer toward in situ characterization.

3.3 STATUS OF MONITORING IN MEMBRANE SEPARATION PROCESSES

3.3.1 INDUSTRIAL PRACTICE

3.3.1.1 Water Quality Monitoring

Most modern RO/NF systems are equipped with flow meters, pressure transmitters, temperature transmitters, conductivity sensors, oxidation-reduction potential sensors, and pH sensors to monitor performance. In RO systems, the feedwater quality is the primary indicator of fouling. For this reason, routine tests include conductivity, pH, turbidity, dissolved oxygen, silt density index (SDI), and total organic carbon (TOC). The placement and frequency of these tests along the membrane feed channel are important in ensuring accuracy and reliability.

Common water quality sensors to determine fouling propensity include SDI, turbidity meter, and particle counter. The insensitivity of these methods for certain particle ranges has directed research to membrane-based fouling indices to predict fouling propensity. These include MFI0.45 and MFI-UF. Since then, alternative indices such as Crossflow Sampler Modified Fouling Index Ultrafiltration (CFS-MFIUF) and feed fouling monitor (FFM) have been developed to account for cake-enhanced osmotic pressure in NF and RO (Lee Nuang Sim et al., 2010; Taheri et al., 2015).

Both SDI and MFI0.45 are based on filtering water at constant pressure through a 0.45 μm microfiltration membrane (Boerlage et al., 2003). The $MFI_{0.45}$ is a membrane-based method to predict rate of flux decline due to particulate fouling in the form of cake filtration. It has been found to be more sensitive to changes in water quality than SDI (Salinas Rodriguez et al., 2019). Despite its use in active ASTM standards (International, 2019c), $MFI_{0.45}$ values are not able to explain practical flux decline rates (Schippers et al., 1981).

3.3.1.2 Membrane Integrity

Membrane integrity measurements can be performed on the membrane when it is offline or can correlate filtrate quality with the extent of loss of membrane integrity (Antony et al., 2012). ASTM D7601–19 Standard Practice for Pressure Driven Membrane Separation Element/Bundle Evaluation outlines the industry standards for inspection, performance testing, and laboratory analyses of pressure-driven membrane separation elements (International, 2019b).

1. *Inspection of structural integrity*: Methods for inspecting structural integrity include non-destructive visual testing, and membrane autopsy, which can only be carried out once the membrane element is dismantled.
2. *Performance testing*: For performance testing, defective membrane elements are primarily identified by examining abnormal or discontinuous changes in outlet conductivity using probes. Performance can also be verified by removing an individual membrane and wet testing it using the membrane manufacturer's product data sheets; data can be compared to as-new performance using standardizing procedures for ultrafiltration (International, 2021) or reverse osmosis (International, 2019a), depending on the process.
3. *Integrity of membranes*: These are determined by using air-based tests (pressure decay and vacuum hold), soluble dye, continuous monitoring particulate light scatter techniques, and TOC monitoring tests for rejecting particles and microbes. Of these, TOC and dye tests are limited to denser membranes, i.e. NF, RO, and tight UF membranes only, while the rest can be applied to MF, UF, NF, and RO (International, 2014, 2017). A pressure decay test (PDT), first developed in 1997 for automatic monitoring of membrane

integrity (Johnson, 1997), is now widely used and accepted in the water industry (Tng et al., 2015).

4. Another standard practice specifically for detecting leaks in NF and RO was approved in 1980; it involves a tube sheet and O-ring tests for hollow fiber elements, a vacuum test for spiral wound elements, and a dye test for spiral wound and tubular elements (Antony et al., 2012; International, 2014). Online TOC monitoring has been found to be more sensitive for integrity monitoring than online particle counting or conductivity monitoring, although researchers suggest that the use of more than one technique may be necessary for reliable operation (Adham et al., 1998).

3.3.1.3 Fouling Monitoring

Performance monitoring uses normalized flux decline characteristics for synthetic feeds in the lab, as established by ASTM D-4516 (International, 2019a). Using this standard method, fouling monitoring is reliant on measuring a decrease in flux or increase in pressure drop on a far from reality dead-end filtration system (Filloux et al., 2012; Nguyen et al., 2011; Sari Erkan et al., 2018). Nevertheless, the three primary indicators of performance issues in reverse osmosis systems are (1) loss of normalized permeate flow, (2) loss of normalized salt rejection, and (3) increase in system differential pressure (Zaidi & Saleem, 2022). Factors such as feedwater composition and feed pressure, temperature, and recovery also affect system performance. For example, a feed temperature drop of 4 °C causes reduced permeate flow by approximately 10% (DOW, 2008). To separate these effects from fouling effects, performance indicators should be normalized with reference to initial system performance. Regularly recorded normalized data is then used to identify potential faults. Table 3.1 shows the symptoms, causes, and corrective measures applied in industrial RO systems (DOW, 2008).

To determine the specific fouling issue, membrane manufacturers use a combination of methods: plant performance data, feedwater analysis, SDI or MFI0.45 tests, and inspection of the membrane element (DOW, 2008). However, none of these methods allow real-time non-invasive detection, nor are they able to accurately determine extent, type, and specific location of fouling on the membrane. Membrane autopsy carried out on a membrane after testing using analytical and microscopic techniques (W. Huang et al., 2021; P. Xu et al., 2010) has long been believed to be the only reliable method of identifying the type of fouling (Darton et al., 2004). Methods that yield any information on the extent and type of fouling accurately are currently available only as offline methods.

Corrective cleaning is typically applied when a 15% deviation from initial value of pressure drop is observed (Amin Saad, 2004). In the case of RO, Dow Water and Process Solutions recommends cleaning the membrane elements when one or more of the following conditions are met (DOW, 2008):

- 15% loss of permeate flow
- 5–10% increase in salt passage (for RO membranes)
- 10–15% increase in normalized pressure drop (feed pressure minus concentrate pressure)

TABLE 3.1

Symptoms, Probable Causes, and Corrective Measures in Reverse Osmosis Systems (DOW, 2008).

Permeate Flow	Salt Passage	Differential Pressure	Direct Cause	Indirect Cause	Corrective Measure
Increasing	Increasing (main symptom)	No changing	Oxidation damage	Free chlorine ozone, $KMnO_4$	Replace element
Increasing	Increasing (main symptom)	No changing	Membrane leak	Permeate backpressure; abrasion	Replace element; improve cartridge filter
Increasing	Increasing (main symptom)	No change	O-ring leak	Improper installation	Replace O-ring
Increasing	Increasing (main symptom)	No change	Leaking product tube	Damaged during element loading	Replace element
Decreasing (main symptom)	Increasing	Increasing	Scaling	Insufficient scale control	Cleaning, scale control
Decreasing (main symptom)	Increasing	Increasing	Colloidal fouling	Insufficient pretreatment	Cleaning; improve pretreatment
Decreasing	No change	Increasing (main symptom)	Biofouling	Contaminated raw water; insufficient pretreatment	Cleaning; disinfection; improve pretreatment
Decreasing (main symptom)	No change	No change	Organic fouling	Oil; cationic polyelectrolytes	Cleaning; improve pretreatment

Observation of changes in these parameters is slow and ineffective; it does not support early detection. Oftentimes, fouling has already progressed in severity and may be irreversible before substantial changes in TMP or flux can be detected. Delayed cleaning also makes it difficult to remove foulants completely (Košutić & Kunst, 2002). This is detrimental to the membrane integrity and risks shortening membrane lifetime. Furthermore, other closely related phenomenon, such as concentration polarization in RO, may have similar effects on permeate flux or transmembrane pressure, making it difficult to distinguish what is actually occurring at the membrane surface, which can result in misinterpretation of data (Rudolph et al., 2019). For example, the "symptoms" of scaling and colloidal fouling are similar in terms of the primary indicators as shown in Table 3.1, yet the cleaning methods and chemicals used will be different. Such misinterpretation translates into delayed correction or choice of unsuitable correction measures, which may in fact damage the membrane

(Vigneswaran, 2009). Thus new and improved industry standards for early detection real-time techniques to characterize fouling and related phenomena during operation are a pressing need. Similar performance indicators, i.e. permeate flux and filtrate water quality, are used to monitor fouling in MF and UF (W. Huang et al., 2021; J. Zhang et al., 2006).

3.3.2 Recent Developments in in Situ Monitoring Techniques for Fouling and Related Phenomena

Many countries have focused research efforts in the area of in situ monitoring of fouling and related phenomena. Figure 3.2 shows the number of publications by country. In total, China has produced the greatest number of publications in this niche area. An interesting observation is the Middle East and North America (MENA) region's interest in monitoring techniques for membrane distillation—a process that is yet to be commercialized but that is very much in line with the region's continued reliance on thermal processes for industrial desalination (Lange, 2019). The higher salinity of the Arabian Gulf compared to that of seawater elsewhere may make the use of membrane distillation systems feasible in the near future.

3.3.2.1 Fouling Monitoring in Membrane Separation Processes

Efforts in online non-invasive monitoring techniques gained momentum at the turn of the millennium. Many experts have cited certain attributes to desirable monitoring systems. Saad points out that early detection warning systems should be functional, affordable, verifiable, reproducible, reliable, simple, comprehensive, universal, and robust (Amin Saad, 2004). Online monitoring tools should be able to provide dynamic information on membrane fouling (Rudolph et al., 2019). According to Güell, monitoring systems should ideally be real-time, non-invasive, and at a molecular scale, and they should make use of pattern recognition (Güell et al., 2009). They recognize that industry is in need of on/inline monitoring and characterization tools.

Both optical and non-optical techniques have been suggested for online detection of fouling and related phenomena (V. Chen et al., 2004). In 1998, Tamachkiarowa and Flemming first proposed the use of a fiber-optical sensor integrated in a membrane module as an online monitoring device to detect deposit formation in real time; their focus was on biofouling (H.-C. Flemming et al., 1998). A fiber-optical device is based on light scattering by deposited material on the tip of the optical fiber. When applied to the water system of a brewery for a period of 2 years, the device was reliable for detecting microorganism colonization $\geq 10^5$ cells cm^{-2} (Tamachkiarow & Flemming, 2003). They also proposed the use of a Fourier transform infrared (FTIR) flow cell to elucidate the chemical nature of the foulant and identification of biological material in real time but acknowledged that this required a high level of technical effort (H.-C. Flemming et al., 1998; H. C. Flemming et al., 1997). Many years later in 2020, Liu et al. reported the probing of a protein-induced membrane fouling by in situ attenuated total reflectance—FTIR (ATR-FTIR) spectroscopy—and revealed a two-step protein-induced membrane fouling process: unfolding of proteins to a more open, less structured state possibly

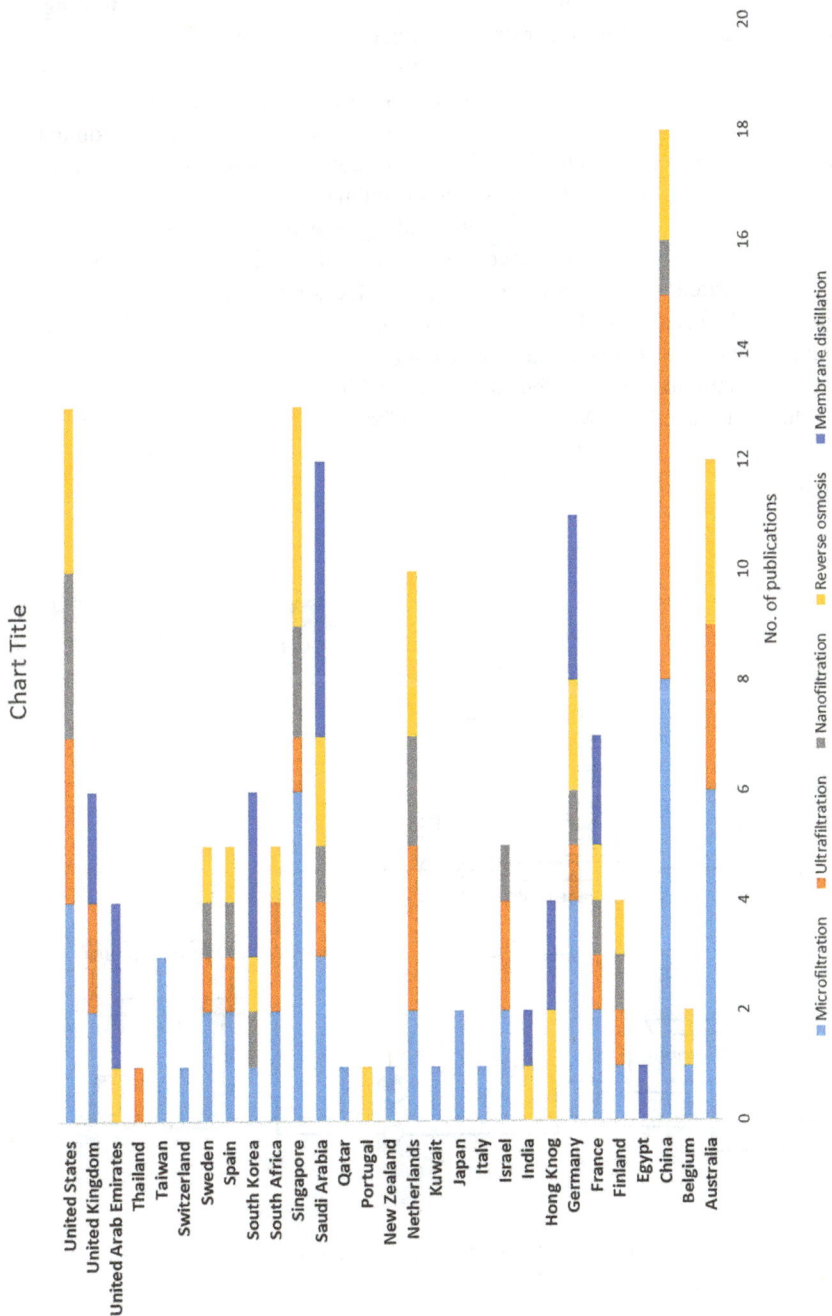

FIGURE 3.2 Publications on in situ monitoring of fouling and related phenomena by country.

Source: Scopus.

due to the hydrophobic surface of the PVDF membrane, followed by aggregation of the proteins to strengthen fouling (X.-Y. Liu et al., 2020). In situ probing techniques not only provide structural information on internal and external fouling, but they can also be used to deconstruct interactions between foulant and membrane surface. Gao et al. developed a fiber-optic reflectance UV-vis spectrometry (FORUS) method for real-time and in situ monitoring of organic fouling using BSA and humic acid (HA) (F. Gao et al., 2020). They measured the surface foulant concentration in UF after establishing a linear correlation between surface concentration and UV-vis absorbance for each type of foulant.

In the early days of research in in situ fouling detection methods, during the same period as Tamachkiarowa's discovery, Li et al. devised a direct observation technique for particle deposition on membrane surfaces (H. Li et al., 1998), using yeast and latex beads as model fouling particles in microfiltration. The technique is based on a microscope attached to a video camera and a module made of Perspex to allow light transmission from the feed side (Figure 3.3). However, direct observation through the membrane (DOTM) is only suitable for observation of two-dimensional foulant distribution; it does not yield cross-sectional profiles or any information on foulants within pores (Meng et al., 2010).

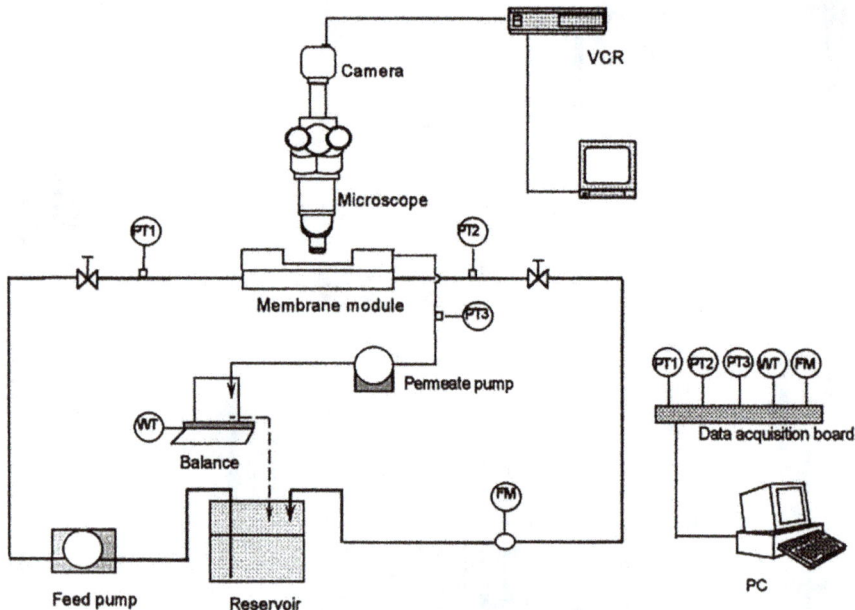

FIGURE 3.3 Experimental setup of cross-flow microfiltration rig with direct observation through the membrane.

Source: Li, H., Fane, A. G., Coster, H. G. L., & Vigneswaran, S. (1998). Direct observation of particle deposition on the membrane surface during crossflow microfiltration. *Journal of Membrane Science, 149*(1), 83–97. Reprinted with permission.

3.3.2.1.1 Neutron Scattering

Su et al. applied small-angle neutron scattering (SANS) for the detection of protein fouling on ceramic ultrafiltration membranes. SANS is a coherent scattering technique used to study structures and inhomogeneities in the range of 1–1000 nm by observing interference effects at very small scattering angles (Kockelmann & Godfrey, 2021; Xiao & Fu, 2021). This technique uses the study of diffraction in the regime where the magnitude of momentum transfer is small compared to the highest d-spacing Bragg peak (Hannon, 2017). SANS is an important technique used in polymer characterization as it allows analyzing the molecular structure as well as motions (Richter et al., 2012; Sanjeeva Murthy, 2013). In polymers, it can be used to investigate chain conformation during phase change.

Exploiting the sensitivity of SANS to the structural composition of the membrane pores, they correlated scattering intensity with time to the dynamic buildup of deposit inside the membrane pores during filtration (Su et al., 1999). This technique is limited to porous and transparent membranes, which casts doubt on its commercial viability (J. C. Chen et al., 2004).

Su et al. used SANS and simultaneous monitoring of flux to study the fouling of ceramic UF membranes by BSA and human serum aluminum (HSA) (Su et al., 2000). Neutrons were fired perpendicularly to the membrane surface. SANS is sensitive to the gradual buildup of a foulant layer inside the membrane flux, while permeate flux measurements only indicate overall blockage of the membrane. Using both methods together allows the location of protein fouling to be determined. They used SANS to elucidate the fouling mechanism at different feed pH levels. SANS has also been applied to polymeric UF membranes in dead-end filtration such that neutrons are fired tangentially to the membrane surface (Pignon et al., 2000). Through SANS, Pignon et al. determined the structural characteristics of protein deposits on polymeric membranes.

3.3.2.1.2 Ultrasonic Time Domain Reflectometry

Mairal et al. compared the use of ultrasonic time-domain reflectometry (UTDR) with traditional fouling indicators for the measurement of inorganic fouling in commercial RO desalination membranes (Mairal et al., 1999). Ultrasonic waves can be used to provide information on the media through which they are travelling as they propagate via cycles of compression and rarefaction. The velocity with which ultrasonic waves travel depends on the material. In UTDR, a transducer transmits ultrasonic wave pulses that propagate through the module and interact with interfaces such as the fouling layer. The amplitude of reflected waves depends on the physical characteristics of the medium (Z. X. Zhang et al., 2003). The acoustic impedance difference between the media on either side of the interface determines the magnitude of the reflected and transmitted waves. The acoustic impedance changes with changes on the membrane or with the development of a new foulant layer. The amplitude of the peaks of the reflected signals will change due to changes to the membrane, and new peaks will appear as fouling develops (Lai et al., 2020). UTDR for fouling detection is based on monitoring the changes in amplitude of reflected ultrasound signals.

Prior to Mairal's work, UTDR had been used for structural characterization of polymeric films (Kools et al., 1998), including in situ measurement of membrane compaction (Peterson et al., 1998; Reinsch et al., 2000). Sanderson's group at the UNESCO Associated Center for Macromolecules and Materials in South Africa has since applied the UTDR technique to characterize inorganic fouling in MF (J. Li et al., 2002a), UF (J. Li et al., 2005; J. Li et al., 2002), NF (X. Li et al., 2015; Z. Zhang et al., 2006), and RO (Sanderson et al., 2002) systems, comparing quantification with traditional indicators of flux decline and membrane autopsy using SEM analysis. They also extended the use of UTDR for membrane cleaning (Sanderson et al., 2002). Soon after, the same group used ultrasonic reflection modeling to predict cake compressibility during cross-flow MF of paper mill effluent (J. Li et al., 2002). Subsequent studies extended the application of UTDR for fouling characterization beyond flat-sheet membranes, in commercial spiral wound (Z.-X. Zhang et al., 2003), and tubular (J.-X. Li et al., 2006) membranes and in hollow fiber membranes (X. Xu et al., 2009). These studies helped develop a signal analysis protocol to represent systemic changes (in terms of shift factors) in the entire acoustic spectrum as a function of module operation time. In testing with spiral wound membranes, three cycles of operation, each consisting of pure water equilibration—fouling and cleaning, regular measurements of permeate flow, acoustic amplitude and arrival time—were taken. The extent of $CaSO_4$ fouling was verified by gravimetric, microscopic, and spectroscopic measurements at the end of the experiments. In the case of hollow fiber membranes, UTDR is effective in distinguishing acoustic response signals generated from different surfaces of the hollow fiber membrane (X. Xu et al., 2009).

Li et al. also studied the effect of pH on the development of BSA gel layers on the membrane using UTDR to characterize the deposited BSA layers (J. Li et al., 2005). They found that the deposit layer was thicker but less compressible at pH 6.9 than at pH 4.9. Kujundzic identified threshold values above which UTDR was difficult to use for biofilm characterization due to decreases in the reflection amplitude at high amounts of exocellular polysaccharides bound to the substrate surface (Kujundzic et al., 2007). They found a correlation between UTDR response and biofilm growth for surface-averaged masses of the biofilms of ≤ 150 µg cm^{-2}. The application of UTDR to detect biofouling is more challenging due to the small difference in acoustic properties between the biofouling layer and water on a membrane (S. Sim et al., 2011). Another group tackled this issue by introducing periodic silica dosing as an "acoustic enhancer" to apply UTDR to detect biofouling for both low and high pressure membrane processes, i.e. flat-sheet polyethersulfone UF and TFC polyamide RO membranes under high pressure operation (S. T. V. Sim et al., 2013). They also concurrently corroborated changes in the UTDR waveform response to biofilm development and membrane performance parameters. Periodic dosing and flushing of the colloidal silica enables repetitive measurement of the fouling layer thickness (S. T. V. Sim et al., 2013). Wang et al. have also applied UTDR to detect the different stages of biofilm formation, namely initial adherence, reversible adhesion, and irreversible adhesion, on a membrane during wastewater treatment, correlating the UTDR response with membrane autopsy (J. Wang et al., 2018). More recently, Tung et al. used a high frequency 50 MHz ultrasound system to measure the fouling distribution in spiral wound UF and RO membrane modules. The high frequency increases the spatial resolution of the fouling deposition; their findings show that in porous membranes, the voltage decreases with increasing fouling, whereas in dense membranes, they observed the opposite (Tung et al., 2015).

3.3.2.1.3 Optical Coherence Tomography

Optical coherence tomography (OCT) is a non-invasive imaging technique commonly used in the medical field to perform high resolution, cross-sectional imaging of biological tissues (Lamirel, 2014; H. Wang & Evans, 2016). The in-depth (axial) resolution possible with OCT (1–10 µm) makes it a suitable technique to inspect the internal structure of objects and films (J. Fujimoto et al., 1998; Targowski & Iwanicka, 2012). OCT allows for three-dimensional visualization of tissue microstructure and has been used in dermatology (Olsen et al., 2018; Sattler et al., 2013; Welzel, 2001), in ophthalmology (Drexler et al., 2001; J. G. Fujimoto et al., 2009), in dentistry (Hsieh et al., 2013; Otis et al., 2000), in clinical trials for multiple sclerosis (MS) (Frohman et al., 2008; Petzold et al., 2010; Siger et al., 2008), as well as in the examination of cultural heritage objects (Targowski & Iwanicka, 2012). In OCT, a low-coherence light beam with a penetration depth of several hundred microns is directed to the membrane, and the backscattered signal is measured using an interferometric setup (Aumann et al., 2019).

To date, there have been eight studies on the use of OCT to characterize fouling in membrane processes. Optical coherence tomography was first used for in situ fouling characterization in membrane filtration by Gao et al., who demonstrated real-time observation of the depth profiles of the cake layer as well as the concentration field of the suspension particles through structural imaging (Y. Gao et al., 2014). OCT acted as a dual-function characterization as Doppler images obtained from the same interference signals were used to provide velocity profiles of the fluid, thus correlating the evolution of fouling with the microhydrodynamic environment (Y. Gao et al., 2014). Bentonite microparticles with a size range of 1–10 µm were used as a model foulant. Deposition was more concentrated toward the upstream end as fouling progressed, as shown in the binary images obtained via digital conversion of the structural images (Figure 3.4) (Y. Gao et al., 2014).

Subsequently, West et al. tested OCT for the visualization of feed spacer biofouling in membrane fouling simulators that mimic the feed spacer channel of spiral wound RO membrane modules (West et al., 2016). They correlated the amount of biofilm to the feed channel pressure drop using river water and wastewater effluent as feed solutions. OCT can distinguish specific deposit (cake volume/membrane surface area) and surface coverage with time for foulants with different particle sizes, which would help in real-time identification of specific particles in the cake layer, as shown by Li et al., who compared fouling evolution between silica nanoparticles and bentonite particles in UF (W. Li et al., 2016). Rapid surface blockage to 95% was observed for the larger bentonite particles (flake-like particles ~10 µ in length) in ~20 minutes, whereas it took silica nanoparticles (spheres with diameter ~10 nm) double the time to reach the same coverage (Figure 3.5) (W. Li et al., 2016). The difference in dimensions could explain the higher blockage rate in the case of bentonite particles.

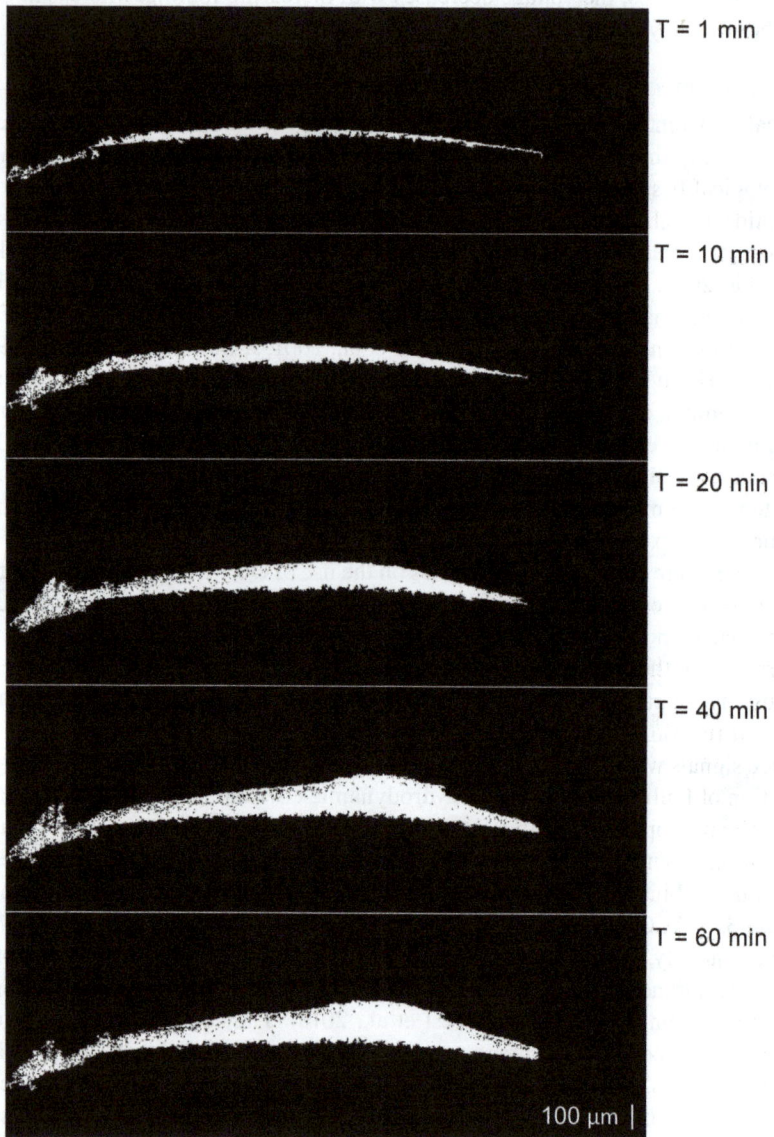

FIGURE 3.4 Binary images of the foulant cake layer formation (white) at different filtration times.

Source: Reprinted with permission from *Environ. Sci. Technol.* 2014, 48, 24, 14273–14281. https://doi.org/10.1021/es503326y. Copyright © 2014 American Chemical Society.

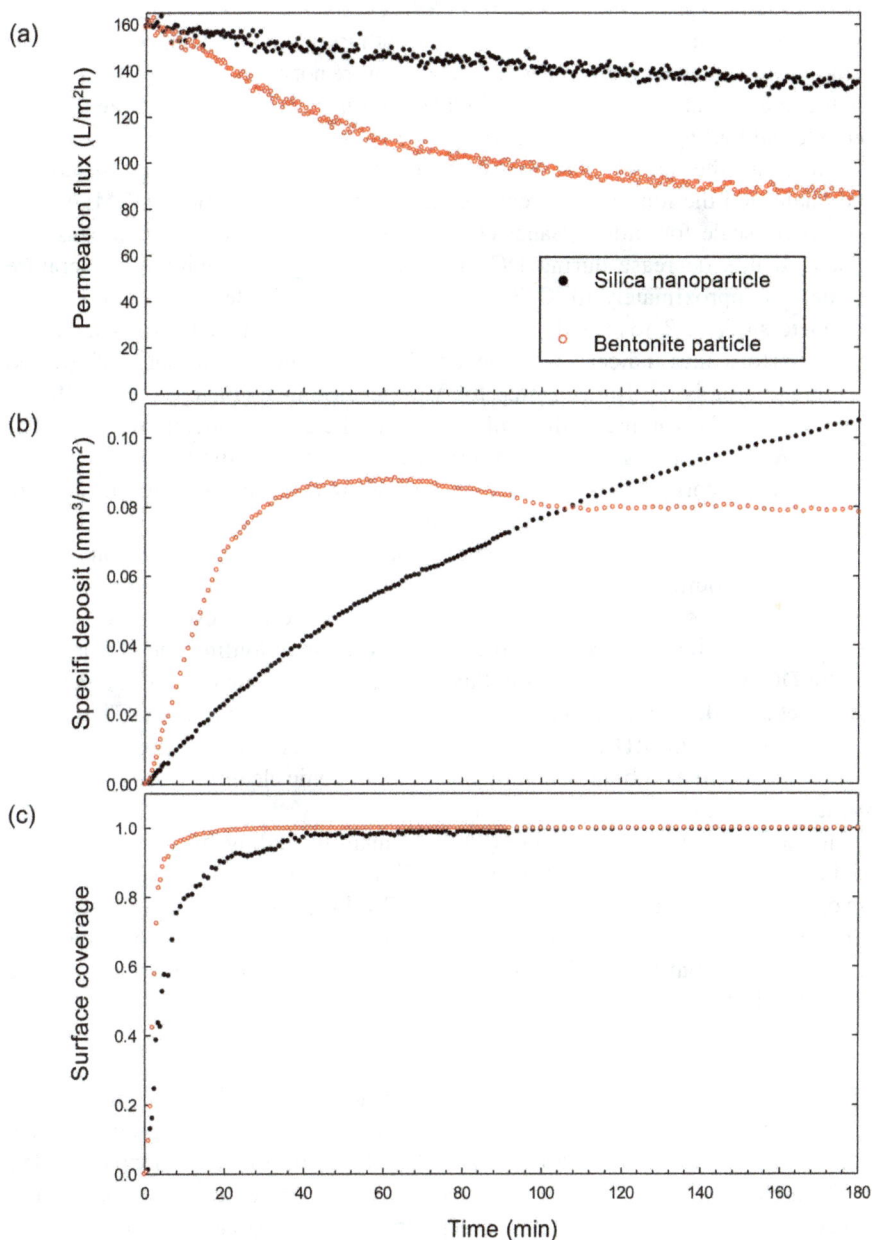

FIGURE 3.5   Fouling evolution using silica nanoparticles (red) and bentonite particles (blue) in terms of (a) permeation flux, (b) specific deposit, and (c) surface coverage.

Source: Reprinted with permission from *Environ. Sci. Technol.* 2016, 50, 13, 6930–6939. https://doi.org/10.1021/acs.est.6b00418. Copyright © 2016 American Chemical Society.

Trinh et al. extended the use of OCT to visualize both external and internal fouling by oil emulsions (Trinh et al., 2018). Silica glass beads were used as a control foulant for external fouling. According to the authors, boundaries between the membrane surface and the oil droplets or glass beads were not distinct in the OCT images, but the intensity at each layer due to foulant deposition helped characterize external and internal fouling (Trinh et al., 2018).

Extending the use of OCT to membrane distillation, Bauer et al. visualized and quantified the fouling in direct contact membrane distillation (DCMD), with a focus on scale formation (Bauer et al., 2019). Fouling layer coverage was correlated to flux decrease during DCMD of hot spring water with a temperature gradient of approximately 10 °C. Their results included the development of scaling parameters R_s and R_c to provide a correlation between scaling behavior and macroscopic performance indices such as flux decline, showing that the share of covered membrane area is the limiting step for MD performance (Bauer et al., 2019). R_c is the ratio of the volume of the fouling layer to the covered membrane area. The smaller R_c values point to more uniform distribution of fouling layer. The other parameter, R_s, correlates the volume of the fouling layer at a selected location in the feed channel and the monitored membrane area. A large R_s value points to a high volumetric amount of fouling on the visualized membrane area (Bauer et al., 2019). These fouling parameters yield information about local scaling structure, which is different and independent from visually based macroscopic assessment. A key result from this work is that the early prediction of fouling and performance loss in DCMD, independently from flux decrease, is possible using OCT methods (Bauer et al., 2019). Alicia Kyoungjin also led projects on the use of OCT for fouling visualization in MD treating industrial textile wastewater, using commercial PTFE and PVDF membranes, combined with ultrasonic cleaning (Guo et al., 2022; Wong et al., 2021).

In another study, Liu et al. used OCT to analyze fouling by clay suspensions during forward osmosis (FO) using a loose NF-like membrane (X. Liu et al., 2020). As opposed to flux-decline measurements, OCT allowed for numerical evaluation of surface coverage and specific deposit, confirming that the back diffusion of the draw solute affects initial foulant deposition and could favor the slow growth of the cake layer (X. Liu et al., 2020).

3.3.2.1.4 Multiphoton Microscopy

Multiphoton microscopy (MPM) is a fluorescence-based laser scanning confocal microscopy technique in which two or more photons combine to produce three-dimensional images of microscopic samples (Erickson-Bhatt & Boppart, 2015). MPM uses localized non-linear excitation to excite fluorescence within a plane. Components of a multiphoton microscope include a laser-scanning confocal that reflects near IR, a pulsed laser, and a few multiphoton peripherals. Compared to other similar optical techniques, MPM has improved depth penetration and less photodamage (A. M. Larson, 2011) and is widely used to examine biological samples involving thick tissue and live animals (Corbin et al., 2014). Although the penetration depth depends strongly on the refractive index and scattering properties of the specimen, it allows penetration 2–3 times deeper into thick specimens than confocal microscopy (Paddock & Eliceiri, 2014). MPM has emerged as a promising technique for imaging of physiology, morphology, and cell–cell interactions of biological samples. The resolution depends on many factors such as the number of photons collected per pixel (Zipfel et al., 2003). In membrane filtration, only one group at the University of Oxford has applied multiphoton microscopy to in situ visualization of fouling by protein aggregates (Field et al., 2009; D. Hughes et al., 2006; D. Hughes et al., 2006; Hughes et al., 2006, 2007). In one of these studies, MPM was used for the in situ 3D characterization of protein fouling on both the surface and within the membrane pores, and the dominant fouling mechanism was identified by combining 3D MPM images with flux decline with time. Using track-etched polycarbonate (PC) membranes, they established that fouling is at first internally dominated but later becomes externally dominated, in agreement with two-stage protein fouling models (D. Hughes, U. K. Tirlapur et al., 2006). It is unclear why further research in using MPM to study membrane fouling, particularly biofouling, has not been carried out. The cost and complexity of the pulsed lasers involved have greatly limited the integration of MPM with membrane systems.

3.3.2.1.5 Fluorescence Spectroscopy

The use of fluorescence spectroscopy in water quality analysis to determine the fouling potential of the feedwater is well documented (Henderson et al., 2009). However, studies of fluorescent techniques on solid phase analysis has been fairly limited. Yamamura et al. applied three-dimensional solid phase fluorescence excitation emission matrix (EEM) spectroscopy) to observe small amounts of fluorescent NOM on the membrane surface without any prior preparation by irradiating the sample with excitation light at 45° (Yamamura et al., 2019). Chen et al. apply multicolor light sheet fluorescence imaging as a visualization technique to differentiate foulants and their temporal–spatial distributions in and around the membrane matrix using BSA and dextran foulants (L. Chen et al., 2022). They found that inner adsorption plays a more critical role for dextran fouling compared to BSA fouling on polymeric membranes. During UF of BSA-dextran mixture solutions, BSA was observed as the dominant contributor to fouling behavior. In another study, Cui et al. applied UV spectrum to in situ monitoring and changes in absorbance curves to detect the extent of yeast fouling in membrane bioreactors (Cui et al., 2022).

Rudolph and coworkers used quartz crystal microbalance with dissipation monitoring (QCM-D) to study adsorptive fouling in situ during microfiltration in lignocellulosic biorefineries (Rudolph et al., 2021). QCM-D is an acoustic sensing tool used to measure mass and viscosity in processes occurring at near surfaces or within thin films (Tonda-Turo et al., 2018). QCM-D is sensitive to small chemical, mechanical, and electrical changes that are converted into electrical signals to study dynamic processes (Tonda-Turo et al., 2018). Although the authors describe their use of QCM-D as in situ, the detection is carried out on the feed, retentate, and permeate streams after microfiltration (Rudolph et al., 2021).

3.3.2.1.6 Small-Angle X-Ray Scattering

X-ray scattering is used in many fields to examine materials at nanometer scales under realistic conditions in real time. Small-angle X-ray scattering (SAXS) has been used for decades to study nanoparticles to determine size, shape, and morphology. In SAXS, the scattering angle (2θ) is less than 5°. SAXS provides low-resolution information on the shape, conformation, and assembly state of biological macromolecules in solution (Kikhney & Svergun, 2015). An advantage of SAXS is that it is non-destructive and provides structural data in native solution environments. It can be used to obtain structural information for nanostructured systems (T. Li et al., 2016). Doudiès et al. used this technique to study external membrane fouling during cross-flow ultrafiltration of milk protein dispersions by measuring the concentration distribution of milk protein in the concentration polarization layer (Doudiès et al., 2021). SAXS captures the evolution of particle deposit and quantifies it with respect to scattering intensities, providing information on thickness and compactness of particle deposits (J. C. Chen et al., 2004). Pignon et al. used SAXS to characterize the structural organization of colloidal fouling during UF (Pignon et al., 2003).

Mollahosseini et al. applied a synchrotron-based X-ray microtomography (SR-μCT) to assess ceramic membrane fouling in dairy stream filtration through porosity variation through the different layers (Mollahosseini et al., 2021).

3.3.3 Role of Electrochemical Impedance Spectroscopy in Membrane Processes

Electrochemical impedance spectroscopy (EIS) is a powerful non-destructive technique used to determine the impedance characteristics of an interface and to gain insight into the interactions between a surface and its environment (Gilbert, 2017; Nossol et al., 2021). In this section, we explore how EIS is used along non-conductive membrane systems for detection. The use of EIS with electrically conductive membrane systems is further elaborated in Chapter 7. Traditionally used for studying materials for energy applications, paintings, and coatings, the use of EIS has gradually expanded to membrane separation over the last 20 years. Figure 3.6 shows the upward trend in research articles with the keywords "impedance spectroscopy" and "membrane". This interest spirals from the ability of EIS to be used in situ as a sensitive tool for real-time monitoring. As a frequency-domain measurement, impedance spectra also carry more structural information than simple time-domain electrochemical measurements that could help elucidate process kinetics.

In EIS, a very small AC potential ≤100 mV is applied to an electrochemical cell to prompt a pseudolinear current response. In a linear or pseudolinear system, the current response to a sinusoidal voltage input will also be a sinusoid at the same frequency but with a phase shift.

As the fouling begins, the electrical properties of the surface change. EIS allows us to monitor these changes. In the case of a non-conducting membrane, the filtration unit must be equipped with a four-electrode system—one pair to inject an AC voltage and another to sense the current response—so that these changes can be monitored (S. Jiang et al., 2017). Nyquist plots and variations in capacitance and conductance of the membrane system are then analyzed and studied to monitor changes within the membrane or at its surface.

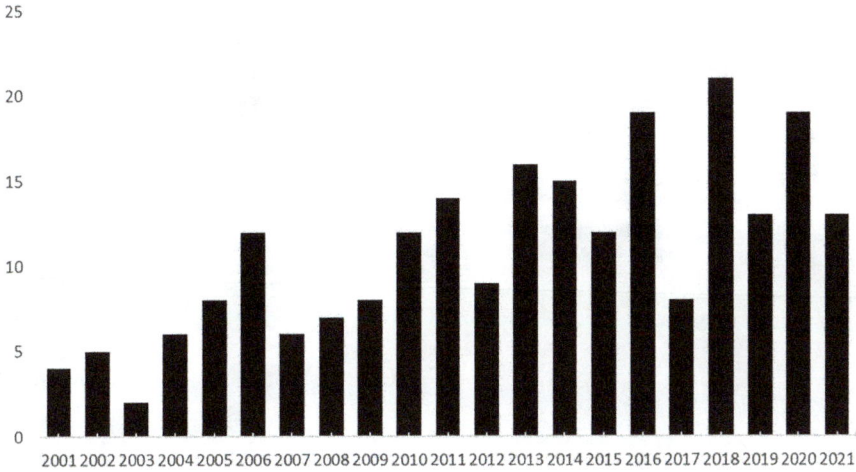

FIGURE 3.6   Number of publications with keywords "impedance spectroscopy", "membrane" in title.

Source: Scopus.

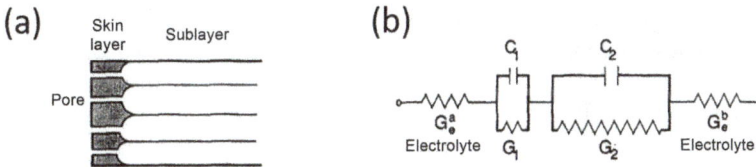

FIGURE 3.7   Skin layer and support layer of an ultrafiltration membrane (a), each represented by a parallel combination of conductance G and capacitance C (b).

Source: Coster, H. G. L., Kim, K. J., Dahlan, K., Smith, J. R., & Fell, C. J. D. (1992). Characterisation of ultrafiltration membranes by impedance spectroscopy. I. Determination of the separate electrical parameters and porosity of the skin and sublayers. *Journal of Membrane Science, 66*(1), 19–26. Reprinted with permission.

3.3.3.1 Structural Characterization of Membranes Using EIS

Coster et al. (Coster et al., 1992) were among the first to distinguish the electrical properties of an asymmetric ultrafiltration membrane and characterize the porosity of the polysulfone active layer using EIS. In fact, over the years, the Coster group at the University of Sydney have spearheaded research in the application of EIS to monitoring in membrane processes, with Singapore now a close second with respect to number of publications.

A schematic of the two membrane layers and corresponding equivalent circuit is shown in Figure 3.7. Each of the two layers, namely the low porosity thin skin layer and the more porous support layer, is represented by a parallel combination of capacitance and conductance elements.

If the dielectric constant of water ϵ_p and that of the membrane material ϵ_m are known, along with thickness of the active layer d, the capacitance obtained from EIS data can be used to determine the porosity of the active layer.

$$C_p = \frac{pA\epsilon_p\epsilon_0}{d};$$

Equation 3.1

$$C_s = \frac{(1-p)A\epsilon_m\epsilon_0}{d}$$

Equation 3.3

$$C = C_s + C_p$$

Equation 3.2

$$p = \frac{\dfrac{Cd}{\epsilon_0 A} - \epsilon_m}{\epsilon_w - \epsilon_m}$$

Equation 3.4

where p is the porosity, A is the membrane area, ϵ_0 is the permittivity of free space, and C is the capacitance of the active layer, which is the sum of the capacitance of the pores and capacitance of the skin. Using EIS, they found that the commercial PSf membranes used have a porosity of 2–5%. Another group used EIS to measure the active layer thickness of a polyethersulfone NF membrane using different electrolyte solutions (see Table 3.2). Y. Xu et al. (2011) used EIS to measure the active layer thickness of a polyethersulfone (PES) commercial nanofiltration membrane in various electrolyte solutions, using Equation 3.5:

$$d_k = \frac{\epsilon_k\epsilon_0}{C_k}$$

Equation 3.5

where d_k is the thickness of the kth layer, ϵ_k is the dielectric constant of the kth layer, and c_k is the capacitance of the kth layer. The thickness obtained from EIS was larger than that observed in SEM images due to swelling of the sulfonated PES layer in electrolyte solutions. The study stressed the importance of measuring thickness in the working state using advanced techniques such as EIS.

TABLE 3.2

Thickness of Active Layer as Measured by SEM and by EIS in Different Electrolytes (Y. Xu et al., 2011).

	SEM	NaCl	MgCl$_2$	LaCl$_3$	KCl	K$_2$SO$_4$	K$_3$PO$_4$
d (μm)	0.150	0.2739	0.2218	0.2501	0.2035	0.2188	0.2096

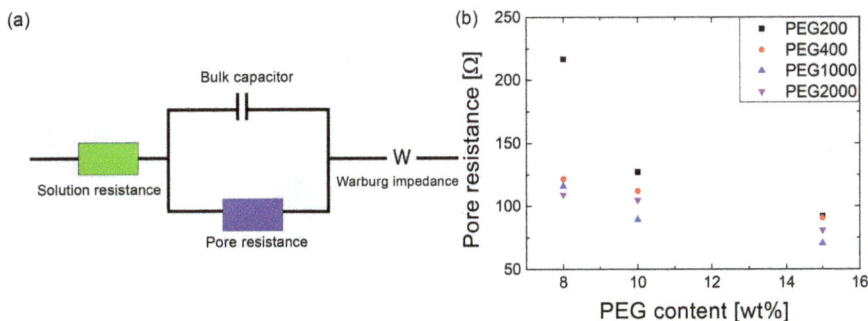

FIGURE 3.8 (a) Equivalent circuit of the membrane–solution system and (b) pore resistance vs. PEG content for different molecular weight PEGs.

Source: Yin, C., Wang, S., Zhang, Y., Chen, Z., Lin, Z., Fu, P., & Yao, L. (2017). Correlation between the pore resistance and water flux of the cellulose acetate membrane. *Environmental Science: Water Research & Technology, 3*(6), 1037–1041. Reprinted with permission.

Yin et al. (Yin et al., 2017) used EIS data to predict water flux for a membrane. Their work relied on the principle that more ions will pass through larger pores and thus increase conductance. Polyethylene glycol (PEG) of different molecular weights was added with varying concentrations to cellulose acetate solutions for phase inversion. PEG acted as a pore forming additive. Figure 3.8a shows the equivalent circuit of the membrane–solution system, and Figure 3.8b shows how the measured pore resistance varied with PEG content.

Their findings revealed that increasing the amount of pore-forming agent reduces pore resistance as more electrolyte solution is flowing through the pores. Using the Hagen–Poiseuille model for water flux through cylindrical pores and an equation for the resistance of a material, they found an inverse relationship between water flux and pore resistance:

$$J = \frac{\rho \Delta P}{8 \pi \eta} \times \frac{1}{R_p} \qquad \text{Equation 3.6}$$

where J is the water flux through the membrane, ρ is the resistivity and is related to the characteristic of the electrolyte, ΔP is the pressure difference across the membrane, η is the viscosity of the electrolyte solution, and R_p is the pore resistance.

Lara and Benavante applied EIS to determine the electrical resistance of a non-conducting membrane for BSA-induced fouling through polymeric PSf and ceramic (ZrO_2/Al_2O_3) MF membranes (de Lara & Benavante, 2009). They also correlated EIS resistance measurements to porosity reduction.

3.3.3.2 Monitoring of Membrane Processes Using EIS

In earlier sections, we overviewed some of the common techniques that are used to characterize the fouling layer on a given membrane. In industrial practice, the

(a)

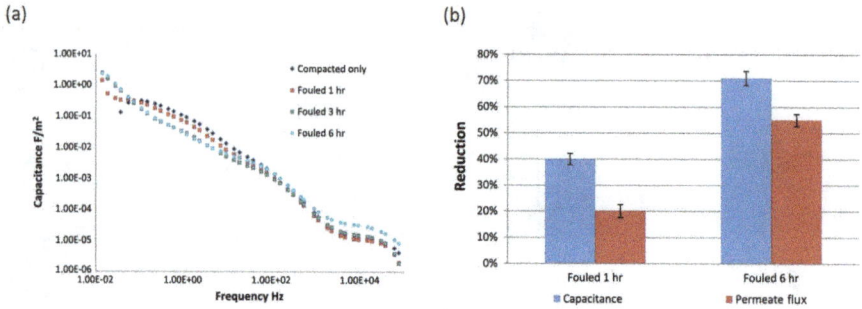

(b)

FIGURE 3.9 (a) Capacitance spectra for clean and fouled membrane; (b) percentage reduction in capacitance (measured at 1 Hz) and flux for different fouling times.

Source: Cen, J., Kavanagh, J., Coster, H., & Barton, G. (2013). Fouling of reverse osmosis membranes by cane molasses fermentation wastewater: Detection by electrical impedance spectroscopy techniques. *Desalination and Water Treatment, 51*(4–6), 969–975. Reprinted with permission.

techniques used are often applied after the membrane has already been destroyed. In recent years, EIS has emerged as a non-invasive, early warning tool for monitoring fouling.

Coster's group studied the fouling of RO membranes with humic and fulvic acids in wastewater (Cen et al., 2013). Impedance spectra for membranes fouled for different times were taken, and capacitance data showed that fouling was reflected in changes at the diffusion polarization (DP) layer (at frequencies <100 Hz), as shown in Figure 3.9a. The DP layer forms when ions near the membrane surface are accumulated and depleted under the effect of an AC potential (Ho et al., 2017). The effects of the DP layer are pronounced at low testing frequencies when ions have sufficient time to respond to the change in bias (Figure 3.10), as opposed to at high frequencies when the polarity of the applied potential changes too fast to notice.

Changes in the DP layer help evaluate fouling as it is formed at the membrane–solution interface. Figure 3.9b shows the percentage reduction in capacitance and permeate flux for the membranes fouled for different durations. It confirms that capacitance measurements are more sensitive to changes due to fouling. To ensure that membrane compaction does not contribute to EIS measurements, membranes should be compacted prior to fouling.

3.3.3.2.1 Wetting in Membrane Distillation

Pore wetting is a challenge that has hindered the widespread use of membrane distillation for desalination and water treatment. The mechanisms, stages, and contributing factors of wetting in MD have been discussed in detail in Section 2.4. Recently, some work has been carried out to detect membrane wetting during MD using EIS and comparing resulting data with flux decline.

Chen et al. (Y. Chen et al., 2017) applied EIS to the early detection of pore wetting during MD. They found that EIS is much quicker to respond to pore wetting than

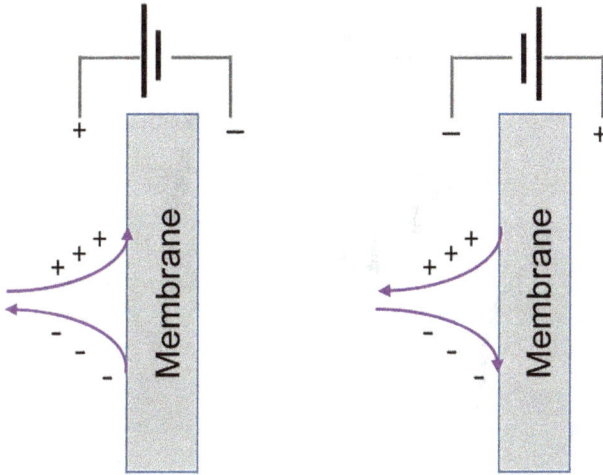

FIGURE 3.10 Schematic showing the effect of alternating polarity of an applied voltage across a membrane on electrically driven transport.

salt rejection measurements. This is because EIS detected early wetting, whereas salt rejection only detected complete wetting. The complexity of employing polymer membranes in canary cells for EIS of membrane wetting has led to interest in electrically conducting membrane materials for wetting detection with MD; this is elaborated in Chapters 5 and 7 of this text.

3.3.3.2.2 Type of Foulant

EIS may also be used to investigate different types of foulants. Coster's group investigated the effect of $CaCO_3$ fouling of RO membranes (Antony et al., 2013; Kavanagh et al., 2009). They found that the impedance response to scaling was different when no pressure was applied vs. when fouling was carried out in a realistic situation under pressure. Their findings suggest that simulating process conditions closely is important. As we will see later, this is an advantage of using electrically conducting membranes where no bypass streams are needed and fouling can be detected in real conditions.

Sim et al. went a step further and applied EIS not just to detect fouling but to detect the mechanism of fouling through EIS using inorganic and organic foulants (L. N. Sim et al., 2013). During the early stages of fouling with silica, the Nyquist plot shifts to the right, and conductance (Figure 3.11a–b) decreases as a flowing layer of non-conductive silica starts to build up on the membrane surface. Later on, conductance decreases as a stagnant layer of silica densifies at the membrane surface and restricts back diffusion of NaCl from the surface to the bulk solution, which increases conductivity at the membrane surface. This can be observed because a cake-enhanced osmotic pressure increases TMP (Figure 3.11c). For BSA fouling, early foulant contact reflected as a shift in conductance toward lower values and conductance decreased as fouling evolved. BSA likely formed a compact film early

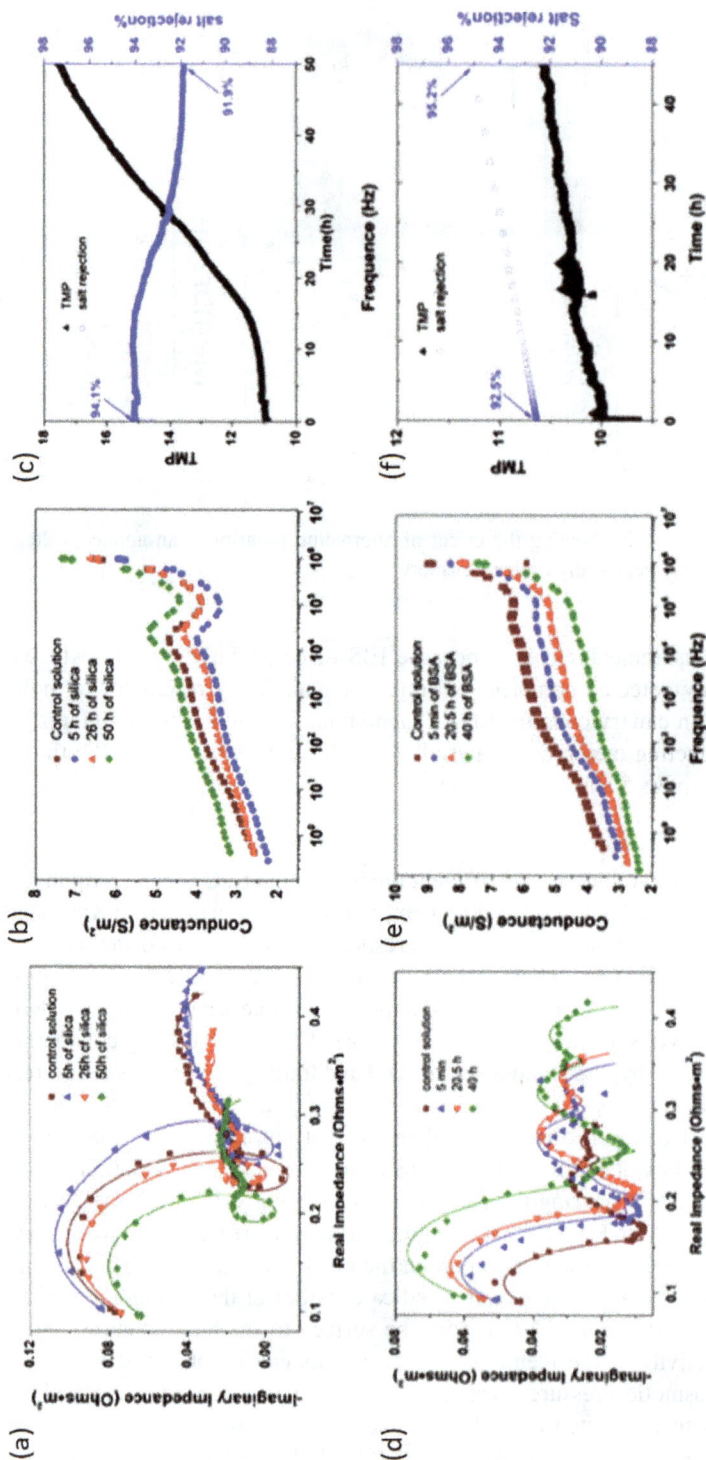

FIGURE 3.11 (a–c) Nyquist plot, conductance, TMP, and salt rejection for fouling with silica; (d–f) Nyquist plot, conductance, TMP, and salt rejection for fouling with BSA.

Source: Sim, L. N., Wang, Z. J., Gu, J., Coster, H. G. L., & Fane, A. G. (2013). Detection of reverse osmosis membrane fouling with silica, bovine serum albumin and their mixture using in-situ electrical impedance spectroscopy. *Journal of Membrane Science, 443*, 45–53. Reprinted with permission.

on, which prevented salts from diffusing from the bulk to the membrane surface, which is verified by the increased salt rejection (SR) shown in Figure 3.11f. The two different fouling types cause different packing and foulant layer structures as silica colloids are soft spheres and tend to foul via cake filtration, whereas BSA would form a film (L. N. Sim et al., 2013).

3.3.3.3 Online Monitoring in Plants

Field trials using EIS have been limited as it is still an emerging technique for membrane separation. Additionally, the lack of electrically conducting membranes means that the setup for in situ electrochemical studies can be complicated. The NEWater RO plant in Singapore is the first field trial to incorporate EIS to characterize fouling (Lee Nuang Sim et al., 2016) over a period of eight months. As the membranes are commercial non-conducting membranes, EIS measurements were carried out in a separate sidestream cell known as a canary cell in which process conditions are mimicked (Ho et al., 2017). The canary cell contains the membrane and spacer materials used in the actual module. Figure 3.12 shows a schematic of the NEWater RO process depicting the canary cell used for impedance measurements.

The study consisted of measuring EIS data at a specified frequency and fitting the data to an equivalent circuit consisting of the membrane layers. Each of the layers has a capacitive and a resistive component. Figure 3.13a shows the Nyquist plot from the canary cell. The low frequency region (<10 Hz) represents the DP layer at the membrane–solution interface, the mid-frequency region (10 Hz—100 kHz) represents the layers of the membrane itself, and the high frequency regions (>100 kHz) represent the solution on either side of the membrane. The real component of impedance in the DP layer, $Z_{real\text{-}DP}$ helps evaluate fouling. Variation in normalized $Z_{real\text{-}DP}$ over time is shown in Figure 3.13b together with plant shutdown and cleaning-in-place (CIP)

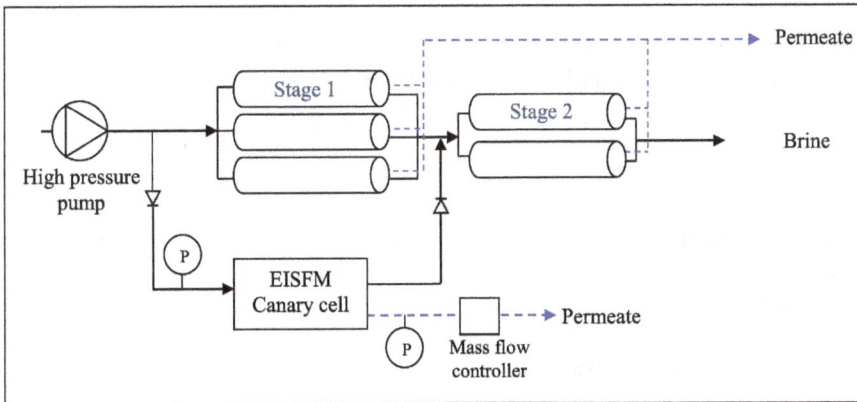

FIGURE 3.12   RO process in NEWater with location of canary cell for EIS measurements.

Source: Ho, J. S., Sim, L. N., Webster, R. D., Viswanath, B., Coster, H. G. L., & Fane, A. G. (2017). Monitoring fouling behavior of reverse osmosis membranes using electrical impedance spectroscopy: A field trial study. *Desalination, 407,* 75–84. Reprinted with permission.

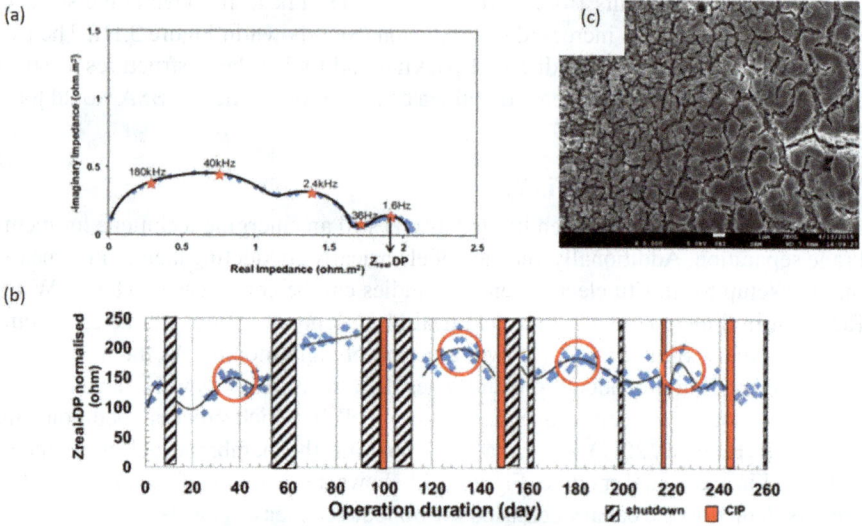

FIGURE 3.13 (a) Nyquist plot from canary cell; full curve depicts fit of the equivalent circuit; (b) normalized $Z_{real-DP}$ over time; (c) SEM image of fouled membrane after 8 months.

Source: Ho, J. S., Sim, L. N., Webster, R. D., Viswanath, B., Coster, H. G. L., & Fane, A. G. (2017). Monitoring fouling behavior of reverse osmosis membranes using electrical impedance spectroscopy: A field trial study. *Desalination*, *407*, 75–84. Reprinted with permission.

events. Before each round of chemical cleaning, a maxima in $Z_{real-DP}$ was observed due to the formation of a loose cake layer of the foulant material. At first, impedance increases after which a denser layer prevents salt ions from diffusing back and impedance decreases, similar to another lab-scale report. Based on their trend, they deduced that inorganic colloidal matter is the dominant foulant here, which was confirmed by SEM imaging during membrane autopsy after it was removed. The major foulant type was 1 μm size particles of $Ca_3(PO_4)_2$. Figure 3.13c shows an SEM image of the fouled membrane. Although UF pretreatment may have removed micron-sized particles, nanosized particles that reached the RO membranes could have coagulated and formed a foulant layer during RO (Ho et al., 2017).

This field study was largely successful and points out the need for tools such as EIS for early detection so that fouling control measures can be strategized efficiently. Such efforts would reduce the frequency of CIP and overall operating cost. Table 3.3 summarizes selected studies of fouling monitoring using EIS. While some useful patterns can be deduced, several contradicting results still necessitate the need for standardized operating conditions and long-term studies to evaluate electrical properties.

It is clear that EIS has a bright potential in membrane technology for online early characterization of fouling and structural changes. However, some challenges still limit real-life implementation, and it is possible that new configurations and materials will further advance the use of EIS for in situ fouling detection. More work is

TABLE 3.3

Summary of EIS Studies Carried Out for Different Foulant Types and Processes.

Fouling Agent	In Situ/Ex Situ	Process	Conductance	Capacitance	Reference
Humic and fulvic acids	ex situ	Reverse osmosis	—	Decreases	(Cen et al., 2013)
CaCO$_3$	in situ	Reverse osmosis	Increases	—	(Antony et al., 2013)
SDS	ex situ	Electrodialysis	Decreases	—	(Zhao et al., 2017)
Alginate	in situ	Reverse osmosis	Decreases, then increases as fouling progresses	Decreases	(Lee Nuang Sim et al., 2016)
Silica	in situ	Reverse osmosis	Decreases, then increases as fouling progresses	Decreases	(Lee Nuang Sim et al., 2016)
Silica	in situ	Reverse osmosis	Decreases, then increases	No significant change	(L. N. Sim et al., 2013)
BSA	in situ	Reverse osmosis	Decreases	Decreases	(L. N. Sim et al., 2013)
BSA	ex situ	Ultrafiltration	Decreases	Decreases	(Sengur-Tasdemir et al., 2018)
CaCO$_3$	in situ	Reverse osmosis	Decreases	Increases	(Kavanagh et al., 2009)

needed using realistic and complex foulant solutions to better understand this technique for evaluating fouling with impedance measurements. As these issues are tackled, it is hoped that large-scale implementation of EIS in membrane separation will only grow in the near future.

3.4 STATUS OF MEMBRANE CLEANING AND CONTROL OF FOULING AND RELATED PHENOMENA

Periodic cleaning must be conducted to prevent severe and irreversible membrane fouling (J. Paul Chen et al., 2003). Inappropriate cleaning methods may worsen fouling and deteriorate performance, thus causing a shorter membrane life and increasing operating and maintenance costs.

3.4.1 PRESSURE-DRIVEN PROCESSES

Cleaning methods can be divided into physical cleaning and chemical cleaning. Physical cleaning methods use physical/mechanical forces to remove foulants from the membrane surface to mitigate the effects of reversible fouling (J Paul Chen et al., 2003). Examples of physical cleaning methods include backwashing, forward flushing, air flushing, as well as the use of turbulence promoters and vibration/ultrasound (Abdel-Karim et al., 2021). In forward flushing, high cross-flow velocities are applied at the feed side that help loosen and remove foulants. It is often combined with reverse flushing for a short amount of time to improve cleaning effectiveness.

3.4.1.1 Industrial Practice

For low pressure processes such as ultrafiltration, there are two common routes to cleaning: (1) backwashing followed by air scrubbing and (2) chemical cleaning (Ali, 2010; H. Chang et al., 2017; Park et al., 2018). Backwashing and air scrubbing should be periodic and automatic, and their frequency depends on raw water quality (Gu et al., 2018; Toray Industries, 2019; Vial & Doussau, 2003; Yang et al., 2021). For surface water filtration, they recommend cleaning with backwash and air scrubbing every 30 minutes (Toray Industries, 2019). For ultrafiltration systems, Toray Industries recommend that chemical cleaning should be carried out before the transmembrane pressure reaches 200 kPa to avoid membrane damage (Toray Industries, 2019).

For high pressure NF/RO membranes, standard cleaners are either acidic or alkaline. It is important to use cleaning solutions appropriate for the type of cleaning, at the correct pH and temperature range to optimize membrane regeneration and prevent fouling from worsening. Acid cleaning solutions are used to remove inorganic precipitates, while alkaline cleaning solutions are used to remove organic fouling and biological matter (DOW, 2008). Sulfuric acid poses the risk of calcium sulfate fouling and therefore should not be used for cleaning. Dow FilmTec recommends alkaline cleaning as the first step unless only calcium carbonate or iron oxide(hydroxide) is confirmed as foulant. This is because acid cleaners react with organics and biofilms and further deteriorate membrane performance, which may be difficult to recover without extreme and damaging cleaning measures (DOW, 2008). Zaidi and Haleema describe cleaning procedures in RO plants in more detail (Zaidi & Saleem, 2022).

3.4.1.2 Recent Developments in Membrane Cleaning and Fouling Control

3.4.1.2.1 Ultrasound

Ultrasound is an acoustic wave that oscillates at frequencies greater than 20 kHz (Gallo et al., 2018; Siegel & Luo, 2008). Ultrasound energy for cleaning first emerged in the 1950s (Cronshaw, 1956; L. G. Larson & Berglund, 1959; McKenna, 1955). Following the development of membrane technology, it was then proposed as a regeneration method for RO and UF membranes very early on (Deqian, 1987). However, ultrasonic cleaning systems back then operated at low frequencies, in the range of 20–40 kHz (Mason, 2016). The very first patent applying ultrasonic vibration to prevent membrane clogging for reverse osmosis was authored by Richard Harvey in 1962 (Harvey, 1965), just four years after the Loeb–Sourirajan membrane was developed and reverse osmosis became a practical reality.

In 1986, McQueen showed that applying a higher frequency increased decontamination of submicroscopic contaminants but that a lower frequency was more suitable for other materials (McQueen, 1986). Timothy Mason points out that high frequency ultrasonic cleaning at around 1 MHz can minimize cavitational damage (Mason, 2016). Ultrasound cleaning relies on the physical effects generated by ultrasound and acoustic cavitation (Yusof et al., 2016). Ultrasonic cavitation is when liquid hollow bubbles grow and collapse during the propagation of ultrasound waves through an elastic medium, as they alternate between compression and rarefaction phases (Aghapour Aktij et al., 2020; Vajnhandl & Majcen Le Marechal, 2005; Z. Wang et al., 2020). Figure 3.14 shows the phenomenon of ultrasonic cavitation. Shchukin et al. describe the mechanism of ultrasonic cavitation near solid surfaces in detail (Shchukin et al., 2011). Compression cycles reduce intermolecular distance, while rarefaction cycles increase molecular distance (Pilli et al., 2011). Cavitation bubbles, or vapor-filled cavities, are generated when the pressure amplitude is greater than the tensile strength, after which microbubbles grow and compress in each alternating phase. When the bubble diameter reaches a critical value, the compression cycle causes it to collapse. Ultrasound in a liquid medium may induce physical and chemical phenomena such as heating or the formation of radicals (Flores et al., 2021). Shock waves from cavity collapse near a solid surface can generate a jet of liquid with a speed on the order of 110 m s⁻¹, which induces a cleaning action (Vajnhandl & Majcen Le Marechal, 2005). Although liquid microjet is the major mechanism of foulant detachment, other cavitational mechanisms such as microsteaming also assist in cleaning (Kyllönen et al., 2005).

In fouling control, ultrasound energy can break the concentration polarization and cake layer at the membrane surface, which helps to regenerate the flux (Kyllönen et al., 2005).

Ultrasonic cleaning of fouled membranes can be either through ex situ ultrasonic transducers or with in situ ultrasonicators (Aghapour Aktij et al., 2020; Arefi-Oskoui

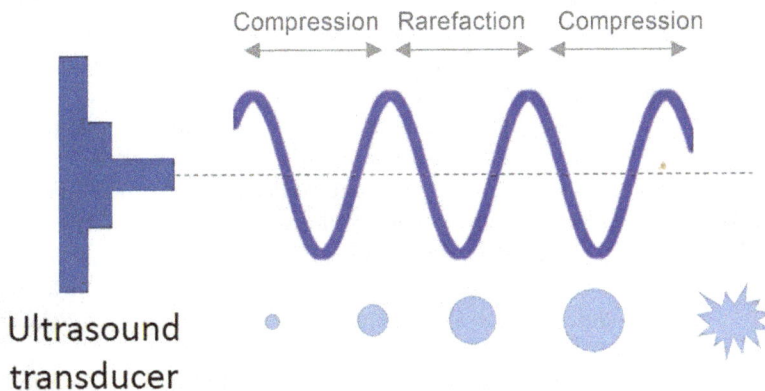

FIGURE 3.14 Schematic of the ultrasound cavitation phenomenon, which involves the formation, growth, and collapse of bubbles.

et al., 2019). In situ methods for membrane cleaning are preferred since ex situ cleaning requires interruption of the filtration and removal of the membrane element from the module and then insertion into a sonication bath (Sari Erkan et al., 2018). The energy released from cavitation physically causes foulants to detach from the membrane surface (Aghapour Aktij et al., 2020; Arefi-Oskoui et al., 2019; Qasim et al., 2018). Ultrasound can be applied in conjunction with other cleaning methods to improve efficiency. Alternatively, ultrasound energy can complement other pretreatment techniques such as ozonation to reduce the amount of organics in the feedwater and thus delay membrane fouling (Arefi-Oskoui et al., 2019; Lozier & Sierka, 1985). Over the years, the amplitude and frequency for various membranes and foulant types have been explored to optimize ultrasonic fouling control. Here we review some recent developments in the use of ultrasonic energy for fouling control in membrane processes.

Some of the most well-known work in the area of ultrasonic cleaning of membranes was carried out in the early 2000s. Lamminen et al. investigated the possible mechanisms for cleaning ceramic membranes fouled by sulfate polystyrene latex particles with ultrasound (Lamminen et al., 2004). Increased power intensity at a lower frequency favored particle removal. While cavitation is responsible for detaching foulant from the membrane, turbulence from ultrasound (i.e. acoustic steaming) helps transport detached particles away from the surface. Membrane surface did not show any visible damage at 20 W cm^{-2}, 20 kHz. An advantage of using ultrasound for fouling control is that it does not affect intrinsic membrane permeability, but depending on operating conditions, there can be some damage due to ultrasound irradiation (Kyllönen et al., 2005). In an early review, Kyllönen et al. attribute the non-commercialization of ultrasonically enhanced membrane filtration to the lack of transducer technology development for membrane filtration (Kyllönen et al., 2005).

Li et al. combined ultrasound with forward flushing to recover permeate flux of flat-sheet nylon MF membranes fouled by Kraft paper mill effluent (J. Li et al., 2002b). They found that a high forward flush velocity and low cleaning solution (water) temperature, together with ultrasound energy at a frequency of 20 kHz and a power of 375 W, was more efficient in cleaning (J. Li et al., 2002b). Kobayashi also applied ultrasound cleaning to MF and UF membranes for the treatment of peptone and milk aqueous solutions (Kobayashi et al., 2003). Muthukumaran et al. also applied ultrasound to dairy-fouled UF membranes (Muthukumaran et al., 2005).

Feng et al. applied online ultrasonic cleaning to remove fouling and to regenerate flux of polyamide reverse osmosis membranes during cross-flow filtration of wastewater effluents containing $CaSO_4$, Fe^{3+} and carboxyl cellulose solutions (Feng et al., 2006). Camara et al. reported a low frequency ultrasound cleaning technique to control fouling during UF of simulated latex paint effluent using a flat-sheet membrane (Figure 3.15) (Camara et al., 2020). When a 20 kHz and 0.29 W cm^{-2} was applied to a feed solution with 0.075 wt% solid, permeate flux increased by 19.7% (Camara et al., 2020).

Recently, interest has moved to modeling the effects of ultrasound on membrane performance, i.e. cleaning efficiency (Luo & Wang, 2022) and feed channel temperature (Horrigan & Freire-Gormaly, 2022). Figure 3.16 shows the effect of ultrasound power and temperature on flux recovery.

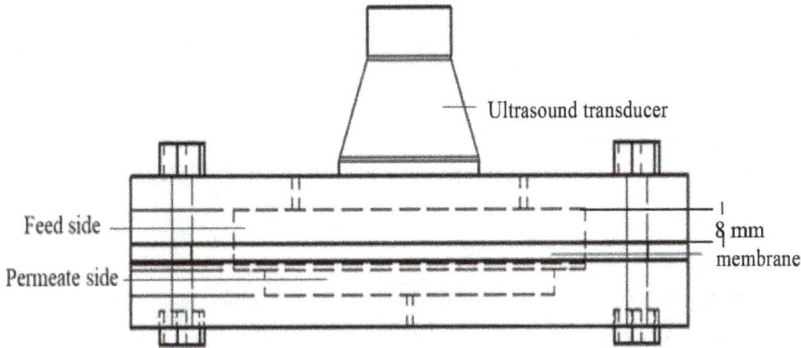

FIGURE 3.15 Schematic of UF membrane module used for ultrasound fouling control.

Source: Camara, H. W. D., Doan, H., & Lohi, A. (2020). In-situ ultrasound-assisted control of polymeric membrane fouling. *Ultrasonics*, *108*, 106206. Reprinted with permission.

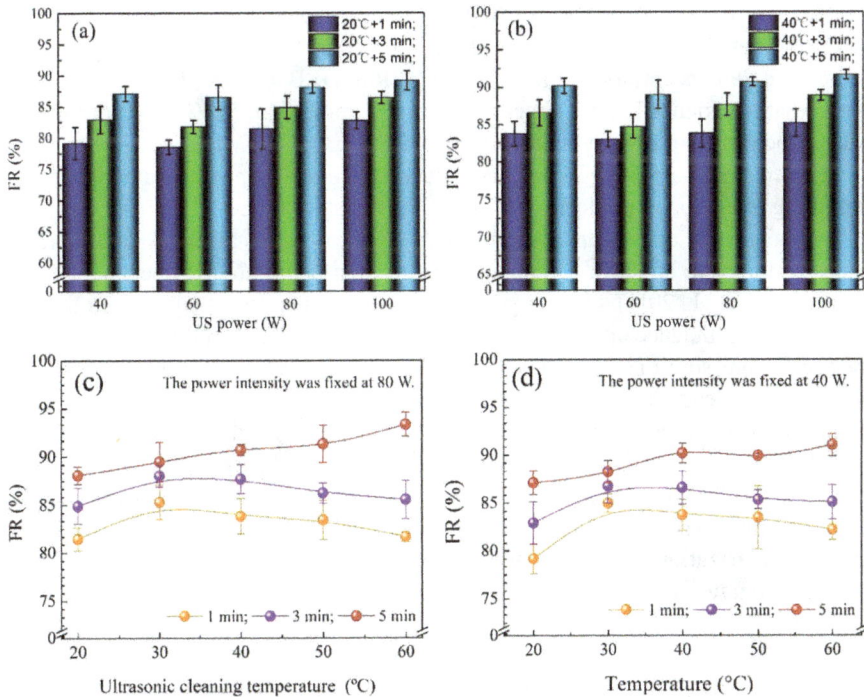

FIGURE 3.16 Effect of ultrasound power intensity on flux recovery.

Source: Luo, H., & Wang, Z. (2022). A new ultrasonic cleaning model for predicting the flux recovery of the UF membrane fouled with humic acid. *Journal of Environmental Chemical Engineering*, *10*(2), 107156. Reprinted with permission.

Interest in ultrasound cleaning has been steady, with commercial prospects still low. The increasing number of studies (Jin et al., 2022), as well as reviews on this topic (Abdel-Karim et al., 2021; Aghapour Aktij et al., 2020; Córdova et al., 2020; Kyllönen et al., 2005; Naji et al., 2021), points toward an upward trend. While much of the research focuses on correlating ultrasound parameters with cleaning efficiency, there is a need to focus on system design and optimization of integrating ultrasound with filtration cells.

3.4.1.2.2 Turbulence Promoters

Turbulence refers to a type of fluid motion characterized by chaotic changes in pressure and flow velocity (C. Liu et al., 2021). Many studies in the last four decades cite the use of turbulence promoters to mitigate fouling. Turbulence improves mass transfer and prevents foulant layer from being deposited, or it removes any deposition of foulants. Turbulence can be generated in many ways including air sparging, jets, spacers, etc. Liu et al. based the effect of turbulence promoters on the characteristics of the cake layer formed on the membrane due to fouling (Y. Liu et al., 2012). Turbulence promoters reduce thickness and increase porosity of the cake layer formed on the membrane due to fouling. At the same time, they reduce the particle size of the cake, which increases specific resistance. However, these effects combine to lower the overall cake resistance and enhance permeate flux.

De Boer et al. used fluidized beds of steel and lead beds to reduce fouling through turbulence for concentration of food liquids by RO (De Boer et al., 1980). Krstić et al. found that turbulence improved permeate flux by more than 500% when turbulence was introduced in the form of static mixing in a cross-flow MF setup (Krstić et al., 2002). Yeo et al. correlated turbulence induced by air sparging around hollow fibers to fouling reduction in the form of TMP rise at constant flux.

Pourbozorg et al. set up a turbulence generator using a vibrating perforated plate to study the effect of turbulence on filtration using hollow fiber membranes (Pourbozorg et al., 2016). Using 4 g L^{-1} yeast feed suspensions, they showed that the presence of turbulence controlled membrane fouling and lowered the rate of transmembrane pressure (TMP) increase. In a later study, Cho et al. used an orifice to generate turbulence to mitigate membrane fouling in microalgae harvesting, attributing flux regeneration to the detachment of the cake layer due to turbulence (Cho et al., 2020). Fouling control through turbulence allowed increased productivity of microalgae harvesting by a factor of almost 3 (Cho et al., 2020). Spacers are also a kind of turbulence promoter, often used on the feed side of the membrane, that mitigate polarization effects and assist in fouling mitigation. The role of spacers in fouling mitigation is discussed in Chapter 6.

3.4.1.2.3 Surface Modification

Membrane fouling in pressure-driven filtration processes depends strongly on surface characteristics such as roughness and hydrophilicity. Membranes with smooth and hydrophilic surfaces exhibit lower fouling tendency than those with surfaces that are rough and hydrophobic. Although fouling can also be hydrophilic, in which case hydrophilic membranes may have a pronounced effect, it is generally interactions between hydrophobic substances and the hydrophobic membrane that cause or worsen fouling. Surface modification is one route through which surface properties such as charge, roughness, and hydrophilicity can be altered to generate low-fouling or antifouling surfaces. Surface modification includes physical and chemical methods i.e. surface grafting. During chemical surface modification, chemical reactions produce a grafted layer with strong adhesion to the membrane's surface. On the other hand, physical modification such as adsorption or coating introduces the desired property on the membrane surface. For an in-depth review of surface modification techniques for fouling mitigation in pressure-driven processes, the reader is referred to Choudhury et al., 2018; Kochkodan et al., 2014; Remanan et al., 2018; Saget et al., 2021. Strategies and advances in surface modification specifically for membrane spacers are discussed in Section 6.3.1.

3.4.2 MEMBRANE DISTILLATION

3.4.2.1 Chemical Cleaning

There has recently been an interest in optimizing chemical cleaning strategies for MD as it inches toward full-scale implementation, partly because chemical cleaning strategies are already widely applied and established in RO desalination processes. Peng et al. studied the effect of four antiscaling and five cleaning agents on $CaSO_4$ scaling in DCMD for RO brine treatment (Peng et al., 2015). They found that an effective antiscaling agent reduces scaling and wetting; a 2% ethylene diamine tetraacetic acid (EDTA)-4Na solution recovered 92.8% flux at 77 °C (Peng et al., 2015). Charfi et al. also sought to define an optimal cleaning strategy based on deionized (DI) water flushing, sodium hypochlorite (NaOCl), and citric acid by varying cleaning frequencies, durations, and cross-flow velocities for fouling control in DCMD for real wastewater (Charfi et al., 2021; Kim et al., 2017). They correlated the effect of each component to flux regeneration. In another study, Jia et al. compared five cleaning methods including hydrogen chloride (HCl)–sodium hydroxide (NaOH), EDTA–NaOH, citric acid, sodium hypochlorite (NaClO), and sodium dodecyl sulphate (SDS) to clean membranes fouled during DCMD to treat RO brine (Jia et al., 2021). They found that 3 wt% SDS was the most effective in recovering membrane performance.

Guillen-Burrieza et al. studied membrane fouling and subsequent chemical cleaning approaches to mitigate the effects of fouling and wetting in long-term plant-scale solar-powered MD operation between 2010 and 2013 (Guillen-Burrieza et al., 2014). They found that the poor flow distribution inside the pilot-sized module reduced cleaning effectiveness and required large amounts of water. However, inactive "dry out" periods in plant operation favored wetting and offset any improvements from cleaning. Fouling, wetting, and subsequent cleaning procedures in pilot-scale long-term studies need more attention and should be included in future MD research as the technology looks toward commercial use. With the development of new cleaning methods, it is important to optimize cleaning strategy for the given configuration, feedwater, cross-flow velocity in order to lower the energy costs associated with MD. Different chemicals are effective for different foulant types. Alkaline solutions help remove organic foulants and possibly biofoulants, but frequent and prolonged exposure may damage the membrane surface and change structure, i.e. "ageing" (Puspitasari et al., 2010). Surfactant-based cleaning agents work by forming a micellar bubble around the hydrophobic organic foulant, helping it detach from the membrane surface. Acidic solutions may be useful in removing scaling agents such as $CaCO_3$ (Abdel-Karim et al., 2021; Guillen-Burrieza et al., 2014). As Peng points out, effective control of organic and inorganic scaling in MD membrane modules are key to its industrialization (Peng et al., 2015). Nevertheless, future directions must emphasize green fouling control strategies for MD (Tijing et al., 2015).

3.4.2.2 Physical Cleaning

3.4.2.2.1 Aeration and Bubbling

Physical cleaning methods such as air sparging (Bhoumick et al., 2021) and gas bubbling have been applied to various configurations of MD (Ding et al., 2011; Ye et al., 2019; W. Zhang et al., 2022) as chemical-free alternatives to recover flux. Air sparging increased flux by 10% for MD of flue gas desulfurized water containing saturated $CaSO_4$ (Bhoumick et al., 2021). Ding et al. studied fouling control through intermittent gas bubbling in concentrating the extract of traditional Chinese medicine by DCMD (Ding et al., 2011). Cleaning efficiency increases with increased gas flow rate, bubbling duration, and MD duration. Microbubble aeration also enhances vacuum membrane distillation (VMD) performance by the mitigation of scaling and concentration polarization, with improved specific energy consumption (Ye et al., 2019; W. Zhang et al., 2022). One must be cautious when selecting process parameters, since a high gas flow rate may promote growth of small crystals and hinder the bubble scouring effect (Y. S. Chang et al., 2021).

3.4.2.2.2 Vibration

Some physical methods work better as preventative measures against fouling than for cleaning. Huang et al. showed that applying a mechanical vibration of 42.5 Hz was suitable as a preventative measure when initiated at the onset of DCMD operation but did not improve flux if applied after fouling had already been observed in the form of 16% flux decline (F. Y. C. Huang et al., 2019). Ultrasonic cleaning has also recently been explored as a technique to enhance flux in MD by up to 100% for various foulants: BSA (Hou et al., 2017), silica (Hou et al., 2016), and calcium (Naji et al., 2020). The high flux enhancement results from a combination of cleaning and improved mass transfer effects (Naji et al., 2020). The efficient use of ultrasound energy and practical industrial-scale applications remain challenges that can be tackled by further insight into theoretical modeling and optimization of process configuration (Naji et al., 2020).

3.4.2.2.3 UV and Visible Light Irradiation

Photocatalytic materials present an exciting ongoing development in MD fouling control. UV and visible light exposure can recover up to 90% of the permeate flux. Hamzah and Leo first reported the incorporation of TiO_2 nanoparticles in PVDF membranes prepared via non-solvent-induced phase separation (NIPS) for self-cleaning under UV irradiation (Hamzah & Leo, 2017); the cleaning effect was demonstrated

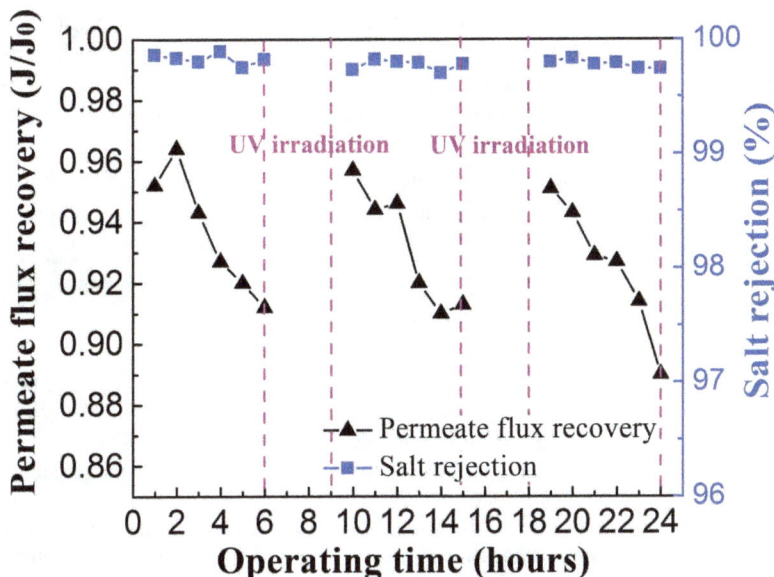

FIGURE 3.17 Performance of self-cleaning PTFE/ZnO membrane for VMD of wastewater effluent with three times UV irradiation.

Source: Huang, Q.-L., Huang, Y., Xiao, C.-F., You, Y.-W., & Zhang, C.-X. (2017). Electrospun ultrafine fibrous PTFE-supported ZnO porous membrane with self-cleaning function for vacuum membrane distillation. *Journal of Membrane Science, 534*, 73–82. Reprinted with permission.

outside the MD cell. Electrospinning is an attractive technique for fabricating high surface area nanofibrous membranes embedded with functional materials such as photocatalysts. Building on Hamzah's work, two groups developed polymer nanofiber membranes with either ZnO or Ag nanoparticles and achieved in situ membrane regeneration >90% under UV or visible light (Guo et al., 2019; Q.-L. Huang et al., 2017). Figure 3.17 shows the flux recovery and salt rejection of electrospun PTFE/ZnO membrane for VMD of dye-containing wastewater.

While MD is not as prone to fouling as pressure-driven membrane processes, the closely related problem of wetting has severely hampered commercialization of the process.

3.5 CONCLUSION

Continuous monitoring of membrane processes is recognized as essential to efficient industrial operation and upscaling of newer technologies. Currently, industrially adopted methods are few and not without challenges: It remains difficult to differentiate different types of foulants during operation, determine the exact location of fouling, and gain a deeper insight into the structural evolution of foulants at the

membrane surface. This chapter considered process monitoring in membrane separation processes and overviewed recent developments in techniques for membrane
monitoring and fouling mitigation. Special attention is paid to the role of electrochemical impedance spectroscopy as a non-invasive tool for the characterization of
fouling and related phenomena. Complex designs and high-cost materials have thus
far limited the widespread use of powerful non-invasive tools for in situ characterization. This is accompanied by a general resistance to transformative integration of
new technologies in the water sector.

BIBLIOGRAPHY

Abdel-Karim, A., Leaper, S., Skuse, C., Zaragoza, G., Gryta, M., & Gorgojo, P. (2021). Membrane cleaning and pretreatments in membrane distillation—a review. *Chemical Engineering Journal, 422,* 129696. https://doi.org/10.1016/j.cej.2021.129696

Adeel, Z. (2017). A renewed focus on water security within the 2030 agenda for sustainable
development. *Sustainability Science, 12*(6), 891–894.

Adham, S., Gagliardo, P., Smith, D., Ross, D., Gramith, K., & Trussell, R. (1998). Monitoring
the integrity of reverse osmosis membranes. *Desalination, 119*(1), 143–150. https://doi.
org/10.1016/S0011-9164(98)00134-9

Aghapour Aktij, S., Taghipour, A., Rahimpour, A., Mollahosseini, A., & Tiraferri, A. (2020).
A critical review on ultrasonic-assisted fouling control and cleaning of fouled membranes. *Ultrasonics, 108,* 106228. https://doi.org/10.1016/j.ultras.2020.106228

Ali, C. (2010). *Chemical cleaning MF/UF systems.* Paper presented at the AMTA Technology
Transfer Workshop.

Amin Saad, M. (2004). Early discovery of RO membrane fouling and real-time monitoring of
plant performance for optimizing cost of water. *Desalination, 165,* 183–191. https://doi.
org/10.1016/j.desal.2004.06.021

Antony, A., Blackbeard, J., & Leslie, G. (2012). Removal efficiency and integrity monitoring
techniques for virus removal by membrane processes. *Critical Reviews in Environmental
Science and Technology, 42*(9), 891–933.

Antony, A., Chilcott, T., Coster, H., & Leslie, G. (2013). In situ structural and functional characterization of reverse osmosis membranes using electrical impedance spectroscopy.
Journal of Membrane Science, 425–426(Suppl C), 89–97. https://doi.org/10.1016/j.
memsci.2012.09.028

Arefi-Oskoui, S., Khataee, A., Safarpour, M., Orooji, Y., & Vatanpour, V. (2019). A review on
the applications of ultrasonic technology in membrane bioreactors. *Ultrasonics Sonochemistry, 58,* 104633. https://doi.org/10.1016/j.ultsonch.2019.104633

Aumann, S., Donner, S., Fischer, J., & Müller, F. (2019). Optical coherence tomography
(OCT): Principle and technical realization. *High Resolution Imaging in Microscopy and
Ophthalmology, 59–85.*

Bauer, A., Wagner, M., Saravia, F., Bartl, S., Hilgenfeldt, V., & Horn, H. (2019). In-situ monitoring and quantification of fouling development in membrane distillation by means of
optical coherence tomography. *Journal of Membrane Science, 577,* 145–152. https://doi.
org/10.1016/j.memsci.2019.02.006

Bhoumick, M. C., Roy, S., & Mitra, S. (2021). Synergistic effect of air sparging in direct contact membrane distillation to control membrane fouling and enhancing flux. *Separation
and Purification Technology, 272,* 118681. https://doi.org/10.1016/j.seppur.2021.118681

Boerlage, S. F. E., Kennedy, M., Aniye, M. P., & Schippers, J. C. (2003). Applications of the
MFI-UF to measure and predict particulate fouling in RO systems. *Journal of Membrane
Science, 220*(1), 97–116. https://doi.org/10.1016/S0376-7388(03)00222-9

Camara, H. W. D., Doan, H., & Lohi, A. (2020). In-situ ultrasound-assisted control of polymeric membrane fouling. *Ultrasonics, 108*, 106206. https://doi.org/10.1016/j.ultras.2020.106206

Cen, J., Kavanagh, J., Coster, H., & Barton, G. (2013). Fouling of reverse osmosis membranes by cane molasses fermentation wastewater: Detection by electrical impedance spectroscopy techniques. *Desalination and Water Treatment, 51*(4–6), 969–975. doi:10.1080/19443994.2012.714657

Chang, H., Liang, H., Qu, F., Liu, B., Yu, H., Du, X., . . . Snyder, S. A. (2017). Hydraulic backwashing for low-pressure membranes in drinking water treatment: A review. *Journal of Membrane Science, 540*, 362–380.

Chang, Y. S., Ooi, B. S., Ahmad, A. L., Leo, C. P., Lyly Leow, H. T., Abdullah, M. Z., & Aziz, N. A. (2021). Correlating scalants characteristic and air bubbling rate in submerged vacuum membrane distillation: A fouling control strategy. *Journal of Membrane Science, 621*, 118991. https://doi.org/10.1016/j.memsci.2020.118991

Charfi, A., Kim, S., Yoon, Y., & Cho, J. (2021). Optimal cleaning strategy to alleviate fouling in membrane distillation process to treat anaerobic digestate. *Chemosphere, 279*, 130524. https://doi.org/10.1016/j.chemosphere.2021.130524

Chen, J. C., Li, Q., & Elimelech, M. (2004). In situ monitoring techniques for concentration polarization and fouling phenomena in membrane filtration. *Advances in Colloid and Interface Science, 107*(2), 83–108. https://doi.org/10.1016/j.cis.2003.10.018

Chen, J. P., Kim, S., & Ting, Y. (2003). Optimization of membrane physical and chemical cleaning by a statistically designed approach. *Journal of Membrane Science, 219*(1–2), 27–45.

Chen, J. P., Mou, H., Wang, L. K., Matsuura, T., & Wei, Y. (2011). Membrane separation: Basics and applications. In L. K. Wang, J. P. Chen, Y.-T. Hung, & N. K. Shammas (Eds.), *Membrane and desalination technologies* (pp. 271–332). Humana Press.

Chen, L. P., Zhang, Y., Li, R., Xu, Y., Zhu, H., Zhang, M., & Zhang, H. (2022). In situ visualization of combined membrane fouling behaviors using multi-color light sheet fluorescence imaging: A study with BSA and dextran mixture. *Journal of Membrane Science, 649*, 120385. https://doi.org/10.1016/j.memsci.2022.120385

Chen, V., Li, H., & Fane, A. G. (2004). Non-invasive observation of synthetic membrane processes—a review of methods. *Journal of Membrane Science, 241*(1), 23–44. https://doi.org/10.1016/j.memsci.2004.04.029

Chen, Y., Wang, Z., Jennings, G. K., & Lin, S. (2017). Probing pore wetting in membrane distillation using impedance: Early detection and mechanism of surfactant-induced wetting. *Environmental Science & Technology Letters, 4*(11), 505–510. doi:10.1021/acs.estlett.7b00372

Cho, H., Mushtaq, A., Hwang, T., Kim, H.-S., & Han, J.-I. (2020). Orifice-based membrane fouling inhibition employing in-situ turbulence for efficient microalgae harvesting. *Separation and Purification Technology, 251*, 117277. https://doi.org/10.1016/j.seppur.2020.117277

Choudhury, R. R., Gohil, J. M., Mohanty, S., & Nayak, S. K. (2018). Antifouling, fouling release and antimicrobial materials for surface modification of reverse osmosis and nanofiltration membranes. *Journal of Materials Chemistry A, 6*(2), 313–333.

Corbin, K., Pinkard, H., Peck, S., Beemiller, P., & Krummel, M. F. (2014). Chapter 8—Assessing and benchmarking multiphoton microscopes for biologists. In J. C. Waters & T. Wittman (Eds.), *Methods in cell biology* (Vol. 123, pp. 135–151). Academic Press.

Córdova, A., Astudillo-Castro, C., Ruby-Figueroa, R., Valencia, P., & Soto, C. (2020). Recent advances and perspectives of ultrasound assisted membrane food processing. *Food Research International, 133*, 109163. https://doi.org/10.1016/j.foodres.2020.109163

Coster, H. G. L., Kim, K. J., Dahlan, K., Smith, J. R., & Fell, C. J. D. (1992). Characterisation of ultrafiltration membranes by impedance spectroscopy. I. Determination of the separate electrical parameters and porosity of the skin and sublayers. *Journal of Membrane Science, 66*(1), 19–26. https://doi.org/10.1016/0376-7388(92)80087-Z

Cronshaw, A. W. (1956). The cleaning of viscose spinning nozzles by means of ultrasonics. *Journal of the Textile Institute Proceedings*, *47*(12), P1015–P1018. doi:10.1080/19447015608665385

Cui, Z., Wang, X., Ngo, H., & Zhu, G. (2022). In-situ monitoring of membrane fouling migration and compression mechanism with improved ultraviolet technique in membrane bioreactors. *Bioresource Technology*, *347*, 126684. https://doi.org/10.1016/j.biortech.2022.126684

Darton, T., Annunziata, U., del Vigo Pisano, F., & Gallego, S. (2004). Membrane autopsy helps to provide solutions to operational problems. *Desalination*, *167*, 239–245. https://doi.org/10.1016/j.desal.2004.06.133

De Boer, R., Zomerman, J. J., Hiddink, J., Aufderheyde, J., Van Swaay, W. P. M., & Smolders, C. A. (1980). Fluidized beds as turbulence promoters in the concentration of food liquids by reverse osmosis. *Journal of Food Science*, *45*(6), 1522–1528. https://doi.org/10.1111/j.1365-2621.1980.tb07554.x

de Lara, R., & Benavente, J. (2009). Use of hydrodynamic and electrical measurements to determine protein fouling mechanisms for microfiltration membranes with different structures and materials. *Separation and Purification Technology*, *66*(3), 517–524. https://doi.org/10.1016/j.seppur.2009.02.003

Deqian, R. (1987). Cleaning and regeneration of membranes. *Desalination*, *62*, 363–371. https://doi.org/10.1016/0011-9164(87)87037-6

Ding, Z., Liu, L., Liu, Z., & Ma, R. (2011). The use of intermittent gas bubbling to control membrane fouling in concentrating TCM extract by membrane distillation. *Journal of Membrane Science*, *372*(1), 172–181. https://doi.org/10.1016/j.memsci.2011.01.063

Doudiès, F., Loginov, M., Hengl, N., Karrouch, M., Leconte, N., Garnier-Lambrouin, F., . . . Gésan-Guiziou, G. (2021). Build-up and relaxation of membrane fouling deposits produced during crossflow ultrafiltration of casein micelle dispersions at 12 °C and 42 °C probed by in situ SAXS. *Journal of Membrane Science*, *618*, 118700. https://doi.org/10.1016/j.memsci.2020.118700

DOW. (2008). *Water & process solutions, FILMTEC reverse osmosis membranes: Technical manual*. Dow Chemical Co.

Drexler, W., Morgner, U., Ghanta, R. K., Kärtner, F. X., Schuman, J. S., & Fujimoto, J. G. (2001). Ultrahigh-resolution ophthalmic optical coherence tomography. *Nature Medicine*, *7*(4), 502–507.

Eke, J., Yusuf, A., Giwa, A., & Sodiq, A. (2020). The global status of desalination: An assessment of current desalination technologies, plants and capacity. *Desalination*, *495*, 114633. https://doi.org/10.1016/j.desal.2020.114633

Erickson-Bhatt, S. J., & Boppart, S. A. (2015). 7—Biophotonics for assessing breast cancer. In I. Meglinski (Ed.), *Biophotonics for medical applications* (pp. 175–214). Woodhead Publishing.

Feng, D., van Deventer, J. S. J., & Aldrich, C. (2006). Ultrasonic defouling of reverse osmosis membranes used to treat wastewater effluents. *Separation and Purification Technology*, *50*(3), 318–323. https://doi.org/10.1016/j.seppur.2005.12.005

Field, R., Hughes, D., Cui, Z., & Tirlapur, U. (2009). In situ characterization of membrane fouling and cleaning using a multiphoton microscope. *Monitoring and Visualizing Membrane-Based Processes*, 151–174.

Filloux, E., Gallard, H., & Croue, J.-P. (2012). Identification of effluent organic matter fractions responsible for low-pressure membrane fouling. *Water Research*, *46*(17), 5531–5540. https://doi.org/10.1016/j.watres.2012.07.034

Flemming, H. C., Schaule, G., Griebe, T., Schmitt, J., & Tamachkiarowa, A. (1997). Biofouling—the Achilles heel of membrane processes. *Desalination*, *113*(2), 215–225. https://doi.org/10.1016/S0011-9164(97)00132-X

Flemming, H.-C., Tamachkiarowa, A., Klahre, J., & Schmitt, J. (1998). Monitoring of fouling and biofouling in technical systems. *Water Science and Technology*, *38*(8), 291–298. https://doi.org/10.1016/S0273-1223(98)00704-5

Flores, E. M. M., Cravotto, G., Bizzi, C. A., Santos, D., & Iop, G. D. (2021). Ultrasound-assisted biomass valorization to industrial interesting products: State-of-the-art, perspectives and challenges. *Ultrasonics Sonochemistry*, *72*, 105455. https://doi.org/10.1016/j.ultsonch.2020.105455

Frohman, E. M., Fujimoto, J. G., Frohman, T. C., Calabresi, P. A., Cutter, G., & Balcer, L. J. (2008). Optical coherence tomography: A window into the mechanisms of multiple sclerosis. *Nature clinical practice Neurology*, *4*(12), 664–675.

Fujimoto, J. G., Bouma, B., Tearney, G., Boppart, S., Pitris, C., Southern, J., & Brezinski, M. E. (1998). New technology for high-speed and high-resolution optical coherence tomography. *Annals of the New York Academy of Sciences*, *838*(1), 95–107.

Fujimoto, J. G., Drexler, W., Schuman, J. S., & Hitzenberger, C. K. (2009). Optical coherence tomography (OCT) in ophthalmology: Introduction. *Optics Express*, *17*(5), 3978–3979.

Gadzalo, Y., Romashchenko, M., & Yatsiuk, M. (2018). Conceptual framework to ensure water security in Ukraine. *Proceedings of the International Association of Hydrological Sciences*, *376*, 63–68.

Gallo, M., Ferrara, L., & Naviglio, D. (2018). Application of ultrasound in food science and technology: A perspective. *Foods (Basel, Switzerland)*, *7*(10), 164. doi:10.3390/foods7100164

Gao, F., Wang, L., Zhang, H., & Wang, J. (2020). Realtime and in-situ monitoring of membrane fouling with fiber-optic reflectance UV-vis spectrophotometry (FORUS). *Chemical Engineering Journal Advances*, *4*, 100058. https://doi.org/10.1016/j.ceja.2020.100058

Gao, Y., Haavisto, S., Li, W., Tang, C. Y., Salmela, J., & Fane, A. G. (2014). Novel approach to characterizing the growth of a fouling layer during membrane filtration via optical coherence tomography. *Environmental Science & Technology*, *48*(24), 14273–14281. doi:10.1021/es503326y

Gilbert, J. L. (2017). 1.2 electrochemical behavior of metals in the biological milieu. In P. Ducheyne (Ed.), *Comprehensive biomaterials II* (pp. 19–49). Elsevier.

Gu, H., Rahardianto, A., Gao, L. X., Caro, X. P., Giralt, J., Rallo, R., . . . Cohen, Y. (2018). Fouling indicators for field monitoring the effectiveness of operational strategies of ultrafiltration as pretreatment for seawater desalination. *Desalination*, *431*, 86–99. https://doi.org/10.1016/j.desal.2017.11.038

Güell, C., Ferrando, M., & López, F. (2009). *Monitoring and visualizing membrane-based processes*. Wiley-VCH.

Guillen-Burrieza, E., Ruiz-Aguirre, A., Zaragoza, G., & Arafat, H. A. (2014). Membrane fouling and cleaning in long term plant-scale membrane distillation operations. *Journal of Membrane Science*, *468*, 360–372.

Guo, J., Wong, P. W., Deka, B. J., Zhang, B., Jeong, S., & An, A. K. (2022). Investigation of fouling mechanism in membrane distillation using in-situ optical coherence tomography with green regeneration of fouled membrane. *Journal of Membrane Science*, *641*, 119894. https://doi.org/10.1016/j.memsci.2021.119894

Guo, J., Yan, D. Y. S., Lam, F. L. Y., Deka, B. J., Lv, X., Ng, Y. H., & An, A. K. (2019). Self-cleaning BiOBr/Ag photocatalytic membrane for membrane regeneration under visible light in membrane distillation. *Chemical Engineering Journal*, *378*, 122137. https://doi.org/10.1016/j.cej.2019.122137

Hamzah, N., & Leo, C. P. (2017). Membrane distillation of saline with phenolic compound using superhydrophobic PVDF membrane incorporated with TiO2 nanoparticles: Separation, fouling and self-cleaning evaluation. *Desalination*, *418*, 79–88. https://doi.org/10.1016/j.desal.2017.05.029

Hannon, A. C. (2017). Neutron diffraction, theory. In J. C. Lindon, G. E. Tranter, & D. W. Koppenaal (Eds.), *Encyclopedia of spectroscopy and spectrometry* (3rd ed., pp. 88–97). Academic Press.

Harvey, R. F. (1965). *Cavitational reverse osmotic separation of water from saline solutions.* Google Patents.

Hejnaes, K. R., & Ransohoff, T. C. (2018). Chapter 50—chemistry, manufacture and control. In G. Jagschies, E. Lindskog, K. Łącki, & P. Galliher (Eds.), *Biopharmaceutical processing* (pp. 1105–1136). Elsevier.

Henderson, R. K., Baker, A., Murphy, K., Hambly, A., Stuetz, R., & Khan, S. (2009). Fluorescence as a potential monitoring tool for recycled water systems: A review. *Water Research, 43*(4), 863–881.

Ho, J. S., Sim, L. N., Webster, R. D., Viswanath, B., Coster, H. G. L., & Fane, A. G. (2017). Monitoring fouling behavior of reverse osmosis membranes using electrical impedance spectroscopy: A field trial study. *Desalination, 407*(Suppl C), 75–84. https://doi.org/10.1016/j.desal.2016.12.012

Horrigan, L., & Freire-Gormaly, M. (2022). Modelling the effects of ultrasonic sonification on reverse osmosis feed channel temperature. *Desalination, 521*, 115332. https://doi.org/10.1016/j.desal.2021.115332

Hou, D., Lin, D., Zhao, C., Wang, J., & Fu, C. (2017). Control of protein (BSA) fouling by ultrasonic irradiation during membrane distillation process. *Separation and Purification Technology, 175*, 287–297. https://doi.org/10.1016/j.seppur.2016.11.047

Hou, D., Zhang, L., Zhao, C., Fan, H., Wang, J., & Huang, H. (2016). Ultrasonic irradiation control of silica fouling during membrane distillation process. *Desalination, 386*, 48–57. https://doi.org/10.1016/j.desal.2016.02.032

Hsieh, Y.-S., Ho, Y.-C., Lee, S.-Y., Chuang, C.-C., Tsai, J.-c., Lin, K.-F., & Sun, C.-W. (2013). Dental optical coherence tomography. *Sensors, 13*(7), 8928–8949.

Huang, F. Y. C., Medin, C., & Arning, A. (2019). Mechanical vibration for the control of membrane fouling in direct contact membrane distillation. *Symmetry, 11*(2), 126.

Huang, Q.-L., Huang, Y., Xiao, C.-F., You, Y.-W., & Zhang, C.-X. (2017). Electrospun ultrafine fibrous PTFE-supported ZnO porous membrane with self-cleaning function for vacuum membrane distillation. *Journal of Membrane Science, 534*, 73–82. https://doi.org/10.1016/j.memsci.2017.04.015

Huang, W., Zhu, Y., Dong, B., Lv, W., Yuan, Q., Zhou, W., & Lv, W. (2021). Investigation of membrane fouling mechanism of intracellular organic matter during ultrafiltration. *Scientific Reports, 11*(1), 1012. doi:10.1038/s41598-020-79272-4

Hughes, D. J., Cui, Z., Field, R. W., & Tirlapur, U. K. (2006). In situ three-dimensional characterization of membrane fouling by protein suspensions using multiphoton microscopy. *Langmuir, 22*(14), 6266–6272. doi:10.1021/la053388q

Hughes, D. J., Cui, Z., Field, R. W., & Tirlapur, U. K. (2007). Membrane fouling by cell-protein mixtures: In situ characterisation using multi-photon microscopy. *Biotechnology and Bioengineering, 96*(6), 1083–1091. https://doi.org/10.1002/bit.21113

Hughes, D. J., Tirlapur, U. K., Field, R., & Cui, Z. (2006a). Multiphoton microscopy—new insights into membrane fouling. *Desalination, 199*(1), 23–25. https://doi.org/10.1016/j.desal.2006.03.135

Hughes, D. J., Tirlapur, U. K., Field, R., & Cui, Z. (2006b). In situ 3D characterization of membrane fouling by yeast suspensions using two-photon femtosecond near infrared non-linear optical imaging. *Journal of Membrane Science, 280*(1), 124–133. https://doi.org/10.1016/j.memsci.2006.01.017

International, A. (2014). *Standard practices for detecting leaks in reverse osmosis and nanofiltration devices.* ASTM D3923–08(2014).

International, A. (2017). *Standard practice for integrity testing of water filtration membrane systems.* ASTM D6908–06(2017).

International, A. (2019a). *Standard practice for standardizing reverse osmosis performance data.* ASTM D4516–19a.

International, A. (2019b). *Standard practice for pressure driven membrane separation element/bundle evaluation.* ASTM D7601–19.

International, A. (2019c). *Standard test method for modified fouling index (MFI-0.45) of water.* ASTM D8002–15e1.

International, A. (2021). *Standard practice for standardizing ultrafiltration permeate flow performance data.* ASTM D5090–20.

Isermann, R. (2011). Supervision, fault-detection and diagnosis methods—a short introduction. In R. Isermann (Ed.), *Fault-diagnosis applications: Model-based condition monitoring: Actuators, drives, machinery, plants, sensors, and fault-tolerant systems* (pp. 11–45). Springer Berlin Heidelberg.

Jevons, K., & Awe, M. (2010). Economic benefits of membrane technology vs. Evaporator. *Desalination, 250*(3), 961–963. https://doi.org/10.1016/j.desal.2009.09.081

Jia, X., Li, K., Wang, B., Zhao, Z., Hou, D., & Wang, J. (2021). Membrane cleaning in membrane distillation of reverse osmosis concentrate generated in landfill leachate treatment. *Water Science and Technology, 85*(1), 244–256. doi:10.2166/wst.2021.614

Jiang, S., Li, Y., & Ladewig, B. P. (2017). A review of reverse osmosis membrane fouling and control strategies. *Science of the Total Environment, 595*(Supplement C), 567–583. https://doi.org/10.1016/j.scitotenv.2017.03.235

Jiang, X., & Foster, C. (2014). *Plant performance monitoring and diagnostics: Remote, real-time and automation.* Paper presented at the ASME Turbo Expo 2014: Turbine Technical Conference and Exposition.

Jin, N., Zhang, F., Cui, Y., Sun, L., Gao, H., Pu, Z., & Yang, W. (2022). Environment-friendly surface cleaning using micro-nano bubbles. *Particuology, 66*, 1–9. https://doi.org/10.1016/j.partic.2021.07.008

Joe Qin, S. (1998). Control performance monitoring—a review and assessment. *Computers & Chemical Engineering, 23*(2), 173–186. https://doi.org/10.1016/S0098-1354(98)00259-2

Johnson, W. T. (1997). Automatic monitoring of membrane integrity in microfiltration systems. *Desalination, 113*(2–3), 303–307. doi:10.1016/S0011-9164(97)00146-X

Kavanagh, J. M., Hussain, S., Chilcott, T. C., & Coster, H. G. L. (2009). Fouling of reverse osmosis membranes using electrical impedance spectroscopy: Measurements and simulations. *Desalination, 236*(1), 187–193. https://doi.org/10.1016/j.desal.2007.10.066

Kikhney, A. G., & Svergun, D. I. (2015). A practical guide to small angle X-ray scattering (SAXS) of flexible and intrinsically disordered proteins. *FEBS Letters, 589*(19, Part A), 2570–2577. https://doi.org/10.1016/j.febslet.2015.08.027

Kim, S., Park, K. Y., & Cho, J. (2017). Evaluation of the efficiency of cleaning method in direct contact membrane distillation of digested livestock wastewater. *Membrane and Water Treatment, 8*(2), 113–123.

Kobayashi, T., Kobayashi, T., Hosaka, Y., & Fujii, N. (2003). Ultrasound-enhanced membrane-cleaning processes applied water treatments: Influence of sonic frequency on filtration treatments. *Ultrasonics, 41*(3), 185–190. https://doi.org/10.1016/S0041-624X(02)00462-6

Kochkodan, V., Johnson, D. J., & Hilal, N. (2014). Polymeric membranes: Surface modification for minimizing (bio) colloidal fouling. *Advances in Colloid and Interface Science, 206*, 116–140.

Kockelmann, W., & Godfrey, E. (2021). Chapter 8—neutron diffraction. In M. Adriaens & M. Dowsett (Eds.), *Spectroscopy, diffraction and tomography in art and heritage science* (pp. 253–286). Elsevier.

Kools, W., Konagurthu, S., Greenberg, A., Bond, L., Krantz, W., Van den Boomgaard, T., & Strathmann, H. (1998). Use of ultrasonic time-domain reflectometry for real-time measurement of thickness changes during evaporative casting of polymeric films. *Journal of Applied Polymer Science, 69*(10), 2013–2019.

Košutić, K., & Kunst, B. (2002). RO and NF membrane fouling and cleaning and pore size distribution variations. *Desalination, 150*(2), 113–120. https://doi.org/10.1016/S0011-9164(02)00936-0

Krstić, D. M., Tekić, M. N., Carić, M. Đ., & Milanović, S. D. (2002). The effect of turbulence promoter on cross-flow microfiltration of skim milk. *Journal of Membrane Science, 208*(1–2), 303–314.

Kujundzic, E., Cristina Fonseca, A., Evans, E. A., Peterson, M., Greenberg, A. R., & Hernandez, M. (2007). Ultrasonic monitoring of earlystage biofilm growth on polymeric surfaces. *Journal of Microbiological Methods, 68*(3), 458–467. https://doi.org/10.1016/j.mimet.2006.10.005

Kyllönen, H. M., Pirkonen, P., & Nyström, M. (2005). Membrane filtration enhanced by ultrasound: A review. *Desalination, 181*(1), 319–335. https://doi.org/10.1016/j.desal.2005.06.003

Lai, L., Sim, L. N., Krantz, W. B., & Chong, T. H. (2020). Characterization of colloidal fouling in forward osmosis via ultrasonic time- (UTDR) and frequency-domain reflectometry (UFDR). *Journal of Membrane Science, 602*, 117969. https://doi.org/10.1016/j.memsci.2020.117969

Lamirel, C. (2014). Optical coherence tomography. In M. J. Aminoff & R. B. Daroff (Eds.), *Encyclopedia of the neurological sciences* (2nd ed., pp. 660–668). Academic Press.

Lamminen, M. O., Walker, H. W., & Weavers, L. K. (2004). Mechanisms and factors influencing the ultrasonic cleaning of particle-fouled ceramic membranes. *Journal of Membrane Science, 237*(1), 213–223. https://doi.org/10.1016/j.memsci.2004.02.031

Lange, M. A. (2019). Impacts of climate change on the Eastern Mediterranean and the Middle East and North Africa region and the water–energy nexus. *Atmosphere, 10*(8), 455.

Larson, A. M. (2011). Multiphoton microscopy. *Nature Photonics, 5*(1), 1–1. doi:10.1038/nphoton.an.2010.2

Larson, L. G., & Berglund, E. D. (1959). Effect of gas content in media on ultrasonic cleaning. *The Journal of the Acoustical Society of America, 31*(2), 247–248. doi:10.1121/1.1907704

Li, H., Fane, A. G., Coster, H. G. L., & Vigneswaran, S. (1998). Direct observation of particle deposition on the membrane surface during crossflow microfiltration. *Journal of Membrane Science, 149*(1), 83–97. https://doi.org/10.1016/S0376-7388(98)00181-1

Li, J., Hallbauer-Zadorozhnaya, V. Y., Hallbauer, D. K., & Sanderson, R. D. (2002). Cake-layer deposition, growth, and compressibility during microfiltration measured and modeled using a noninvasive ultrasonic technique. *Industrial & Engineering Chemistry Research, 41*(16), 4106–4115. doi:10.1021/ie020142x

Li, J., Sanderson, R., Chai, G., & Hallbauer, D. (2005). Development of an ultrasonic technique for in situ investigating the properties of deposited protein during crossflow ultrafiltration. *Journal of Colloid and Interface Science, 284*(1), 228–238.

Li, J., Sanderson, R. D., Hallbauer, D., & Hallbauer-Zadorozhnaya, V. (2002). Measurement and modelling of organic fouling deposition in ultrafiltration by ultrasonic transfer signals and reflections. *Desalination, 146*(1–3), 177–185.

Li, J., Sanderson, R. D., & Jacobs, E. P. (2002a). Non-invasive visualization of the fouling of microfiltration membranes by ultrasonic time-domain reflectometry. *Journal of Membrane Science, 201*(1), 17–29. https://doi.org/10.1016/S0376-7388(01)00664-0

Li, J., Sanderson, R. D., & Jacobs, E. P. (2002b). Ultrasonic cleaning of nylon microfiltration membranes fouled by Kraft paper mill effluent. *Journal of Membrane Science, 205*(1), 247–257. https://doi.org/10.1016/S0376-7388(02)00121-7

Li, J.-X., Sanderson, R. D., & Chai, G. Y. (2006). A focused ultrasonic sensor for in situ detection of protein fouling on tubular ultrafiltration membranes. *Sensors and Actuators B: Chemical, 114*(1), 182–191. https://doi.org/10.1016/j.snb.2005.04.041

Li, T., Senesi, A. J., & Lee, B. (2016). Small angle X-ray scattering for nanoparticle research. *Chemical Reviews, 116*(18), 11128–11180. doi:10.1021/acs.chemrev.5b00690

Li, W., Liu, X., Wang, Y.-N., Chong, T. H., Tang, C. Y., & Fane, A. G. (2016). Analyzing the evolution of membrane fouling via a novel method based on 3D optical coherence tomography imaging. *Environmental Science & Technology, 50*(13), 6930–6939. doi:10.1021/acs.est.6b00418

Li, X., Zhang, H., Hou, Y., Gao, Y., Li, J., Guo, W., & Ngo, H. H. (2015). In situ investigation of combined organic and colloidal fouling for nanofiltration membrane using ultrasonic time domain reflectometry. *Desalination, 362*, 43–51. https://doi.org/10.1016/j.desal.2015.02.005

Liu, C., Xu, H., Cai, X., & Gao, Y. (2021). Chapter 8—Liutex similarity, structure, and asymmetry in turbulent boundary layer. In C. Liu, H. Xu, X. Cai, & Y. Gao (Eds.), *Liutex and its applications in turbulence research* (pp. 199–225). Academic Press.

Liu, X.-Y., Chen, G., Tu, G., Li, Z., Deng, B., & Li, W. (2020). Membrane fouling by clay suspensions during NF-like forward osmosis: Characterization via optical coherence tomography. *Journal of Membrane Science, 602*, 117965. https://doi.org/10.1016/j.memsci.2020.117965

Liu, X.-Y., Chen, W., & Yu, H.-Q. (2020). Probing protein-induced membrane fouling with in-situ attenuated total reflectance Fourier transform infrared spectroscopy and multivariate curve resolution-alternating least squares. *Water Research, 183*, 116052. https://doi.org/10.1016/j.watres.2020.116052

Liu, Y., He, G., Li, B., Hu, Z., & Ju, J. (2012). A comparison of cake properties in traditional and turbulence promoter assisted microfiltration of particulate suspensions. *Water Research, 46*(8), 2535–2544. https://doi.org/10.1016/j.watres.2012.02.002

Lorenzo, G. D., Araneo, R., Mitolo, M., Niccolai, A., & Grimaccia, F. (2020). Review of O&M practices in PV plants: Failures, solutions, remote control, and monitoring tools. *IEEE Journal of Photovoltaics, 10*(4), 914–926. doi:10.1109/JPHOTOV.2020.2994531

Lozier, J. C., & Sierka, R. A. (1985). Using ozone and ultrasound to reduce RO membrane fouling. *Journal (American Water Works Association), 77*(8), 60–65.

Luo, H., & Wang, Z. (2022). A new ultrasonic cleaning model for predicting the flux recovery of the UF membrane fouled with humic acid. *Journal of Environmental Chemical Engineering, 10*(2), 107156. https://doi.org/10.1016/j.jece.2022.107156

Mairal, A. P., Greenberg, A. R., Krantz, W. B., & Bond, L. J. (1999). Real-time measurement of inorganic fouling of RO desalination membranes using ultrasonic time-domain reflectometry. *Journal of Membrane Science, 159*(1), 185–196. https://doi.org/10.1016/S0376-7388(99)00058-7

Mannan, S. (2014). Chapter 20—computer aids and expert systems. In S. Mannan (Ed.), *Lees' process safety essentials* (pp. 383–401). Butterworth-Heinemann.

Mason, T. J. (2016). Ultrasonic cleaning: An historical perspective. *Ultrasonics Sonochemistry, 29*, 519–523. https://doi.org/10.1016/j.ultsonch.2015.05.004

McKenna, Q. C. (1955). Ultrasonic cleaning of miniature devices. *IRE Transactions on Ultrasonic Engineering, PGUE-3*, 16–22. doi:10.1109/T-PGUE.1955.29214

McQueen, D. (1986). Frequency dependence of ultrasonic cleaning. *Ultrasonics, 24*(5), 273–280.

Mendret, J., Guigui, C., Schmitz, P., & Cabassud, C. (2009). Optical and acoustic methods for in situ characterization of membrane fouling. *Monitoring and Visualizing Membrane-Based Processes*, 229–251.

Meng, F., Liao, B., Liang, S., Yang, F., Zhang, H., & Song, L. (2010). Morphological visualization, componential characterization and microbiological identification of membrane fouling in membrane bioreactors (MBRs). *Journal of Membrane Science, 361*(1), 1–14. https://doi.org/10.1016/j.memsci.2010.06.006

Mollahosseini, A., Min Lee, K., Abdelrasoul, A., Doan, H., & Zhu, N. (2021). Innovative in situ investigations using synchrotron-based micro tomography and molecular dynamics simulation for fouling assessment in ceramic membranes for dairy and food industry. *International Journal of Applied Ceramic Technology, 18*(6), 2143–2157. https://doi.org/10.1111/ijac.13824

Muller, M., Schreiner, B., Smith, L., van Koppen, B., Sally, H., Aliber, M., . . . Karar, E. (2009). Water security in South Africa. *Development Planning Division. Working Paper Series, 12.*

Muthukumaran, S., Kentish, S., Lalchandani, S., Ashokkumar, M., Mawson, R., Stevens, G. W., & Grieser, F. (2005). The optimisation of ultrasonic cleaning procedures for dairy fouled ultrafiltration membranes. *Ultrasonics Sonochemistry, 12*(1), 29–35. https://doi.org/10.1016/j.ultsonch.2004.05.007

Naji, O., Al-juboori, R. A., Bowtell, L., Alpatova, A., & Ghaffour, N. (2020). Direct contact ultrasound for fouling control and flux enhancement in air-gap membrane distillation. *Ultrasonics Sonochemistry, 61*, 104816. https://doi.org/10.1016/j.ultsonch.2019.104816

Naji, O., Al-juboori, R. A., Khan, A., Yadav, S., Altaee, A., Alpatova, A., . . . Ghaffour, N. (2021). Ultrasound-assisted membrane technologies for fouling control and performance improvement: A review. *Journal of Water Process Engineering, 43*, 102268. https://doi.org/10.1016/j.jwpe.2021.102268

Nguyen, A. H., Tobiason, J. E., & Howe, K. J. (2011). Fouling indices for low pressure hollow fiber membrane performance assessment. *Water Research, 45*(8), 2627–2637. https://doi.org/10.1016/j.watres.2011.02.020

Nossol, E., Muñoz, R. A. A., Richter, E. M., de Souza Borges, P. H., Silva, S. C., & Rocha, D. P. (2021). Sensing materials: Graphene. In *Reference module in biomedical sciences.* Elsevier.

Ochando-Pulido, J. (2016). The use of membranes in olive mill wastewater treatment: How to control dynamic fouling. *Polymer Science, 2*, 1–7.

Olsen, J., Holmes, J., & Jemec, G. B. (2018). Advances in optical coherence tomography in dermatology—a review. *Journal of Biomedical Optics, 23*(4), 040901.

Otis, L. L., Everett, M. J., Sathyam, U. S., & Colston Jr, B. W. (2000). Optical coherence tomography: A new imaging: Technology for dentistry. *The Journal of the American Dental Association, 131*(4), 511–514.

Paddock, S. W., & Eliceiri, K. W. (2014). Laser scanning confocal microscopy: History, applications, and related optical sectioning techniques. In S. W. Paddock (Ed.), *Confocal microscopy: Methods and protocols* (pp. 9–47). Springer.

Park, S., Kang, J.-S., Lee, J. J., Vo, T.-K.-Q., & Kim, H.-S. (2018). Application of physical and chemical enhanced backwashing to reduce membrane fouling in the water treatment process using ceramic membranes. *Membranes, 8*(4), 110. doi:10.3390/membranes8040110

Peng, Y., Ge, J., Li, Z., & Wang, S. (2015). Effects of anti-scaling and cleaning chemicals on membrane scale in direct contact membrane distillation process for RO brine concentrate. *Separation and Purification Technology, 154*, 22–26. https://doi.org/10.1016/j.seppur.2015.09.007

Peterson, R., Greenberg, A., Bond, L., & Krantz, W. (1998). Use of ultrasonic TDR for real-time noninvasive measurement of compressive strain during membrane compaction. *Desalination, 116*(2–3), 115–122.

Petzold, A., de Boer, J. F., Schippling, S., Vermersch, P., Kardon, R., Green, A., . . . Polman, C. (2010). Optical coherence tomography in multiple sclerosis: A systematic review and meta-analysis. *The Lancet Neurology, 9*(9), 921–932.

Pignon, F., Alemdar, A., Magnin, A., & Narayanan, T. (2003). Small-angle X-ray scattering studies of Fe-montmorillonite deposits during ultrafiltration in a magnetic field. *Langmuir, 19*(21), 8638–8645. doi:10.1021/la030020p

Pignon, F., Magnin, A., Piau, J.-M., Cabane, B., Aimar, P., Meireles, M., & Lindner, P. (2000). Structural characterisation of deposits formed during frontal filtration. *Journal of Membrane Science, 174*(2), 189–204. https://doi.org/10.1016/S0376-7388(00)00394-X

Pilli, S., Bhunia, P., Yan, S., LeBlanc, R. J., Tyagi, R. D., & Surampalli, R. Y. (2011). Ultrasonic pretreatment of sludge: A review. *Ultrasonics Sonochemistry, 18*(1), 1–18. https://doi.org/10.1016/j.ultsonch.2010.02.014

Pourbozorg, M., Li, T., & Law, A. W. K. (2016). Effect of turbulence on fouling control of submerged hollow fibre membrane filtration. *Water Research*, *99*, 101–111. https://doi.org/10.1016/j.watres.2016.04.045

Puspitasari, V., Granville, A., Le-Clech, P., & Chen, V. (2010). Cleaning and ageing effect of sodium hypochlorite on polyvinylidene fluoride (PVDF) membrane. *Separation and Purification Technology*, *72*(3), 301–308. https://doi.org/10.1016/j.seppur.2010.03.001

Qasim, M., Darwish, N. N., Mhiyo, S., Darwish, N. A., & Hilal, N. (2018). The use of ultrasound to mitigate membrane fouling in desalination and water treatment. *Desalination*, *443*, 143–164. https://doi.org/10.1016/j.desal.2018.04.007

Reinsch, V. E., Greenberg, A. R., Kelley, S. S., Peterson, R., & Bond, L. J. (2000). A new technique for the simultaneous, real-time measurement of membrane compaction and performance during exposure to high-pressure gas. *Journal of Membrane Science*, *171*(2), 217–228.

Remanan, S., Sharma, M., Bose, S., & Das, N. C. (2018). Recent advances in preparation of porous polymeric membranes by unique techniques and mitigation of fouling through surface modification. *ChemistrySelect*, *3*(2), 609–633.

Richter, D., Monkenbusch, M., & Schwahn, D. (2012). 2.11—neutron scattering. In K. Matyjaszewski & M. Möller (Eds.), *Polymer science: A comprehensive reference* (pp. 331–361). Elsevier.

Rudolph, G., Hermansson, A., Jönsson, A.-S., & Lipnizki, F. (2021). In situ real-time investigations on adsorptive membrane fouling by thermomechanical pulping process water with quartz crystal microbalance with dissipation monitoring (QCM-D). *Separation and Purification Technology*, *254*, 117578. https://doi.org/10.1016/j.seppur.2020.117578

Rudolph, G., Virtanen, T., Ferrando, M., Güell, C., Lipnizki, F., & Kallioinen, M. (2019). A review of in situ real-time monitoring techniques for membrane fouling in the biotechnology, biorefinery and food sectors. *Journal of Membrane Science*, *588*, 117221. https://doi.org/10.1016/j.memsci.2019.117221

Saget, M., de Almeida, C. F., Fierro, V., Celzard, A., Delaplace, G., Thomy, V., . . . Jimenez, M. (2021). A critical review on surface modifications mitigating dairy fouling. *Comprehensive Reviews in Food Science and Food Safety*, *20*(5), 4324–4366.

Salinas Rodriguez, S. G., Sithole, N., Dhakal, N., Olive, M., Schippers, J. C., & Kennedy, M. D. (2019). Monitoring particulate fouling of North Sea water with SDI and new ASTM MFI0.45 test. *Desalination*, *454*, 10–19. https://doi.org/10.1016/j.desal.2018.12.006

Sanderson, R., Li, J., Koen, L., & Lorenzen, L. (2002). Ultrasonic time-domain reflectometry as a non-destructive instrumental visualization technique to monitor inorganic fouling and cleaning on reverse osmosis membranes. *Journal of Membrane Science*, *207*(1), 105–117.

Sanjeeva Murthy, N. (2013). 2—Scattering techniques for structural analysis of biomaterials. In M. Jaffe, W. Hammond, P. Tolias, & T. Arinzeh (Eds.), *Characterization of biomaterials* (pp. 34–72). Woodhead Publishing.

Sari Erkan, H., Bakaraki Turan, N., & Önkal Engin, G. (2018). Chapter five—membrane bioreactors for wastewater treatment. In D. S. Chormey, S. Bakırdere, N. B. Turan, & G. Ö. Engin (Eds.), *Comprehensive analytical chemistry* (Vol. 81, pp. 151–200). Elsevier.

Sattler, E. C., Kästle, R., & Welzel, J. (2013). Optical coherence tomography in dermatology. *Journal of Biomedical Optics*, *18*(6), 061224.

Schippers, J., Hanemaayer, J., Smolders, C., & Kostense, A. (1981). Predicting flux decline of reverse osmosis membranes. *Desalination*, *38*, 339–348.

Sengur-Tasdemir, R., Guler-Gokce, Z., Sezai Sarac, A., & Koyuncu, I. (2018). Determination of membrane protein fouling by UV spectroscopy and electrochemical impedance spectroscopy. *Polymer-Plastics Technology and Engineering*, *57*(2), 59–69. doi:10.1080/03602559.2017.1300816

Shchukin, D. G., Skorb, E., Belova, V., & Möhwald, H. (2011). Ultrasonic cavitation at solid surfaces. *Advanced Materials*, *23*(17), 1922–1934. https://doi.org/10.1002/adma.201004494

Siegel, R. J., & Luo, H. (2008). Ultrasound thrombolysis. *Ultrasonics, 48*(4), 312–320.

Siger, M., Dzięgielewski, K., Jasek, L., Bieniek, M., Nicpan, A., Nawrocki, J., & Selmaj, K. (2008). Optical coherence tomography in multiple sclerosis. *Journal of Neurology, 255*(10), 1555–1560.

Sim, L. N., Gu, J., Coster, H. G. L., & Fane, A. G. (2016). Quantitative determination of the electrical properties of RO membranes during fouling and cleaning processes using electrical impedance spectroscopy. *Desalination, 379*(Suppl C), 126–136. https://doi.org/10.1016/j.desal.2015.11.006

Sim, L. N., Wang, Z. J., Gu, J., Coster, H. G. L., & Fane, A. G. (2013). Detection of reverse osmosis membrane fouling with silica, bovine serum albumin and their mixture using in-situ electrical impedance spectroscopy. *Journal of Membrane Science, 443*(Suppl C), 45–53. https://doi.org/10.1016/j.memsci.2013.04.047

Sim, L. N., Ye, Y., Chen, V., & Fane, A. G. (2010). Crossflow sampler modified fouling index ultrafiltration (CFS-MFIUF)—an alternative fouling index. *Journal of Membrane Science, 360*(1), 174–184. https://doi.org/10.1016/j.memsci.2010.05.010

Sim, S. T. V., Suwarno, S. R., Chong, T. H., Krantz, W. B., & Fane, A. G. (2013). Monitoring membrane biofouling via ultrasonic time-domain reflectometry enhanced by silica dosing. *Journal of Membrane Science, 428*, 24–37. https://doi.org/10.1016/j.memsci.2012.10.032

Sim, S. T. V., Suwarno, S. R., Chong, T. H., Yeo, P., Krantz, W., & Fane, A. (2011). Enhanced acoustic diagnostic tool for fouling in membrane processes. *US Provisional Patent, 61*(553,637).

Su, T. J., Lu, J. R., Cui, Z. F., Bellhouse, B. J., Thomas, R. K., & Heenan, R. K. (1999). Identification of the location of protein fouling on ceramic membranes under dynamic filtration conditions. *Journal of Membrane Science, 163*(2), 265–275. https://doi.org/10.1016/S0376-7388(99)00170-2

Su, T. J., Lu, J. R., Cui, Z. F., & Thomas, R. K. (2000). Fouling of ceramic membranes by albumins under dynamic filtration conditions. *Journal of Membrane Science, 173*(2), 167–178. https://doi.org/10.1016/S0376-7388(00)00370-7

Taheri, A. H., Sim, L. N., Chong, T. H., Krantz, W. B., & Fane, A. G. (2015). Prediction of reverse osmosis fouling using the feed fouling monitor and salt tracer response technique. *Journal of Membrane Science, 475*, 433–444. https://doi.org/10.1016/j.memsci.2014.10.043

Tamachkiarow, A., & Flemming, H.-C. (2003). On-line monitoring of biofilm formation in a brewery water pipeline system with a fibre optical device. *Water Science and Technology, 47*(5), 19–24. doi:10.2166/wst.2003.0270

Targowski, P., & Iwanicka, M. (2012). Optical coherence tomography: Its role in the non-invasive structural examination and conservation of cultural heritage objects—a review. *Applied Physics A, 106*(2), 265–277. doi:10.1007/s00339-011-6687-3

Tijing, L. D., Woo, Y. C., Choi, J.-S., Lee, S., Kim, S.-H., & Shon, H. K. (2015). Fouling and its control in membrane distillation—A review. *Journal of Membrane Science, 475*, 215–244. https://doi.org/10.1016/j.memsci.2014.09.042

Tng, K. H., Antony, A., Wang, Y., & Leslie, G. L. (2015). 11—Membrane ageing during water treatment: Mechanisms, monitoring, and control. In A. Basile, A. Cassano, & N. K. Rastogi (Eds.), *Advances in membrane technologies for water treatment* (pp. 349–378). Woodhead Publishing.

Tonda-Turo, C., Carmagnola, I., & Ciardelli, G. (2018). Quartz crystal microbalance with dissipation monitoring: A powerful method to predict the in vivo behavior of bioengineered surfaces. *Frontiers in Bioengineering and Biotechnology, 6.* doi:10.3389/fbioe.2018.00158

Toray Industries, I. (2019). *TORAY pressurized PVDF hollow fiber membrane module.* "TORAYFIL™" Instruction Manual.

Trinh, T. A., Li, W., Han, Q., Liu, X., Fane, A. G., & Chew, J. W. (2018). Analyzing external and internal membrane fouling by oil emulsions via 3D optical coherence tomography. *Journal of Membrane Science*, *548*, 632–640. https://doi.org/10.1016/j.memsci.2017.10.043

Tung, K.-L., Teoh, H.-C., Lee, C.-W., Chen, C.-H., Li, Y.-L., Lin, Y.-F., . . . Huang, M.-S. (2015). Characterization of membrane fouling distribution in a spiral wound module using high-frequency ultrasound image analysis. *Journal of Membrane Science*, *495*, 489–501. https://doi.org/10.1016/j.memsci.2015.08.035

Tzafestas, S. G., Pouliezos, A. D., & Stavrakakis, G. S. (2010). *Real time fault monitoring of industrial processes*. Springer Netherlands.

Vajnhandl, S., & Majcen Le Marechal, A. (2005). Ultrasound in textile dyeing and the decolouration/mineralization of textile dyes. *Dyes and Pigments*, *65*(2), 89–101. https://doi.org/10.1016/j.dyepig.2004.06.012

Vial, D., & Doussau, G. (2003). The use of microfiltration membranes for seawater pre-treatment prior to reverse osmosis membranes. *Desalination*, *153*(1), 141–147. https://doi.org/10.1016/S0011-9164(02)01115-3

Vigneswaran, S. (2009). *Waste water treatment technologies—volume III*. EOLSS Publ.

Virtanen, T., Reinikainen, S.-P., Lahti, J., Mänttäri, M., & Kallioinen, M. (2018). Visual tool for real-time monitoring of membrane fouling via Raman spectroscopy and process model based on principal component analysis. *Scientific Reports*, *8*(1), 11057. doi:10.1038/s41598-018-29268-y

Wang, H., & Evans, C. L. (2016). Chapter 11—Coherent Raman scattering microscopy in dermatological imaging. In M. R. Hamblin, P. Avci, & G. K. Gupta (Eds.), *Imaging in dermatology* (pp. 103–117). Academic Press.

Wang, J., Ren, H., Li, X., Li, J., Ding, L., Geng, J., . . . Hu, H. (2018). In situ monitoring of wastewater biofilm formation process via ultrasonic time domain reflectometry (UTDR). *Chemical Engineering Journal*, *334*, 2134–2141. https://doi.org/10.1016/j.cej.2017.11.043

Wang, Z., Fang, R., & Guo, H. (2020). Advances in ultrasonic production units for enhanced oil recovery in China. *Ultrasonics Sonochemistry*, *60*, 104791. https://doi.org/10.1016/j.ultsonch.2019.104791

Welzel, J. (2001). Optical coherence tomography in dermatology: A review. *Skin Research and Technology: Review article*, *7*(1), 1–9.

West, S., Wagner, M., Engelke, C., & Horn, H. (2016). Optical coherence tomography for the in situ three-dimensional visualization and quantification of feed spacer channel fouling in reverse osmosis membrane modules. *Journal of Membrane Science*, *498*, 345–352. https://doi.org/10.1016/j.memsci.2015.09.047

Wong, P. W., Guo, J., Khanzada, N. K., Yim, V. M. W., & Kyoungjin, A. (2021). In-situ 3D fouling visualization of membrane distillation treating industrial textile wastewater by optical coherence tomography imaging. *Water Research*, *205*, 117668. https://doi.org/10.1016/j.watres.2021.117668

Xenarios, S., Shenhav, R., Abdullaev, I., & Mastellari, A. (2018). Current and future challenges of water security in Central Asia. In *Global water security* (pp. 117–142). Springer.

Xiao, Y., & Fu, Z. (2021). Nanomaterials characterization by neutron scattering methods. In *Reference module in materials science and materials engineering*. Elsevier.

Xu, P., Bellona, C., & Drewes, J. E. (2010). Fouling of nanofiltration and reverse osmosis membranes during municipal wastewater reclamation: Membrane autopsy results from pilot-scale investigations. *Journal of Membrane Science*, *353*(1), 111–121. https://doi.org/10.1016/j.memsci.2010.02.037

Xu, X., Li, J., Li, H., Cai, Y., Cao, Y., He, B., & Zhang, Y. (2009). Non-invasive monitoring of fouling in hollow fiber membrane via UTDR. *Journal of Membrane Science*, *326*(1), 103–110. https://doi.org/10.1016/j.memsci.2008.09.042

Xu, Y., Wang, M., Ma, Z., & Gao, C. (2011). Electrochemical impedance spectroscopy analysis of sulfonated polyethersulfone nanofiltration membrane. *Desalination*, *271*(1), 29–33. https://doi.org/10.1016/j.desal.2010.12.007

Yamamura, H., Ding, Q., & Watanabe, Y. (2019). Solid-phase fluorescence excitation emission matrix for in-situ monitoring of membrane fouling during microfiltration using a polyvinylidene fluoride hollow fiber membrane. *Water Research*, *164*, 114928. https://doi.org/10.1016/j.watres.2019.114928

Yang, J., Monnot, M., Ercolei, L., & Moulin, P. (2021). Impact of chlorinated-assisted backwash and air backwash on ultrafiltration fouling management for urban wastewater tertiary treatment. *Membranes*, *11*(10), 733.

Ye, Y., Yu, S., Liu, B., Xia, Q., Liu, G., & Li, P. (2019). Microbubble aeration enhances performance of vacuum membrane distillation desalination by alleviating membrane scaling. *Water Research*, *149*, 588–595.

Yin, C., Wang, S., Zhang, Y., Chen, Z., Lin, Z., Fu, P., & Yao, L. (2017). Correlation between the pore resistance and water flux of the cellulose acetate membrane. *Environmental Science: Water Research & Technology*, *3*(6), 1037–1041. doi:10.1039/C7EW00274B

Yusof, N. S. M., Babgi, B., Alghamdi, Y., Aksu, M., Madhavan, J., & Ashokkumar, M. (2016). Physical and chemical effects of acoustic cavitation in selected ultrasonic cleaning applications. *Ultrasonics Sonochemistry*, *29*, 568–576. https://doi.org/10.1016/j.ultsonch.2015.06.013

Zaidi, S. J., & Saleem, H. (2022). Chapter 11—reverse osmosis system troubleshooting. In S. J. Zaidi & H. Saleem (Eds.), *Reverse osmosis systems* (pp. 375–406). Elsevier.

Zhang, J., Liu, Y., Gao, S., Li, C., Zhang, F., Zen, H., & Ye, C. (2006). Pilot testing of outside-in UF pretreatment prior to RO for high turbidity seawater desalination. *Desalination*, *189*(1–3), 269–277.

Zhang, W., Yu, S., Zhao, H., Ji, X., & Ning, R. (2022). Vacuum membrane distillation for seawater concentrate treatment coupled with microbubble aeration cleaning to alleviate membrane fouling. *Separation and Purification Technology*, *290*, 120864. https://doi.org/10.1016/j.seppur.2022.120864

Zhang, Z. X., Bright, V. M., & Greenberg, A. R. (2006). Use of capacitive microsensors and ultrasonic time-domain reflectometry for in-situ quantification of concentration polarization and membrane fouling in pressure-driven membrane filtration. *Sensors and Actuators B: Chemical*, *117*(2), 323–331. https://doi.org/10.1016/j.snb.2005.11.016

Zhang, Z. X., Greenberg, A. R., Krantz, W. B., & Chai, G. Y. (2003). Chapter 4—study of membrane fouling and cleaning in spiral wound modules using ultrasonic time-domain reflectometry. In D. Bhattacharyya & D. A. Butterfield (Eds.), *Membrane science and technology* (Vol. 8, pp. 65–88). Elsevier.

Zhao, Z., Shi, S., Cao, H., & Li, Y. (2017). Electrochemical impedance spectroscopy and surface properties characterization of anion exchange membrane fouled by sodium dodecyl sulfate. *Journal of Membrane Science*, *530*(Suppl C), 220–231. https://doi.org/10.1016/j.memsci.2017.02.037

Zipfel, W. R., Williams, R. M., & Webb, W. W. (2003). Nonlinear magic: Multiphoton microscopy in the biosciences. *Nature Biotechnology*, *21*(11), 1369–1377. doi:10.1038/nbt899

4 Electrical Conductivity in Materials

4.1 INTRODUCTION

Electrical conductivity is a measure of a material's ability to carry an electric current. Historically, electrically conductive materials were seldom explored outside of electronic engineering applications and energy storage devices. Today, they form a new generation of materials for a diverse range of applications: tissue engineering, textile design, and water purification, to name a few. Smart electrically conducting materials that exhibit coupling between multiple physical domains are now integrated into everyday devices and used in sensing, pH-responsive films, separation membranes, antimicrobial applications, desalination, electrocatalysis, capacitive deionization (CDI), actuators, and protective coatings. The same is true for membrane processes that continue to move toward more intelligent systems. Electrically conducting materials are used in spacers and in membranes for electrochemical self-cleaning and electrically enhanced separation. Depending on many factors, electrical simulation impacts the interaction between solutes and the membrane or spacer, which in turn affects performance and fouling propensity. In order to appreciate the role of electrically conductive materials in membrane separation, we must familiarize ourselves with the fundamentals of electrical conductivity in materials, factors affecting electrical and electrochemical properties using a bottom-up approach, and methods commonly used to fabricate electrically conducting materials.

The electrical nature of atoms is a function of the configuration of electrons, in particular the electrons present in the outermost orbit of an atom. These are known as valence electrons. The behavior of solids has been described using electron theories: Drude–Lorentz (classical) theory, free electron theory, energy band theory, and Brillouin zone theory. The first two explain the mechanism of conduction in solids, while the two latter theories explain semiconduction. They have been referred to throughout this text, but for a detailed description of the assumptions and energy equations, the reader is referred to other literature ("Conductive Materials: Electron Theories, Properties and Behaviour," 2015).

Apart from metals and semiconductors, electrically conductive polymers and carbon-based nanomaterials have been at the forefront of recent advances. The prospect of developing materials that combine the flexibility of polymers with the electrical conductivity of metals has fascinated researchers for decades. Figure 4.1 shows the conductivity range for various classes of materials that are discussed in this text. In this chapter, we overview the electronic structure and mechanisms of electrical conductivity in various classes of materials, including polymers, carbon nanotubes,

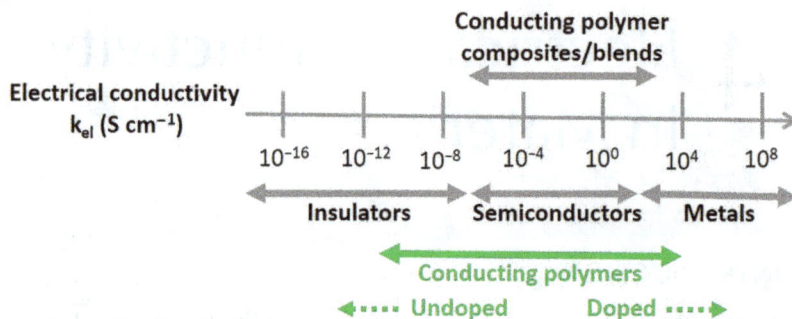

FIGURE 4.1 Typical conductivity range of various classes of materials.

graphene, metals, semiconductors, and polymer nanocomposites. We elucidate connections between structure and electrochemical activity in an attempt to gain insight into their electronic and electrochemical behavior.

Although their electrochemical properties are most commonly exploited for (bio) sensing and energy storage devices (Pumera, 2009), the development of "membrane electrodes" has realized the marriage of electrochemistry with membrane separation for the detection of toxic compounds, fouling and related phenomena, electrolytic self-cleaning, and electrically enhanced performance. We will discuss these aspects of electrically conductive materials in subsequent chapters. In electrically conducting membrane systems, carbonaceous materials can serve as either anodes or cathodes. With a low overpotential for oxygen evolution reaction (OER) of under 0.4 V, carbon anodes are considered "active" (Martínez-Huitle et al., 2015). In aqueous media, for physicochemical-based mechanisms such as electrostatic repulsion, electrophoresis, and direct oxidation/reduction, a voltage lower than the oxygen evolution potential (1.23 V) needs to be applied (Gao et al., 2014; Martínez-Huitle et al., 2015; Zhu et al., 2018). An electrode material should possess high specific surface area, high electrical conductivity, ease of processability, low cost, and chemical and thermal stability.

4.2 POLYMERS

4.2.1 Electrical Conductivity in Conducting Polymers

Conducting polymers, first reported in 1973 (Walatka et al., 1973), are polymers that exhibit a conductivity > 100 S cm^{-1} at room temperature (Brédas, 1987). An early 1980s review ruled that conducting polymers at the time did not possess the properties for a technologically useful material (Street & Clarke, 1981). However, they optimistically concluded that the field of conducting polymers remains young, with increasing opportunity to tailor properties via molecular engineering. More than half a century later, an unprecedented interest in conducting polymers, together with advances in nanotechnology, have proved their statement true. The possibility of combining electrical properties of metals with low density, high plasticity,

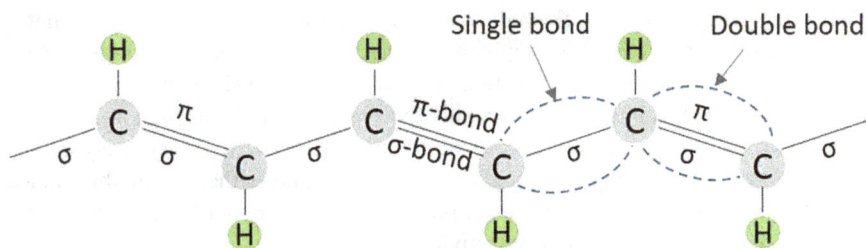

FIGURE 4.2 Schematic showing the conjugated double bonds along the backbone of conductive polymers; σ-bonds ensure the strength of the chain, while π-bonds allow the delocalization of electrons.

low cost, and engineering feasibility of organic polymers makes conducting polymers an exciting class of materials for new-generation membrane systems (Brédas, 1987).

Conducting polymers carry current without possessing a partially empty or partially filled band. When an electron is removed from the valence band by oxidation, a vacancy is created. There are several known conductive polymer systems; of these, conductive polymers such as polypyrrole (PPy), polyacetylene, polyaniline, and polythiophene have been studied the most extensively (D. D. Ateh et al., 2006). Conductivity in polymers arises from the presence of conjugated double bonds along the backbone, as shown in Figure 4.2. The overlapping p-orbitals in the series of π-bonds lead to delocalized π-electrons into a conduction band that enables metallic behavior (Kamath et al., 2021; Wallace & Spinks, 2007). The formation of an energy gap in the electronic spectrum means that dopant ions need to be introduced to overcome the energy gap and impart electrical conductivity. Dopant ions act to neutralize the unstable backbone of the polymer by donating or accepting electrons.

Dopants are added as a random dispersion or aggregation in molar concentrations to improve the conductivity of polymers (Wallace & Spinks, 2007). Carrier concentration and conductivity are proportional to doping concentration (Kumar & Sharma, 1998). Doping introduces a charge carrier by removing or adding electrons to/from the polymer chain, which then leads to the formation of conjugational defects, i.e. polarons or bipolarons. A polaron is a charge that is surrounded by a distorted crystal lattice. The mechanism of conduction is based on the presence of delocalized polarons and bipolarons. When an electrical potential is applied, these polarons and bipolarons, or localized electrons, move in the polymer backbone and allow the charge to conduct through the polymer (Ravichandran et al., 2010).

Common dopants include small salt ions, such as Cl^-, Br^-, and NO_3^-, and larger dopants, such as hyaluronic acid, peptides, or polymers (Kamath et al., 2021). Large dopants are more integrated into the polymer and will not be leached out with time or with the application of an electrical stimulus, granting the polymer greater electrochemical stability, while small dopants can leave and reenter the polymer with electrical stimulation (X. Liu et al., 2009).

Ateh suggests that electrical conductivity in PPy, a widely used conductive material, is a combination of p-type (bipolaron) conduction, interchain hopping of electrons, and the motion of anions or cations within the material (F. Ahmed et al., 2016; D. Ateh et al., 2006). The electrical conductivity of PPy can be as high as 7.5×10^3 S cm^{-1} (Dai, 2004). The affinity of conducting polymers toward many heavy metals (Arora, 2019; Mahmud et al., 2016; Zare et al., 2015) makes them a suitable choice for wastewater treatment. For instance, poly(1,8-diaminophtalene) can sorb copper ions at acidic pH levels with high selectivity.

In non-conducting polymers, the movement of electrons is restricted by directional covalent bonds through which electrons are tightly bound to stable ions (F. Wilson, 2012). The electrical conductivity of polymers can be increased by introducing conductive additives or doping agents to allow electrons or holes to jump freely from valence band to conduction band (Van Krevelen & Te Nijenhuis, 2009). Other factors that contribute to electrical conductivity in polymers include chemical composition, presence of impurities, and type of hardener. Intrinsically conducting polymers (ICPs) are not as mechanically strong as other polymers and suffer from poor processability (Lalegül-Ülker et al., 2018), but their electrical conductivity can be similar to that of metals. They are often used in composites to overcome challenges associated with thermal, chemical, and mechanical stability (Hatchett & Josowicz, 2008).

4.2.2 ELECTROCHEMICAL PROPERTIES OF CONDUCTING POLYMERS

Doping of conducting polymers causes changes in the geometric structure of the polymer. Subsequent reduction of the polymer back to its undoped state enables it to recover its geometric structure. The charge-discharge cycle of doping and undoping is the principle of operation in polymer-based sensors and energy storage devices (Schoetz et al., 2018).

During electro-oxidation of conducting polymers, electrons in the π-bond move through the polymer backbone, and cations from the electrolyte are injected into the polymer chain to neutralize electric charge (Le et al., 2017). On the other hand, electroreduction consists of electrons transported to the backbone while anions intercalate from the electrolyte into the polymer chain (Szumska et al., 2021). During redox reactions, polymer chains can be either positively or negatively charged depending on whether the reaction is reduction or oxidation. Electrochemical studies of conducting polymer highlight macroscopic, near-equilibrium processes that occur when redox reactions are studied at low rates.

In electrochemical applications, conducting polymers that are n- and p-dopable exhibit greater energy density than polymers that can only be p-doped (Rudge et al., 1994). During redox reactions, the insertion/de-insertion of ions into the CP matrix can cause them to swell and de-swell with changes in the redox state. This actuation behavior of CPs is exploited in voltage-gated membranes, as discussed in Chapter 8. During oxidation, i.e. extraction of electrons, charge neutrality necessitates the insertion of small ions into the polymer matrix, forming ionic cross-links between polymer chains and ions and increasing overall volume. When voltages are reversed, these dopant anions are de-inserted during reduction, which causes de-swelling. The expansion depends on electrolyte concentration, physicochemical properties of the CP, size/type of dopant, and electrolyte ions. The electrochemical behavior of CPs is limited by the diffusion rates of dopant species; increasing diffusivities would further strengthen their role in electrochemical applications with greater control over morphology and structure.

If CP electrodes are scanned at potentials greater than an established potential range, they may experience irreversible oxidation and structural degradation. Apart from electrochemical potential, this phenomenon of overoxidation depends on electrolyte pH and ionic concentrations.

Conducting polymers are widely employed in electrochemical applications, especially in capacitors, rechargeable batteries, and fuel cells. One such CP that has garnered attention is polyaniline (PANI), due to ease of synthesis, flexibility, environmental friendliness, high electrical conductivity, and ease of conversion between oxidation states. PANI has been synthesized in various morphologies including nanowires, nanofibers, and nanograins. Its tunable properties have allowed it to be used in anticorrosive coatings, energy storage systems, gas sensing and electrocatalytic and electrochromic devices. PANI contains "n" reduced (benzenoid diamine) and "m" oxidized (quinoid diamine) repeating units (Dhand et al., 2011). It has particularly garnered interest in biosensing applications, which, combined with ease of processability, have raised curiosity in its use as a membrane for desalination and water treatment for fouling mitigation.

Figure 4.3 shows a typical voltammogram for PANI with reversible p-doping (Le et al., 2017). The voltammogram shows a multielectron transfer process with two oxidation peaks and two reduction peaks. The kind of acid electrolyte used changes voltammetric behavior due to differing charges and sizes of $SO2^{-4}$ and CL^- anions. When HCl is used as dopant, PANI has a higher electrical conductivity, which points to the effective migration of dopant ions into the electrode material.

Improvements in exploiting PANI for the desired electrochemical behavior remain limited by its relatively high cost, long-term stability, and the trade-off between electrochemical performance and mechanical properties (Z. Li & Gong, 2020).

PPy has recently attracted attention as a flexible electrode material due to it electrical conductivity, facile fabrication, and suitable redox properties (Çirmi et al., 2022; Fan & Maier, 2006). Recent research has allowed the synthesis of conductive PPy into various nanostructures: nanowires, nanotubes, nanosheets, and hydrogels (Yang Huang et al., 2016). Nanostructured PPy exhibits better energy storage capacity and superior mechanical strength.

Studies on the origins of conductivity and the effect of doping have allowed enhanced control over the kinetics of charge-transfer reactions involving CPs. As discussed, the mobility of charge carriers (polarons, bipolarons, etc.) depends on dopant type, size, temperature, and inherent structure of the CP. Electrical conductivity of CPs can be controlled via electrochemical doping. The poor long-term electrochemical stability and relatively inferior electrical conductivity compared to metals still hinder the use of CPs for electrochemical applications.

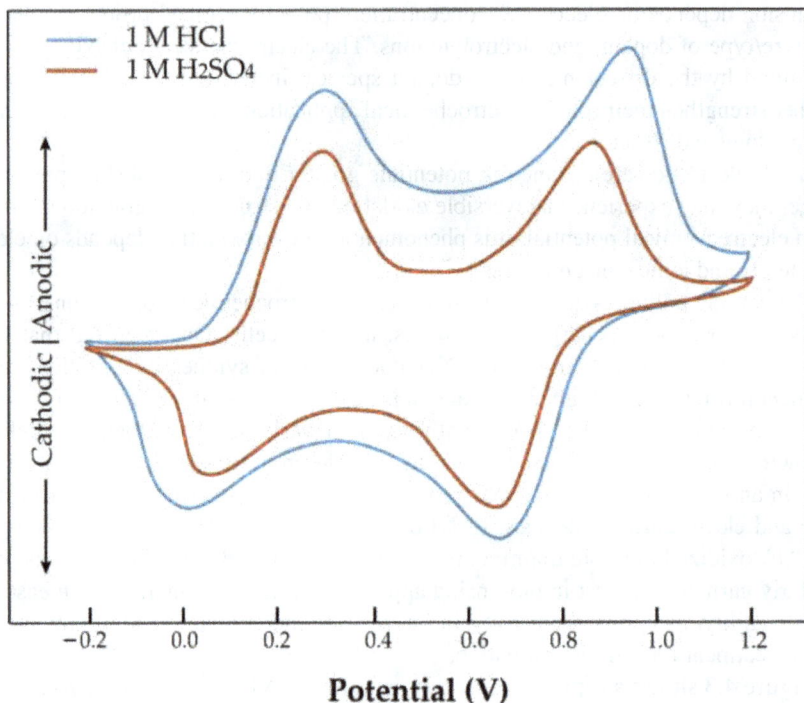

FIGURE 4.3 Cyclic voltammograms of polyaniline (PANI) film in HCl or H$_2$SO$_4$ with a potential scan rate of 50 mV s^{-1}.

Source: Le, T.-H., Kim, Y., & Yoon, H. (2017). Electrical and electrochemical properties of conducting polymers. *Polymers*, *9*(4), 150.

4.3 METALS

Electrical transport in metals is very well understood today, owing to over a century's worth of scientific literature. Thus we will only briefly overview the mechanism and discovery of electrical transport in metallic materials here.

Electrical conductivity is arguably one of the most important characteristics of metals (Quinn & Yi, 2018). Metals have a high density of conduction electrons. One may recall that conductivity is proportional to both carrier concentration and mobility (Cui et al., 2016). Electrical transport in metallic materials is driven by the motion of free electrons moving through a solid (Bardeen, 1940). While earlier models such as the Lorentz and Drude models alluded to the concept of free electrons as the underlying explanation for metallic conduction, it was German scientist Arnold Sommerfield who developed the free electron theory of metals (Hall, 1928). Because he combined the Drude model with quantum mechanical Fermi–Dirac statistics, the free electron model is also known as the Drude–Sommerfeld model. Somerfield's theory of metals provided a model for the behavior of electrons in a solid metallic crystal (Roy, 2019) and remains successful for simple metals whose conduction electrons are donated from *sp*-shells. The model assumes that, in the solid state, valence electrons in a metallic atom become conduction electrons. The Sommerfeld model can successfully explain the approximate temperature dependence and magnitudes of the thermal and electrical conductivities of metals (Singleton, 2001). Roy notes that the free electron theory is not accurate for transition metals such as iron, manganese, and bismuth (Roy, 2019). Identifying the shortcomings of the free electron model, the band theory of solids emerged as a more accurate description of electron transport in metals (Britannica), including transition metals.

The band theory was developed to explain inconsistencies in previous models and to determine electrical and thermal properties in insulators, semiconductors, and metals. It is foundational in the development of solid-state electronics, which uses semiconductor materials extensively. As opposed to electrons in free space, the band theory considers electrons in a periodic lattice and as an extension the electron–lattice interaction (Shinozuka, 2021). Electrons are assigned states with energy values within specific ranges. A solid becomes conductive when electrons have sufficient excitation to jump from valence band to conduction band; this can be through thermal excitation across the bandgap or thermal excitation from localized impurities or dopants (Bube, 2003).

Insulators have wide bandgaps, typically >3–5 eV (Klimm, 2014). Semiconductors are crystalline or amorphous substances in which the valence band is filled, but the conduction band is empty. Most III-V semiconductors have high electron mobilities and a wide range of bandgap values (Lovell et al., 1976). According to the band theory, the high electrical conductivity observed in metals is due to the overlapping of their valence and conduction bands (Bardeen, 1940).

Factors that affect the electrical conductivity of metals and alloys include the following (Taherian & Kausar, 2019):

1. *Microstructure defects such as vacancies, interstitial, or substitutional defects, precipitates and impurities:* Defects cause electrons to scatter, reducing electron mobility and affecting their mean free path.
2. *Work hardening:* Strengthening by plastic deformation can disrupt the atomic network and create dislocations, resulting in barriers for electrons. Annealing also affects electrical conductivity.
3. *Alloying:* Alloying decreases electrical conductivity because the alloying elements act as impurities and cause scattering.

For a more detailed description on theories surrounding electron transport in metals and semiconductors, as well as their historical evolution, the reader is guided to other resources (Altmann, 2013; Bardeen, 1940; Chester & Thellung, 1959; Dugdale, 1995; Jones et al., 1934; Mizutani, 2001; Quinn & Yi, 2018; Schubin et al., 1934; Tosi, 2005).

4.4 CARBON-BASED NANOMATERIALS

While nanotechnology has manifested its potential in the form of massive advances and applications, no group of nanostructured materials has attracted attention in

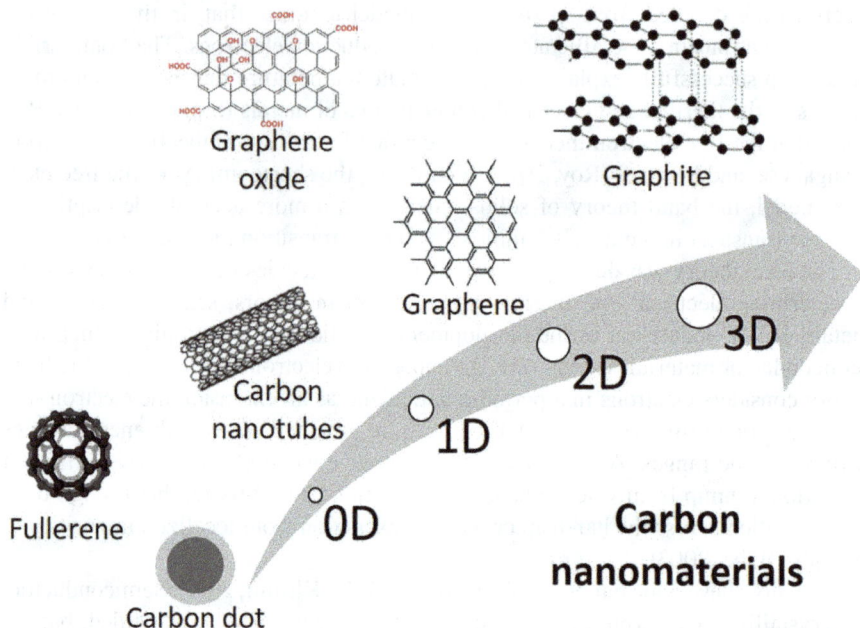

FIGURE 4.4 Carbon-based nanomaterials.

quite the same manner as carbon-based nanomaterials. Adding to the more classical carbon allotropes of "hard" diamond and "soft" graphite (Falcao & Wudl, 2007), discoveries between 1985 and 2010 led to zero-dimensional, one-dimensional, and two-dimensional carbon nanomaterials in the form of fullerene, carbon nanotubes, and graphene, respectively (Jariwala et al., 2013). Figure 4.4 shows a schematic of selected carbon nanomaterials. The term "wonder materials" has repeatedly been used in the literature to refer to carbon nanomaterials, most notably graphene and carbon nanotubes (Baines, 2018; H. Huang et al., 2020; Khan & Ahmad, 2020; Waite & Nazarpour, 2016). Nanoparticles, nanorods, and nanoplates are also carbon nanostructures that have emerged, each offering different properties depending on size, shape, and spatial arrangement (N. Yang et al., 2016). The unique physicochemical properties of graphene and CNTs arise from the formation of sp^2 hybridized carbon atoms in a hexagonal network with in-place σ-bonds and out-of-plane π-orbitals (Georgakilas et al., 2015).

In fact, when considering electrically conducting membranes reported between 2010 and 2020, approximately 60% are prepared using carbon-based materials (M. Sun et al., 2021). The high electrical conductivity in carbon-based materials is a result of sp^2 hybridization and delocalized π–π electrons, as we will see. In this section, we look at CNTs and graphene-based materials, electronic transport in each, and their applicability to water treatment and desalination.

Carbon-based materials such as activated carbon, CNTs, and graphene have been studied extensively as electrode materials due to their high conductivity, wide electrochemical potential window, and strong electrocatalytic behavior for various redox reactions. In addition, certain carbon structures allow chemical modification that enable them to be used for surface reactions. In this section, we review the electrical and electrochemical properties of carbon-based materials with a focus on CNTs and graphene.

4.4.1 Carbon Nanotubes

Perhaps the most widely investigated nanomaterial in polymer composite membrane systems, carbon nanotubes are cylindrical structures consisting rolled-up graphene sheets with hollow centers (Holban et al., 2016). Depending on the number of rolled-up graphene sheets, carbon nanotubes CNTs are identified as single walled carbon nanotubes (SWCNTs) or multiwalled carbon nanotubes (MWCNTs). The cylinders are 2–40 nm in diameter and up to hundreds of µm in length. Pristine CNTs have been measured to be 100 times stronger than steel and six times lighter (Grobert, 2006). Their structure imparts a unique combination of physicochemical properties that makes them a promising material for a wide range of applications. CNTs have high aspect ratios and surface areas (Birch et al., 2013; Zang et al., 2015), are lightweight (Maruyama, 2021), can be functionalized with ease (Chadar et al., 2021), demonstrated antifouling behavior (C. Hu et al., 2007), superior mechanical strength (Norizan et al., 2020), and high temperature stability and distinctive water transport (H. G. Park & Jung, 2014). All of these properties can be exploited to develop high performing materials for desalination and water treatment. This has resulted in fortified interest in CNTs for sustainable membrane separation processes. Every article on CNTs and CNT-based materials touts their exceptional properties.

Trends in understanding electron transport mechanisms in CNTs has more or less complemented the growing interest in CNTs, with approximately 400 publications addressing electron transport exclusively. Of these, over 200 articles were published between 2004 and 2011. Here, we provide a concise review in the understanding of electron transport in CNTs with the aim of developing a greater appreciation of the contributions of CNTs to macroscopic electrically conductive membranes and systems.

4.4.1.1 Individual CNTs

Early studies which sparked interest in CNTs consisted of predictions of their electronic properties, which were said to be controlled by small structural vibrations (Ebbesen et al., 1996). Two years after his original discovery, Sumio Iijima, the inventor of CNTs (Iijima, 1991), used band structure calculations to conclude that SWCNTs should behave as semiconductors (Iijima & Ichihashi, 1993). The following year, Chintamani Rao's group in India carried out the first study to address the electron transport properties of CNTs (Seshadri et al., 1994). Based on tunneling conductance measurements on bulk samples of CNTs, their findings suggested the presence of a bandgap, the width of which is inversely proportional to tube diameter. However, bandgap width differed significantly between resistivity measurements (1 eV) and tunneling conductance measurements (146 meV) for the same nanotube (9.7 nm), which they attributed to Coulomb blockades in tunneling measurements.

The electrical properties of individual CNTs were first determined in 1996 (Ebbesen et al., 1996; Langer et al., 1996). Over the years, various techniques have been applied to better understand electrical transport in CNTs.

Nanotube diameter and atomic structure of nanotubes, often described by their helicity or tube chirality, determine whether the nanotube exhibits metallic or semiconducting behavior (Charoenpakdee et al., 2020; Jorio & Dresselhaus, 2008; Jiangtao Wang et al., 2018). Studies have shown ballistic electron transport in MWCNTs at room temperature with mean free paths on the order of tens of microns (Berger et al., 2003; Poncharal et al., 2002). Defects in the nanotube structure, however, can considerably reduce electron transport in nanotubes. Defects may cause a disordered structure that can increase inhibition of electron transport with increasing CNT length (Algharagholy, 2019). Low concentrations of dopants, such as boron or nitrogen, to induce holes or electrons can improve electronic conductance without affecting mechanical properties (Terrones et al., 2004). Doping also enables a more reactive CNT surface.

4.4.1.2 Buckypaper

In water purification, CNTs are often present as mats or networks either on their own, also known as buckypaper, or as fillers in polymer nanocomposites. Such network nanostructures form freestanding CNT membranes, often of entangled and randomly aligned CNTs held together by weak van der Waal forces at interconnections (Q. Liu et al., 2013). Buckypapers make excellent candidates for membrane materials. Networks of CNTs can show interesting in-plane electrical conductivity behavior. The formation of a network influences electrical conductivity in buckypapers as factors such as misalignment, agglomeration, and junctions come into play as compared to individual CNTs.

Electrical conduction in CNT networks is thus of great interest in the use of electrically conductive CNT-based membranes for desalination and water treatment. In these complex structures, conduction occurs in regions of metallic conduction combined with hopping or tunneling through various types of defects or electrical barriers (Kaiser et al., 1999). In 2022, Shuang Tang used Seebeck coefficient measurements to confirm metallic transport behavior in a network of semiconducting CNTs at room temperature (S. Tang, 2022). Intratube ballistic electron transport is the dominating contributor to conduction in CNT networks as compared to diffusive scattering by long-range order or to quantum hopping resistance at nanotube contact points (S. Tang, 2022). Improved electron transfer in longer CNTs also enhances electrical conductivity in millimeter-scale-long CNTs (Sakurai et al., 2013).

4.4.1.3 Electrochemical Activity in CNTs

CNTs can be used in electrodes as freestanding buckypaper or through modification of other materials via aligned or randomly oriented nanotube forests (Merkoçi et al., 2005). Several methods have been used to fabricate CNT electrodes: vacuum filtration (S. Liu et al., 2013), electrophoretic deposition (Du & Pan, 2006), drop-dying (Y. Song et al., 2016), and r (J. H. Kim et al., 2006). The performance of CNT electrodes depends on several factors such as aggregation, CNT diameter, presence of defects and impurities, and number of walls (Sgobba & Guldi, 2009).

When the diameter is greater than 10 nm, the behavior of CNTs is very close to that of graphite as shown by Compton et al. They argue that edge-plane sites in graphite and tube ends in CNTs are electrochemically active, whereas the pristine walls of the CNTs are analogous to the basal plane of graphite. Consequently, the tube ends and wall defects exhibit fast electron transfer of edge-plane pyrolytic graphite. Introducing defects in CNT walls can therefore tailor the electrochemical activity of MWCNTs. In SWCNTs, the smaller diameter and greater curvature reduces the overlap between p-orbitals on neighboring carbon atoms involved in π-bonds. However, this does not affect electrochemical activity (Holloway et al., 2008).

The role of oxygen-containing groups on HET is unclear, with contradicting findings on whether carboxylic acid moieties at CNT ends promote faster HET (Chou et al., 2005; Pumera, 2007). Martin Pumera argues that increased HET could be due to an increase in defects during the oxidation process as opposed to the introduction of oxygen-containing groups (Pumera, 2009). The presence of nanographite impurities also contribute to the electrochemical properties of CNTs (Ambrosi & Pumera, 2010).

CNTs are typically prepared via chemical vapor deposition (CVD) or arc discharge methods, and metal catalyst nanoparticles of nickel, cobalt, iron, or molybdenum are commonly used in the process. Often the metal remains at the ends or within the nanotubes. These residual metallic impurities contribute to electrochemical activity. Iron- and copper-based impurities are responsible for the electrocatalytic oxidation of glucose and the reduction of hydrogen peroxide (Pumera & Iwai, 2009). Other impurities such as Co and Mo may not affect the reduction of hydrogen peroxide (Pumera & Miyahara, 2009). One reason for the significant impact of certain metallic nanoparticles is that their diffusion layers, on the order of tens or hundreds of μm, overlap, and the particles within the CNTs act as metallic nanoelectrodes with a distance of up to hundreds of nm between them.

4.4.2 GRAPHENE

Graphene is a two-dimensional material consisting of a single or few layer(s) of carbon atoms packed in a honeycomb lattice. A single-layer graphene sheet has two structural components: the edge, which is the most commonly found linear defect in graphene and consists of dangling bonds and various capping moieties, and the basal plane, which consists of 2D conjugated sp^2 carbon atoms (Yuan et al., 2013).

An advantage of graphene over CNTs is the relatively low-cost production methods available to extract graphene from graphite. These include mechanical stripping and electrochemical exfoliation, among others. Mechanical exfoliation or stripping is a top-down technique that is also one of the first methods discovered for graphene synthesis. In this technique, a longitudinal or transverse stress is generated on the surface of graphite to separate it into mono-atomic thin layers. Simple Scotch Tape is commonly used in what is now known as the "Scotch Tape method" (Randviir et al., 2014), although other agents such as ultrasonication (Ci et al., 2009) or electric field (X. Liang et al., 2009) may also be used for mechanical stripping. In electrochemical exfoliation, an applied voltage causes ionic species to intercalate into graphite such that the formation of gaseous species at the electrode–electrolyte surface causes exfoliation of the graphite into individual graphene sheets (Achee et al., 2018).

Graphene serves as a building block for many other carbon-based materials. Extremely high predicted values for high carrier mobility in graphene sparked interest in graphene for electronics (Allen et al., 2010). A high-quality 2D crystal lattice prevents scattering from electron–phonon interactions and promotes high charge transport. For pure graphene, ballistic electron transport with extremely high carrier mobilities enables its high electrical conductivity. However, the accidental or deliberate occurrence of defects in graphene causes the scattering of electron waves, which can lead to a significant drop in electrical conductivity (Banhart et al., 2011). Conduction in graphene can be tuned through appropriate chemical doping that injects charge into the electron system.

The band structure of graphene contributes to its unique electrical properties. The exceptional electron mobility in graphene is due to the effective masslessness of electrons in the Dirac band structure (M. Wilson, 2006). Although there is no bandgap, graphene is still considered semimetallic as it has zero density of state electrons at the Fermi level. Its valence and conduction bands coincide at the Dirac points, which are six locations in momentum space (Peres et al., 2009). The electronic structure of graphene can be modified through various techniques, such as electric and magnetic fields, chemical intercalation and adsorption, stacking geometry, edge-chiralty, defects, and strain (Zhan et al., 2012).

4.4.2.1 Electrochemical Activity in Graphene

Exploitation of electrically conductive membrane systems can be maximized by hybridizing membrane technologies with electrochemical processes. Graphene particularly is promising due to its large specific surface area for electrochemical reactions compared to other carbon structures. In theory, the surface area of graphene is close to 2600 m^2 g^{-1}, higher than any other carbon-based material (Zhang et al., 2014). In fact, its high surface area is a sought-out characteristic in electrode materials (Mano, 2020). Graphene is a popular material choice for preparing such electroactive membrane systems. Thus we seek to understand the electrochemical behavior in graphene and graphene-based composites as a first step toward realizing their value for mitigation of fouling and related phenomena. This includes parameters such as electrochemical potential window, electron transfer rate, redox potentials, etc.

While bulk properties derived from interior structure are important, the electrochemical behavior of a material is also dictated by surface chemical and electrical properties (McCreery, 2008). Graphene is considered electrochemically stable and has a large potential window, similar to that of graphite. However, the heterogeneous electron transfer (HET) constant for a single-layer defect-free graphene sheet is double that of multilayer graphene. Increasing the number of sheets, however, reduces HET to values similar to those in edge plane pyrolytic graphite.

Combined with low-cost mass production methods, the unique properties of graphene render it an attractive material for electrochemical devices such as batteries and sensors. Opened edges of graphene exhibit faster electron transfer kinetics compared to folded edges (Ambrosi et al., 2011). As previously mentioned, defects alter electron transfer kinetics. Edges are a common linear defect found in graphene, and the literature to date shows contradicting effects of edges on electron transfer kinetics for the same redox reactions (Pavlov et al., 2019). Yuan et al. found that the edge has a specific capacitance that is four orders of magnitude higher, a faster electron transfer rate, and better electrocatalytic activity as compared to the graphene basal plane (Yuan et al., 2013). Point defects also affect electrocatalytic activities: Kislenko et al. found that electron transfer increases by an order of magnitude for monovacancies compared to pristine graphene for redox couples with standard potentials between –0.2 V and 0.3 V (Kislenko et al., 2020).

Electrochemical properties depend on defect type as different defects induce varying changes in the electron structure (Kislenko et al., 2020). Controlled engineering of defects in graphene, such as zigzag or armchair, is an interesting route to tune electron transport in graphene during growth (T. Ma et al., 2013). Goh and Pumera further confirmed that the electrochemical response does not depend on the number of graphene layers (Ambrosi et al., 2014; Goh & Pumera, 2010). However, other studies show a correlation between kinetics and the number of layers (Brownson et al., 2014; Velický et al., 2014). Similarly, Güell et al. showed that the rate of heterogeneous electron transfer increases systematically with the number of graphene layers (Güell et al., 2012).

4.5 POLYMER COMPOSITES

4.5.1 POLYMER-CNT COMPOSITES

Embedding CNTs homogeneously into lightweight, flexible engineered polymers is a common strategy to exploit their superior properties on a macroscopic level (Spitalsky et al., 2010). This holds especially true in membrane technology, which is dominated by polymeric materials such as polyamide, polyethersulfone (PES), cellulose acetate (CA), PVDF, PTFE, polypropylene, etc. (Bera et al., 2022). The large success of polymeric materials in desalination and water treatment results from their ease of processing, excellent film forming ability, high mechanical properties, and, maybe most importantly, ability to tailor polymeric membranes for high selectivity and productivity. Polymeric membranes can be prepared to be non-porous for NF, RO, and FO, such that separation occurs via selective diffusion, or they can be porous with different ranges of pore sizes for MF, UF, and MD, as discussed in Chapter 1. Hence, while each of the conducting polymers and CNT membranes belongs to a niche in the scientific literature on conductive membranes, it is polymer-CNT composite materials that have garnered the most attention and remain the most frequently adopted material type for the development of electrically conductive membrane materials.

4.5.2 POLYMER-GRAPHENE COMPOSITES

Kotov cites graphene as a potential cost-effective material to use as filler in conductive composites (Kotov, 2006). CNTs and graphene are both better candidates for polymer nanocomposites compared to clay or other carbon-based fillers, as composites based on these materials have lower percolation thresholds (H. Kim et al., 2010; Kuilla et al., 2010). The physicochemical properties of polymer-graphene nanocomposites depend on the distribution of graphene layers within the polymer matrix and the interface between graphene and polymer (Kuilla et al., 2010). Few-layer graphene oxide sheets are compatible with organic polymers due to the presence of functional groups at its edges, although homogeneous composites are difficult to prepare using pristine graphene (Kuilla et al., 2010). However, swelling of highly hydrophilic graphene oxide films due to the presence of carbonyl and carboxyl groups (Y. Xu et al., 2008) may cast doubt on their suitability in membrane technology; further treatment may be required to prevent swelling as is commonly done with hydrophilic polymers. Additionally, graphene oxide is electrically non-conductive and requires surface modification or chemical reduction to increase electrical conductivity of graphene-based nanocomposites. Emphasizing the role of interface modification in the structure and functional properties of polymer-graphene composites, Ma and coworkers covalently modified graphene nanoplatelets to enhance the electrical conductivity of nanocomposites by 116% (J. Ma et al., 2013). Interface modification leads to more uniform dispersion and a greater degree of electron transfer across the interface.

Conductive polymer-graphene composites with segregated structures such that the conductive filler is only positioned at the interface between the polymer matrix and bulk have shown ultralow conductivity percolation. As an example, Pang et al. fabricated a polyethylene graphene composite using a water/ethanol-solvent-assisted dispersion with a percolation threshold as low as 0.07 vol. %, owing to the formation of a two-dimensional conductive network (Pang et al., 2010).

The electrical conductivity of polymer-graphene composites and the percolation threshold (see Section 4.5.3) are used to characterize the electrical properties. Superior electrical properties consists of a low percolation threshold and higher conductivity at a lower graphene loading, which would reduce the cost of filler loading and ensure that the composite retains the processability of the matrix material. Factors such as intrinsic electrical conductivity, filler aspect ratio, dispersion state, and contact resistance between graphene all affect the electrical properties of polymer-graphene composites. Thermally reduced graphene oxide (TRGO) is commonly used as conductive filler via melt processing, but its composites often suffer from poor dispersion and loss of functional groups as a result of reduction.

Although Xie et al. have shown that improvement through graphene is more effective than that with competing nanofillers such as CNTs from a theoretical standpoint (S. Xie et al., 2008), comparison between the two types of composites is not straightforward. Conflicting articles report that practical graphene composites have a higher percolation threshold and are less effective in conductivity improvement than the same composite with CNTs. This has been attributed to the tendency of graphene to aggregate because of its larger surface area and plane-to-plane contact area and to the propensity of TRGO to wrinkle and crimp. Additionally, electron transfer in TRGO occurs through various types of contacts: plane-to-plane, edge-to-edge, and edge-to-plane, whereas overlapping contact is the main mechanism in CNTs. It is also more challenging to interlace 2D graphene into a network structure as compared to 1D CNTs. As a result, the electrical properties of fabricated polymer-graphene composites are far from their theoretical predictions. Table 4.1 shows the electrical properties of selected polymer-graphene nanocomposites.

TABLE 4.1

Electrical Properties of Selected Polymer Nanocomposites.

Matrix	Filler	Filler Loading	Process	σ (S cm^{-1}) of Matrix	σ (S cm^{-1}) of Composite	Reference
Carbon nanotubes						
Polyvinylidene fluoride (PVDF)	MWCNTs	0.2 vol. %	Mechanical mixing	10^{-14}	10^{-2}	(Q. Liu et al., 2012)
	MWCNTs	5 wt%	Sonication	—	3.7×10^{-5}	(Gao et al., 2014)
Nafion	CNTs	50 wt%	Solution	—	1.6×10^{1}	(C. Luo et al., 2008)
Polyvinyl alcohol (PVA)	Carboxylated MWCNTs	20 wt%	Sonication and vacuum filtration	—	3.6×10^{1}	(de Lannoy et al., 2012)
	MWCNTs		Spinning		9×10^{2}	(K. Liu et al., 2010)
Graphene						
Epoxy	Reduced GO	0.2 wt%	Sonication	10^{-11}	10^{-8}	(L.-C. Tang et al., 2013)
	Graphene foam	0.2 wt%	Solution	10^{-12}	3×10^{0}	(Jia et al., 2014)
	Graphene nanoplatelets	0.025 vol. %	Solution	10^{-17}	10^{-6}	(King et al., 2013)
Polypropylene (PP)	Graphene nanoplatelets	0.5 wt%	Coating	10^{-16}	10^{-1}	(Imran et al., 2018)
	Graphene	1.85 wt%	Coating	—	4×10^{-2}	(T. Sun et al., 2020)
		2.5 wt%	Melt mixing	10^{-18}	10^{-3}	(Gkourmpis et al., 2019)
PVDF	Reduced GO	0.015 vol. %	Solvothermal reduction	10^{-12}	10^{-6}	(He & Tjong, 2013)
	Functionalized graphene	0.5 wt%	Sonication	10^{-18}	10^{-6}	(Eswaraiah et al., 2011)
	Reduced GO	0.2 vol. %	Sonication	10^{-12}	10^{-5}	(M. Li et al., 2013)
Polysulfone (PS)	Graphene nanoplatelets	10 wt%	Sonication	10^{-12}	10^{-4}	(Abbasi et al., 2020)

4.5.3 PERCOLATION THRESHOLD

The electrical conductivity behavior in polymer-CNT and polymer-graphene composites is linked to understanding percolation theory. Percolation refers to the sharp increase in electrical conductivity of a composite material when the arrangement of the conductive filler forms a conductive network within the polymer matrix.

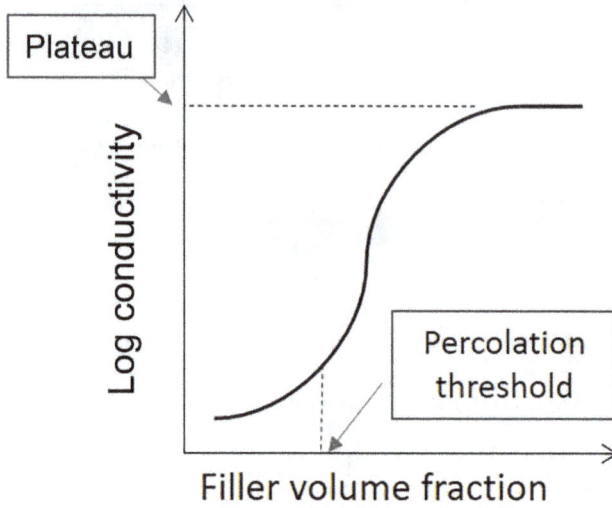

FIGURE 4.5 Electrical conductivity as a function of volume fraction for a non-conducting polymer with a conductive filler.

Percolation is typical in composites with very high contrast constituent properties (Yang Wang & Weng, 2018).

"Percolation" is the term for long-range connectivity, and the filler volume fraction at which this sharp increase occurs is known as the "percolation threshold" (Moniruzzaman & Winey, 2006). Above the percolation threshold, the electrical conductivity can increase by 10^{10}–10^{15} S m^{-1} (Grady, 2011). Figure 4.5 shows a typical conductivity plot with filler for an insulating matrix and conductive filler such as CNTs.

At percolation, the formation of a conductive network enables an open path for electrons to travel macroscopic distances (Grunlan et al., 2001; Shante & Kirkpatrick, 1971). According to the classical percolation theory, the electrical conductivity of a composite is given by (X. Luo et al., 2022):

$$\sigma = \sigma_0 (\Phi - \Phi_c)^t \qquad\qquad \text{Equation 4.1}$$

where σ and σ_0 are the conductivities of the composite and the electrical filler, respectively, Φ is the volume fraction of fillers, Φ_c is the percolation threshold, and t is the critical exponent. Experimental values between 0.7 and 3.1 have been reported for the critical exponent for CNT-polymer composites (Min et al., 2010).

Researchers have focused on determining the factors that affect percolation threshold in an attempt to lower percolation threshold for polymer-CNT composites. A lower percolation threshold retains mechanical properties of the matrix polymer material (He & Tjong, 2013; Nogales et al., 2004). Morphology, size, alignment, polymer matrix properties, interphase depth, and tunneling distance of the filler are all factors that affect percolation threshold in composites (Bauhofer & Kovacs, 2009; Kazemi et al., 2021). For conventional fillers such as microscale metal powder or carbon black, high filler contents of 10–50% are needed to achieve electrical conductivity (Ma et al., 2010). On the other hand, nanoscale filler dimensions can lower the percolation threshold to as low as 0.5 wt% for polymer-CNT composites (Bauhofer & Kovacs, 2009).

It thus makes sense that adequate conductivity at low filler contents can be achieved in a composite material if the percolation threshold is low. Many studies have thus focused on lowering the percolation threshold of various polymer nanocomposites (Al-Saleh, 2017; Nogales et al., 2004; S. B. Park et al., 2014).

4.6 MEASUREMENT OF ELECTRICAL CONDUCTIVITY

4.6.1 FOUR-POINT PROBE

Electrical conductivity of membrane materials is most commonly measured using the four-point probe method involving a direct current (DC). In this technique, a probe with four tungsten carbide needles equidistant from one another and arrayed in a straight line is in contact with the sample, as shown in Figure 4.6.

A constant electrical current I is applied through the outermost needles, and a voltmeter measures the voltage V generated between the two inner needles. For a sample with thickness less than the separation distance d between consecutive probe needles, the resistivity can be calculated using Equation 4.2:

$$\rho = 2\pi d\left(\frac{V}{I}\right)$$

Equation 4.2

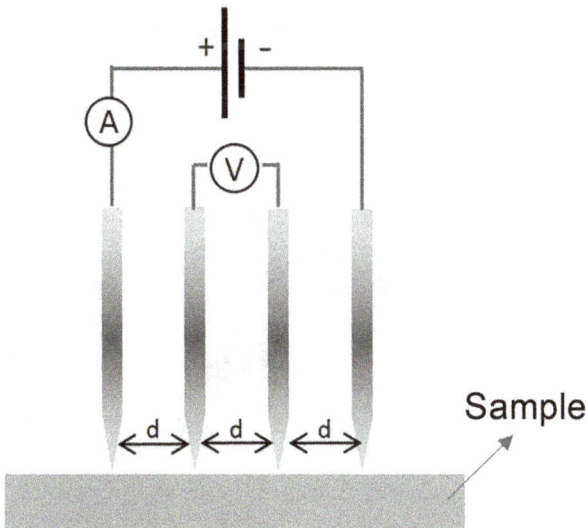

FIGURE 4.6 Schematic of the four-point probe measurement technique for electrical conductivity of thin film samples.

4.7 PREPARATION OF ELECTRICALLY CONDUCTING MEMBRANE SYSTEMS

Electrically conducting membranes can be prepared in several different ways. Some of these are shown schematically in Figure 4.7. Freestanding membrane casting involves the blending of conducting materials with binders such as polymers. Electrospinning of polymer solutions containing electrically conductive fillers is also a growing versatile method of freestanding membrane fabrication.

Another common preparation is method is to modify substrates to improve their electrochemical performance, pore size, hydrophobicity, and permeability in order for them to be employed as membrane electrodes. This can be done using either electrically conducting or non-conducting substrates. Examples of conducting substrates include carbon fiber cloth (Yihua Li et al., 2015), Ti mesh (Hui et al., 2019; L. Xu et al., 2019), stainless steel (Xiaohui Li et al., 2020; Xueye Wang et al., 2017), and PANI (B. Li et al., 2020; L. Xie et al., 2020). Non-conducting substrates can be polymeric (J. Liu et al., 2017; Tan et al., 2019) or ceramic (S. Chen et al., 2020; Geng & Chen, 2017; Xiaoxiong Wang et al., 2020), on which an electrically conducting layer can be deposited by coating, pressing, chemical vapor deposition, physical vapor deposition, etc.

The third method of preparing membrane electrodes is by assembling composite membranes by incorporating mesh electrodes on commercial membranes. Mesh electrodes have the added advantage of acting as spacers to improve flow characteristics.

Some methods commonly used to fabricate electrically conductive membranes are described next.

4.7.1 VACUUM FILTRATION

Vacuum-assisted filtration or vacuum filtration is the process of filtering a suspension of conductive material through a porous substrate to obtain uniform freestanding

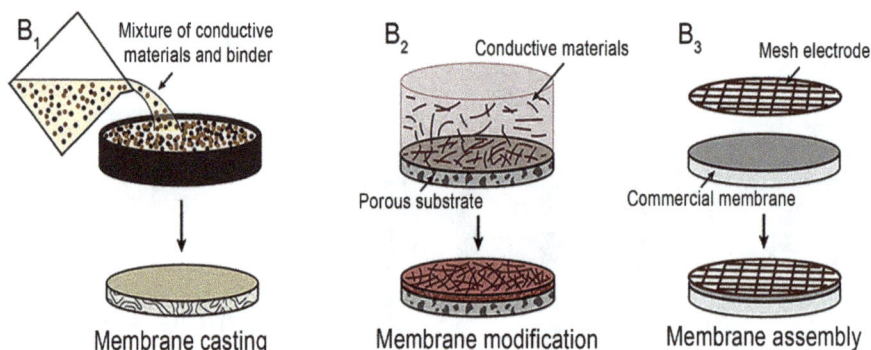

FIGURE 4.7 Electrically conducting membrane materials, preparation, and modules. Schematic illustration of electrically conducting membrane preparation through (a) freestanding membrane casting, (b) membrane modification, and (c) composite membrane assembly.

Source: Reprinted with permission from *ACS EST Engg.* 2021, 1, 4, 725–752. https://doi.org/10.1021/acsestengg.1c00015. Copyright © 2021 American Chemical Society.

membranes. It is the most commonly employed method to prepare freestanding GO (Hou et al., 2018; Yi Huang et al., 2015; Joshi et al., 2014; B. Liang et al., 2015; G. Liu et al., 2016; Mohammed et al., 2022; J. J. Song et al., 2015) and CNT membranes (Rashid et al., 2014; Urper et al., 2018). Peng Wang's group showed that vacuum-filtered GO membranes are in fact asymmetric, with variation in the top and bottom surface pores; their asymmetry should be considered when employing them in separation applications (B. Tang et al., 2016).

Madaria et al. applied vacuum-assisted filtration to fabricate films of a percolating silver nanowire network (Madaria et al., 2010). This technique has also been widely applied to develop cellulosic films (Meng & Manas-Zloczower, 2015; Noh et al., 2021; Xi et al., 2022; Yin et al., 2020). Hu's group compared Fe-based ceramic membranes made via vacuum filtration and electrospinning for the removal of Cd^{2+} and found that the high specific surface area of Fe_3O_4 nanoparticles in the vacuum-filtered membrane resulted in higher adsorption capacity and Cd^{2+} ion removal efficiency (Wu et al., 2019).

GO membranes prepared via VF have demonstrated high water flux and selectivity for organic dyes and salt ions. However, challenges associated with VF include higher pressure and long duration, especially when thicker membranes are required (L. Chen et al., 2018).

4.7.2 ELECTROSPINNING

Electrospinning is an emerging simple technique to prepare nanofibrous membranes that have a large specific surface area, high porosity, and an interconnected pore structure. These features allow electrospun membranes to be suitable for desalination and wastewater treatment (F. E. Ahmed et al., 2015). Electrospinning consists of charging a polymer solution or melt in a syringe/needle via a strong electric field; competing electrostatic and surface forces cause the fluid to form a conical structure when it is ejected and consequently collected on a collector. The versatility of this technique in terms of solution properties, end membrane morphology, operating conditions, etc. makes it promising for membrane fabrication, as is shown in two recent reviews (F. E. Ahmed et al., 2015; Francis et al., 2022). Some challenges related to electrospinning include high electric field and low throughput. However, manufacturers have applied various approaches such as multineedle and needleless spinning, and centrifugal methods to increase production rate to as high as 450 g h^{-1}, which make electrospinning more feasible on a larger scale (Neubert et al., 2011; Omer et al., 2021; Persano et al., 2013). Lv et al. used electrospinning followed by silver mirror reaction to develop silver-nanoparticle-coated acrylonitrile-styrene nanofibers with high thermal and electrical conductivity (Lv et al., 2020). Often, CNTs are blended with the polymer solution prior to electrospinning to yield conductive membranes. Polymers used with CNTs in conductive electrospun membranes include polypropylene (PP) (Cao et al., 2014), polyimide (W. Xu et al., 2014), polyvinylidene fluoride (PVDF) (X. Yang et al., 2019), and polystyrene (Jingwen Wang et al., 2012). Conductive polymers such as polyaniline and PPy have also been used for electrospinning nanofibrous membranes (Kai et al., 2011; Merlini et al., 2015; Neubert et al., 2011; Perdigão et al., 2020). Li et al. carried out polymerization of PPy on electrospun polyacrylonitrile (PAN) templates to prepare conductive nanofibrous membranes (Yifu Li et al., 2022).

Electrospinning and electrospraying are both electrohydrodynamic processes relying on the same principal. However, in electrospraying, an atomization method is used via application of electrical forces to disperse a liquid or aerosol into fine charged droplets (Zong et al., 2018). The liquid at the end of a nozzle is subjected to electrically induced shear stress, and small droplets can be sprayed. Electrospraying allows control of the motion of the charged droplets and can be used for surface coating and thin-film production (Jaworek & Sobczyk, 2008). Electrospraying facilitates the preparation of 3D membrane structures via controlled layer-by-layer deposition, micro/nanodroplet production, and in situ polymerization. Compared to other membrane fabrication processes, the ability to tune droplet size, little or no chemical waste, control over film thickness, the ability to form composite structures, the ability to attain in situ reactions, and the potential to exploit electrochemical activity make electrospraying a promising technique for membrane separation (S. Huang et al., 2022).

Chen et al. used electrospraying to develop a GO NF membrane that was applied to dead-end filtration with a high pure water flux of 11–20 L m^{-2} h^{-1} bar^{-1} under a low pressure of 1 bar (L. Chen et al., 2018). The membrane also demonstrated high organic dye rejection >98.8% through physical size sieving and electrostatic interaction. The membrane was able to reject Na$_2$SO$_4$ with 63% efficiency through Donnan exclusion and steric hindrance. Its smooth and hydrophilic surface enabled ease of flux recovery by DI water washing. Roso and coworkers prepared electrically conductive membranes by combining electrospraying of pristine MWCNTs with simultaneous electrospinning of polyurethane nanofibers (Roso et al., 2016).

4.7.3 Dip Coating

Dip coating is a simple and commercially successful method of depositing onto a substrate. In dip coating, the substrate is dipped into the coating solution and removed (Brinker, 2013; Scriven, 2011). Coating parameters depend on dipping speed, polymer solution concentration, temperature, etc. (Himma et al., 2017). It is a versatile technique in that it can be applied to various membranes. Huang et al. coated PPy on a polypropylene membrane for solar-driven desalination (X. Huang et al., 2017). The PPy thin film attaches tightly to individual PP fibers via this technique. The authors emphasize that dip coating can be used on many types of membranes as it is not sensitive to the wetting property, conductivity, or surface curvature of the substrate. Electrically conductive membranes have been prepared for various applications using this method (Bak et al., 2010; Surana et al., 2015).

Dip coating is used as a facile and scalable technique to develop electrically conducting fibrous membranes. Kang and Jin fabricated electrically conducting silk non-woven membranes by dipping electrospun silk fibroin membranes into an MWCNT dispersion bath, thus adsorbing MWCNTs on the surface of silk nanofibers (Kang & Jin, 2007). Qiian and coworkers attached an ultrathin coating of sulfonated graphene onto electrospun carbon nanofibers via dip coating (Qian et al., 2015). The resulting composite showed a hydrophilic surface, high electrochemical specific capacitance, and lower interfacial charge transfer rate and were used for desalination via capacitive deionization with enhanced electrosorption capacity and charge efficiency. Sun and coworkers fabricated an electrically conductive superwetting mesh film through dip coating of a layer-by-layer graphene assembly, which they then used for adsorptive separation of oil and organic compounds from water. Dip coating has also been used to assemble SWCNTs (Bak et al., 2010) and MWCNTs (H. S. Kim et al., 2006) onto electrospun nylon 6 nanofibrous membranes.

4.7.4 Drop Casting

Drop casting consists of a suspension drop cast on a smooth substrate such as silica or paper and subsequent drying. This is a simple and fast method of generating freestanding uniform conductive films upon separation from the substrate. Drop casting is often used to modify membranes with nanotubes or nanoparticles for electrocatalysis and is a common technique in electrode preparation (Dinh et al., 2022). However, this technique has its own drawbacks: Robinson's group points out that resultant films can be incoherent with voids and cracks (Ha et al., 2012; Kaliyaraj Selva Kumar et al., 2020). Nevertheless, it is widely used in the fabrication of graphene-based films (Tran et al., 2021; Valentini et al., 2013; Zhao et al., 2014), mainly for electronic and biomedical devices. Wang et al. also used drop casting to develop conductive polymer composite strain sensors by using a Ag nanoparticles precursor (L. Wang et al., 2022). Song et al. used a conductive rGO ink for drop casting flexible large-area reduced rGO, into which they also incorporated 1D CNTs and Ag nanowires to construct a 3D network for faster electron transfer (C. Song et al., 2021). Prehn et al. used a modified drop casting technique to fabricate polymer composites using CNTs and sulfonated polyether ether ketone for fuel cells (Prehn et al., 2011). Tenorio et al. used drop casting to fabricate transparent TiO_2 photoanodes based on SWCNTs (Alvarado Tenorio et al., 2011). In another study, a composite was deposited from a graphene-PANI dispersion using drop casting as an ion-to-electron transducer (Boeva & Lindfors, 2016). Wang et al. drop cast graphene nanoplatelet dispersions onto polyethylene terephthalate (PET) substrates, which yielded mechanically flexible films with an electrical conductivity of 65 S m^{-1}. MXene has also been loaded using drop casting onto a clean wiper to achieve high uniformity and fast coating of films with high electrical conductivity of 230 S m^{-1}. The composite exhibits a large specific capacitance, high porosity, and high tensile strength. Choi et al. fabricated counter-electrodes for dye-sensitized solar cells (DSSCs) using graphene-based MWCNTs drop casted onto a SiO_2/Si substrate (Choi et al., 2011).

Drop casting is yet to receive attention in the preparation of conductive membranes for desalination and water treatment, making it a direction that perhaps researchers should consider when developing electrically conductive membranes for desalination and water treatment.

4.7.5 Spin Coating

Spin coating is yet another common method for low-cost deposition of thin films from materials ranging from polymers to functional inorganic films. It is used to optimize coating solutions and processes, although large-scale implementation is not considered appropriate (Shariatinia, 2021). Some advantages of spin coating include controllable thickness, uniformity, short processing time, and low cost (Sahu et al., 2009).

Spin coating consists of pouring a small amount of solution onto a substrate to induce complete wetting. The covered substrate is spun to accelerate solvent evaporation. Factors such as polymer solution concentration and rotating speed affect the thickness of the resulting film (Schubert & Dunkel, 2003). GO nanosheets have been attached to polymeric membranes by spin coating for high salt rejection (Dong et al., 2019; X. Hu et al., 2018). increased organic fouling resistance (Igbinigun et al., 2016), increased chlorine resistance (Shao et al., 2017), and as a pretreatment step before desalination (Abdelkader et al., 2019). Antibacterial (Y. Chen et al., 2016) and anticorrosion (Syed et al., 2015) films have also been developed using spin coating. Antibacterial films involve a silver nanowire film (van Berkel et al., 2021). Spin coated PANI-PAA/PEI composite coatings with a multilayer structure were used to prevent 316 stainless steels from corrosion.

Schmidt et al. investigated the feasibility of replacing indium tin oxide (ITO) electrodes with spin coated polymer-based composite films filled with MWCNTs (Schmidt et al., 2007). Their coating mixture consisted of a solvent with low volatility, a dissolved thermoplastic polymer, and MWCNTs. They attributed high electrical conductivity and low percolation threshold of the composites to the high aspect ratio of MWCNTs. Jo et al. fabricated highly conductive transparent thin films from SWCNTs using spin coating (Jo et al., 2010). Chang et al. reported the synthesis of a new type of graphene composite with graphene and PEDOT:PSS suitable for spin coating into conductive, transparent, and flexible electrodes without the need for high temperature processing (Chang et al., 2010). Wang et al. reduced the surface roughness of copper nanowires by spin coating a PEDOT:PSS solution on their surface (Yaxiong Wang et al., 2019). In another work, electrically conductive composite silica films were reported by incorporation of individual GO sheets into silica sols followed by spin coating, chemical reduction, and thermal curing (Watcharotone et al., 2007). Schlicke and coworkers spin coated freestanding films of cross-linked gold nanoparticles (Schlicke et al., 2011).

4.8 CONCLUSION

This chapter described the electronic structures and resulting electrical and electrochemical properties in several important classes of conductive materials: polymers, metals, carbon-based materials, and polymer composites. In conducting polymers, metallic behavior is enabled by doping polymers with acceptor or donor ions to introduce delocalized charges that move through the polymer backbone. Synthesis routes permit nanostructured conducting polymers whose large surface area imparts superior electrochemical activity.

In CNTs and graphene, the electrochemical behavior is similar to that of graphite, except that it changes with CNT diameter, number of defects, etc. In polymer composites with conductive fillers, the electrical conductivity follows a percolation threshold, which determines the filler content required to form a conductive network through the base material. We have discussed the main routes and their drawbacks for producing electrically conducting membranes.

BIBLIOGRAPHY

Abbasi, H., Antunes, M., & Velasco, J. I. (2020). Electrical conduction behavior of high-performance microcellular nanocomposites made of graphene nanoplatelet-filled polysulfone. *Nanomaterials, 10*(12), 2425.

Abdelkader, B. A., Antar, M. A., Laoui, T., & Khan, Z. (2019). Development of graphene oxide-based membrane as a pretreatment for thermal seawater desalination. *Desalination, 465*, 13–24. https://doi.org/10.1016/j.desal.2019.04.028

Achee, T. C., Sun, W., Hope, J. T., Quitzau, S. G., Sweeney, C. B., Shah, S. A., . . . Green, M. J. (2018). High-yield scalable graphene nanosheet production from compressed graphite using electrochemical exfoliation. *Scientific Reports, 8*(1), 14525. doi:10.1038/s41598-018-32741-3

Ahmed, F. E., Lalia, B. S., & Hashaikeh, R. (2015). A review on electrospinning for membrane fabrication: Challenges and applications. *Desalination, 356*, 15–30.

Ahmed, F. E., Lalia, B. S., Kochkodan, V., Hilal, N., & Hashaikeh, R. (2016). Electrically conductive polymeric membranes for fouling prevention and detection: A review. *Desalination, 391*, 1–15. https://doi.org/10.1016/j.desal.2016.01.030

Al-Saleh, M. H. (2017). Clay/carbon nanotube hybrid mixture to reduce the electrical percolation threshold of polymer nanocomposites. *Composites Science and Technology, 149*, 34–40. https://doi.org/10.1016/j.compscitech.2017.06.009

Algharagholy, L. A. (2019). Defects in carbon nanotubes and their impact on the electronic transport properties. *Journal of Electronic Materials, 48*(4), 2301–2306. doi:10.1007/s11664-019-06955-8

Allen, M. J., Tung, V. C., & Kaner, R. B. (2010). Honeycomb carbon: A review of graphene. *Chemical Reviews, 110*(1), 132–145. doi:10.1021/cr900070d

Altmann, S. L. (2013). *Band theory of metals: The elements*. Elsevier Science.

Alvarado Tenorio, G., Rincón González, M. E., Calva-Yáñez, J. C., & Solís de la Fuente, M. (2011). Photoanodes based on carbon nanotubes deposited by drop casting and filtration methods: Photoelectrochemical characterization. *ECS Transactions, 36*(1), 511–517. doi:10.1149/1.3660646

Ambrosi, A., Bonanni, A., & Pumera, M. (2011). Electrochemistry of folded graphene edges. *Nanoscale, 3*(5), 2256–2260.

Ambrosi, A., Chua, C. K., Bonanni, A., & Pumera, M. (2014). Electrochemistry of graphene and related materials. *Chemical Reviews, 114*(14), 7150–7188. doi:10.1021/cr500023c

Ambrosi, A., & Pumera, M. (2010). Nanographite impurities dominate electrochemistry of carbon nanotubes. *Chemistry—A European Journal, 16*(36), 10946–10949. https://doi.org/10.1002/chem.201001584

Arora, R. (2019). Adsorption of heavy metals–a review. *Materials Today: Proceedings, 18*, 4745–4750.

Ateh, D. D., Navsaria, H. A., & Vadgama, P. (2006). Polypyrrole-based conducting polymers and interactions with biological tissues. *Journal of the Royal Society Interface, 3*(11), 741–752. doi:10.1098/rsif.2006.0141

Baines, S. (2018). From 2D to 3D. *Science Museum Group Journal, 10*(10).

Bak, H., Cho, S. Y., Yun, Y. S., & Jin, H.-J. (2010). Electrically conductive transparent films based on nylon 6 membranes and single-walled carbon nanotubes. *Current Applied Physics, 10*(3, Suppl), S468–S472. https://doi.org/10.1016/j.cap.2010.02.034

Banhart, F., Kotakoski, J., & Krasheninnikov, A. V. (2011). Structural defects in graphene. *ACS Nano, 5*(1), 26–41. doi:10.1021/nn102598m

Bardeen, J. (1940). Electrical conductivity of metals. *Journal of Applied Physics*, *11*(2), 88–111.

Bauhofer, W., & Kovacs, J. Z. (2009). A review and analysis of electrical percolation in carbon nanotube polymer composites. *Composites Science and Technology*, *69*(10), 1486–1498. https://doi.org/10.1016/j.compscitech.2008.06.018

Bera, S. P., Godhaniya, M., & Kothari, C. (2022). Emerging and advanced membrane technology for wastewater treatment: A review. *Journal of Basic Microbiology*, *62*(3–4), 245–259. https://doi.org/10.1002/jobm.202100259

Berger, C., Poncharal, P., Yi, Y., & de Heer, W. (2003). Ballistic conduction in multiwalled carbon nanotubes. *Journal of Nanoscience and Nanotechnology*, *3*(1–2), 171–177.

Birch, M. E., Ruda-Eberenz, T. A., Chai, M., Andrews, R., & Hatfield, R. L. (2013). Properties that influence the specific surface areas of carbon nanotubes and nanofibers. *The Annals of Occupational Hygiene*, *57*(9), 1148–1166. doi:10.1093/annhyg/met042

Boeva, Z. A., & Lindfors, T. (2016). Few-layer graphene and polyaniline composite as ion-to-electron transducer in silicone rubber solid-contact ion-selective electrodes. *Sensors and Actuators B: Chemical*, *224*, 624–631. https://doi.org/10.1016/j.snb.2015.10.054

Brédas, J. (1987). Electronic and transport properties of highly conducting polymers. In *Chemical physics of intercalation* (pp. 253–269). Springer.

Brinker, C. J. (2013). Dip coating. In T. Schneller, R. Waser, M. Kosec, & D. Payne (Eds.), *Chemical solution deposition of functional oxide thin films* (pp. 233–261). Springer.

Britannica, T. & Editors of Encyclopaedia (2019). *Free-electron model of metals*. Encyclopedia Britannica. https://www.britannica.com/science/free-electron-model-of-metals

Brownson, D. A., Varey, S. A., Hussain, F., Haigh, S. J., & Banks, C. E. (2014). Electrochemical properties of CVD grown pristine graphene: Monolayer-vs. quasi-graphene. *Nanoscale*, *6*(3), 1607–1621.

Bube, R. H. (2003). Electrons in solids. In R. A. Meyers (Ed.), *Encyclopedia of physical science and technology* (3rd ed., pp. 307–329). Academic Press.

Cao, L., Su, D., Su, Z., & Chen, X. (2014). Fabrication of multiwalled carbon nanotube/polypropylene conductive fibrous membranes by melt electrospinning. *Industrial & Engineering Chemistry Research*, *53*(6), 2308–2317. doi:10.1021/ie403746p

Chadar, R., Afzal, O., Alqahtani, S. M., & Kesharwani, P. (2021). Carbon nanotubes as an emerging nanocarrier for the delivery of doxorubicin for improved chemotherapy. *Colloids and Surfaces B: Biointerfaces*, *208*, 112044.

Chang, H., Wang, G., Yang, A., Tao, X., Liu, X., Shen, Y., & Zheng, Z. (2010). A transparent, flexible, low-temperature, and solution-processible graphene composite electrode. *Advanced Functional Materials*, *20*(17), 2893–2902. https://doi.org/10.1002/adfm.201000900

Charoenpakdee, J., Suntijitrungruang, O., & Boonchui, S. (2020). Chirality effects on an electron transport in single-walled carbon nanotube. *Scientific Reports*, *10*(1), 18949. doi:10.1038/s41598-020-76047-9

Chen, L., Moon, J.-H., Ma, X., Zhang, L., Chen, Q., Chen, L., . . . Ci, L. (2018). High performance graphene oxide nanofiltration membrane prepared by electrospraying for wastewater purification. *Carbon*, *130*, 487–494. https://doi.org/10.1016/j.carbon.2018.01.062

Chen, S., Wang, G., Li, S., Li, X., Yu, H., & Quan, X. (2020). Porous carbon membrane with enhanced selectivity and antifouling capability for water treatment under electrochemical assistance. *Journal of Colloid and Interface Science*, *560*, 59–68.

Chen, Y., Lan, W., Wang, J., Zhu, R., Yang, Z., Ding, D., . . . Xie, E. (2016). Highly flexible, transparent, conductive and antibacterial films made of spin-coated silver nanowires and a protective ZnO layer. *Physica E: Low-dimensional Systems and Nanostructures*, *76*, 88–94. https://doi.org/10.1016/j.physe.2015.10.009

Chester, G. V., & Thellung, A. (1959). On the electrical conductivity of metals. *Proceedings of the Physical Society*, *73*(5), 745–766. doi:10.1088/0370-1328/73/5/308

Choi, H., Kim, H., Hwang, S., Choi, W., & Jeon, M. (2011). Dye-sensitized solar cells using graphene-based carbon nano composite as counter electrode. *Solar Energy Materials and Solar Cells, 95*(1), 323–325. https://doi.org/10.1016/j.solmat.2010.04.044

Chou, A., Böcking, T., Singh, N. K., & Gooding, J. J. (2005). Demonstration of the importance of oxygenated species at the ends of carbon nanotubes for their favourable electrochemical properties. *Chemical Communications* (7), 842–844.

Ci, L., Song, L., Jariwala, D., Elias, A. L., Gao, W., Terrones, M., & Ajayan, P. M. (2009). Graphene shape control by multistage cutting and transfer. *Advanced Materials, 21*(44), 4487–4491.

Çirmi, D., Karatekin, R. S., Aydın, R., & Köleli, F. (2022). Hydrogen evolution and CO2-reduction on a non-supported polypyrrole electrode. *Synthetic Metals, 289*, 117102.

Cui, Z., Zhou, C., Qiu, S., Chen, Z., Lin, J., Zhao, J., . . . Su, W. (2016). *Printed electronics: Materials, technologies and applications.* Wiley.

Dai, L. (2004). *Intelligent macromolecules for smart devices: From materials synthesis to device applications.* Springer Science & Business Media.

de Lannoy, C. F., Jassby, D., Davis, D. D., & Wiesner, M. R. (2012). A highly electrically conductive polymer–multiwalled carbon nanotube nanocomposite membrane. *Journal of Membrane Science, 415–416*, 718–724. https://doi.org/10.1016/j.memsci.2012.05.061

Dhand, C., Das, M., Datta, M., & Malhotra, B. D. (2011). Recent advances in polyaniline based biosensors. *Biosensors and Bioelectronics, 26*(6), 2811–2821. https://doi.org/10.1016/j.bios.2010.10.017

Dinh, L. N. M., Tran, B. N., Agarwal, V., & Zetterlund, P. B. (2022). Synthesis of highly stretchable and electrically conductive multiwalled carbon nanotube/polymer nanocomposite films. *ACS Applied Polymer Materials, 4*(3), 1867–1877. doi:10.1021/acsapm.1c01738

Dong, Y., Cheng, Y., Xu, G., Cheng, H., Huang, K., Duan, J., . . . Yao, H. (2019). Selectively enhanced ion transport in graphene oxide membrane/PET conical nanopore system. *ACS Applied Materials & Interfaces, 11*(16), 14960–14969. doi:10.1021/acsami.9b01071

Du, C., & Pan, N. (2006). Supercapacitors using carbon nanotubes films by electrophoretic deposition. *Journal of Power Sources, 160*(2), 1487–1494.

Dugdale, J. S. (1995). *Electrical properties of disordered metals.* Cambridge University Press.

Ebbesen, T. W., Lezec, H. J., Hiura, H., Bennett, J. W., Ghaemi, H. F., & Thio, T. (1996). Electrical conductivity of individual carbon nanotubes. *Nature, 382*(6586), 54–56. doi:10.1038/382054a0

Eswaraiah, V., Sankaranarayanan, V., & Ramaprabhu, S. (2011). Functionalized graphene–PVDF foam composites for EMI shielding. *Macromolecular Materials and Engineering, 296*(10), 894–898. https://doi.org/10.1002/mame.201100035

Falcao, E. H., & Wudl, F. (2007). Carbon allotropes: Beyond graphite and diamond. *Journal of Chemical Technology & Biotechnology, 82*(6), 524–531. https://doi.org/10.1002/jctb.1693

Fan, L.-Z., & Maier, J. (2006). High-performance polypyrrole electrode materials for redox supercapacitors. *Electrochemistry Communications, 8*(6), 937–940.

Francis, L., Ahmed, F. E., & Hilal, N. (2022). Electrospun membranes for membrane distillation: The state of play and recent advances. *Desalination, 526*, 115511. https://doi.org/10.1016/j.desal.2021.115511

Gao, G., Zhang, Q., & Vecitis, C. D. (2014). CNT–PVDF composite flow-through electrode for single-pass sequential reduction–oxidation. *Journal of Materials Chemistry A, 2*(17), 6185–6190.

Geng, P., & Chen, G. (2017). Antifouling ceramic membrane electrode modified by Magnéli Ti4O7 for electro-microfiltration of humic acid. *Separation and Purification Technology, 185*, 61–71.

Georgakilas, V., Perman, J. A., Tucek, J., & Zboril, R. (2015). Broad family of carbon nano allotropes: Classification, chemistry, and applications of fullerenes, carbon dots, nanotubes, graphene, nanodiamonds, and combined superstructures. *Chemical Reviews, 115*(11), 4744–4822.

Gkourmpis, T., Gaska, K., Tranchida, D., Gitsas, A., Müller, C., Matic, A., & Kádár, R. (2019). Melt-mixed 3D hierarchical graphene/polypropylene nanocomposites with low electrical percolation threshold. *Nanomaterials*, *9*(12), 1766.

Goh, M. S., & Pumera, M. (2010). The electrochemical response of graphene sheets is independent of the number of layers from a single graphene sheet to multilayer stacked graphene platelets. *Chemistry—An Asian Journal*, *5*(11), 2355–2357. https://doi.org/10.1002/asia.201000437

Grady, B. P. (2011). *Carbon nanotube-polymer composites: Manufacture, properties, and applications*. Wiley.

Grobert, N. (2006). Nanotubes—grow or go? *Materials Today*, *9*(10), 64. https://doi.org/10.1016/S1369-7021(06)71680-7

Grunlan, J. C., Gerberich, W. W., & Francis, L. F. (2001). Lowering the percolation threshold of conductive composites using particulate polymer microstructure. *Journal of Applied Polymer Science*, *80*(4), 692–705. https://doi.org/10.1002/1097-4628(20010425)80:4<692::AID-APP1146>3.0.CO;2-W

Güell, A. G., Ebejer, N., Snowden, M. E., Macpherson, J. V., & Unwin, P. R. (2012). Structural correlations in heterogeneous electron transfer at monolayer and multilayer graphene electrodes. *Journal of the American Chemical Society*, *134*(17), 7258–7261. doi:10.1021/ja3014902

Gupta, K. M., and Gupta, N. (Eds.) (2015). Conductive materials: Electron theories, properties and behaviour. In *Advanced electrical and electronics materials*. Scrivener Publishing LLC. https://doi.org/10.1002/9781118998564.ch4

Ha, D.-H., Islam, M. A., & Robinson, R. D. (2012). Binder-free and carbon-free nanoparticle batteries: A method for nanoparticle electrodes without polymeric binders or carbon black. *Nano Letters*, *12*(10), 5122–5130. doi:10.1021/nl3019559

Hall, E. H. (1928). Sommerfeld's electron-theory of metals. *Proceedings of the National Academy of Sciences of the United States of America*, *14*(5), 370–377.

Hatchett, D. W., & Josowicz, M. (2008). Composites of intrinsically conducting polymers as sensing nanomaterials. *Chemical Reviews*, *108*(2), 746–769. doi:10.1021/cr068112h

He, L., & Tjong, S. C. (2013). Low percolation threshold of graphene/polymer composites prepared by solvothermal reduction of graphene oxide in the polymer solution. *Nanoscale Research Letters*, *8*(1), 132. doi:10.1186/1556-276X-8-132

Himma, N. F., Wardani, A. K., & Wenten, I. G. (2017). Preparation of superhydrophobic polypropylene membrane using dip-coating method: The effects of solution and process parameters. *Polymer-Plastics Technology and Engineering*, *56*(2), 184–194. doi:10.1080/03602559.2016.1185666

Holban, A. M., Grumezescu, A. M., & Andronescu, E. (2016). Chapter 10—inorganic nanoarchitectonics designed for drug delivery and anti-infective surfaces. In A. M. Grumezescu (Ed.), *Surface chemistry of nanobiomaterials* (pp. 301–327). William Andrew Publishing.

Holloway, A. F., Toghill, K., Wildgoose, G. G., Compton, R. G., Ward, M. A. H., Tobias, G., . . . Crossley, A. (2008). Electrochemical opening of single-walled carbon nanotubes filled with metal halides and with closed ends. *The Journal of Physical Chemistry C*, *112*(28), 10389–10397. doi:10.1021/jp802127p

Hou, J., Bao, C., Qu, S., Hu, X., Nair, S., & Chen, Y. (2018). Graphene oxide membranes for ion separation: Detailed studies on the effects of fabricating conditions. *Applied Surface Science*, *459*, 185–193. https://doi.org/10.1016/j.apsusc.2018.07.207

Hu, C., Yang, C., & Hu, S. (2007). Hydrophobic adsorption of surfactants on water-soluble carbon nanotubes: A simple approach to improve sensitivity and antifouling capacity of carbon nanotubes-based electrochemical sensors. *Electrochemistry Communications*, *9*(1), 128–134. doi:10.1016/j.elecom.2006.08.055

Hu, X., Yu, Y., Lin, N., Ren, S., & Zhang, X. (2018). Graphene oxide/Al_2O_3 membrane with efficient salt rejection for water purification. *Water Science & Technology, 18*(6), 2162–2169. https://doi.org/10.2166/ws.2018.037

Huang, H., Shi, H., Das, P., Qin, J., Li, Y., Wang, X., . . . Cheng, H.-M. (2020). The chemistry and promising applications of graphene and porous graphene materials. *Advanced Functional Materials, 30*(41), 1909035. https://doi.org/10.1002/adfm.201909035

Huang, S., Mansouri, J., Le-Clech, P., Leslie, G., Tang, C. Y., & Fane, A. G. (2022). A comprehensive review of electrospray technique for membrane development: Current status, challenges, and opportunities. *Journal of Membrane Science, 646*, 120248. https://doi.org/10.1016/j.memsci.2021.120248

Huang, X., Yu, Y.-H., de Llergo, O. L., Marquez, S. M., & Cheng, Z. (2017). Facile polypyrrole thin film coating on polypropylene membrane for efficient solar-driven interfacial water evaporation. *RSC Advances, 7*(16), 9495–9499.

Huang, Y., Li, H., Wang, L., Qiao, Y., Tang, C., Jung, C., . . . Yu, M. (2015). Ultrafiltration membranes with structure-optimized graphene-oxide coatings for antifouling oil/water separation. *Advanced Materials Interfaces, 2*(2), 1400433. https://doi.org/10.1002/admi.201400433

Huang, Y., Li, H., Wang, Z., Zhu, M., Pei, Z., Xue, Q., . . . Zhi, C. (2016). Nanostructured polypyrrole as a flexible electrode material of supercapacitor. *Nano Energy, 22*, 422–438. https://doi.org/10.1016/j.nanoen.2016.02.047

Hui, H., Wang, H., Mo, Y., Yin, Z., & Li, J. (2019). Optimal design and evaluation of electrocatalytic reactors with nano-MnOx/Ti membrane electrode for wastewater treatment. *Chemical Engineering Journal, 376*, 120190.

Igbinigun, E., Fennell, Y., Malaisamy, R., Jones, K. L., & Morris, V. (2016). Graphene oxide functionalized polyethersulfone membrane to reduce organic fouling. *Journal of Membrane Science, 514*, 518–526. https://doi.org/10.1016/j.memsci.2016.05.024

Iijima, S. (1991). Helical microtubules of graphitic carbon. *Nature, 354*(6348), 56–58. doi:10.1038/354056a0

Iijima, S., & Ichihashi, T. (1993). Single-shell carbon nanotubes of 1-nm diameter. *Nature, 363*(6430), 603–605. doi:10.1038/363603a0

Imran, K. A., Lou, J., & Shivakumar, K. N. (2018). Enhancement of electrical and thermal conductivity of polypropylene by graphene nanoplatelets. *Journal of Applied Polymer Science, 135*(9), 45833. https://doi.org/10.1002/app.45833

Jariwala, D., Sangwan, V. K., Lauhon, L. J., Marks, T. J., & Hersam, M. C. (2013). Carbon nanomaterials for electronics, optoelectronics, photovoltaics, and sensing. *Chemical Society Reviews, 42*(7), 2824–2860.

Jaworek, A., & Sobczyk, A. T. (2008). Electrospraying route to nanotechnology: An overview. *Journal of Electrostatics, 66*(3), 197–219. https://doi.org/10.1016/j.elstat.2007.10.001

Jia, J., Sun, X., Lin, X., Shen, X., Mai, Y.-W., & Kim, J.-K. (2014). Exceptional electrical conductivity and fracture resistance of 3D interconnected graphene foam/epoxy composites. *ACS Nano, 8*(6), 5774–5783. doi:10.1021/nn500590g

Jo, J. W., Jung, J. W., Lee, J. U., & Jo, W. H. (2010). Fabrication of highly conductive and transparent thin films from single-walled carbon nanotubes using a new non-ionic surfactant via spin coating. *ACS Nano, 4*(9), 5382–5388. doi:10.1021/nn1009837

Jones, H., Zener, C., & Lennard-Jones, J. E. (1934). The general proof of certain fundamental equations in the theory of metallic conduction. *Proceedings of the Royal Society of London. Series A, Containing Papers of a Mathematical and Physical Character, 144*(851), 101–117. doi:10.1098/rspa.1934.0036

Jorio, A., & Dresselhaus, G. (2008). *Carbon nanotubes: Advanced topics in the synthesis, structure, properties and applications* (Vol. 111). Springer.

Joshi, R. K., Carbone, P., Wang, F. C., Kravets, V. G., Su, Y., Grigorieva, I. V., . . . Nair, R. R. (2014). Precise and ultrafast molecular sieving through graphene oxide membranes. *Science, 343*(6172), 752–754.

Kai, D., Prabhakaran, M. P., Jin, G., & Ramakrishna, S. (2011). Polypyrrole-contained electrospun conductive nanofibrous membranes for cardiac tissue engineering. *Journal
of Biomedical Materials Research Part A*, *99A*(3), 376–385. https://doi.org/10.1002/
jbm.a.33200

Kaiser, B., Park, Y. W., Kim, G. T., Choi, E. S., Düsberg, G., & Roth, S. (1999). Electronic
transport in carbon nanotube ropes and mats. *Synthetic Metals*, *103*(1), 2547–2550.
https://doi.org/10.1016/S0379-6779(98)00222-7

Kaliyaraj Selva Kumar, A., Zhang, Y., Li, D., & Compton, R. G. (2020). A mini-review: How
reliable is the drop casting technique? *Electrochemistry Communications*, *121*, 106867.
https://doi.org/10.1016/j.elecom.2020.106867

Kamath, S. V., Aruchamy, K., & Sanna Kotrappanavar, N. (2021). 2—Conjugated polymer-based
smart composites for optoelectronics and energy applications. In N. K. Subramani,
H. Siddaramaiah, & J. H. Lee (Eds.), *Polymer-based advanced functional composites for
optoelectronic and energy applications* (pp. 31–49). Elsevier.

Kang, M., & Jin, H.-J. (2007). Electrically conducting electrospun silk membranes fabricated
by adsorption of carbon nanotubes. *Colloid and Polymer Science*, *285*(10), 1163–1167.
doi:10.1007/s00396-007-1668-y

Kazemi, F., Mohammadpour, Z., Naghib, S. M., Zare, Y., & Rhee, K. Y. (2021). Percolation
onset and electrical conductivity for a multiphase system containing carbon nanotubes
and nanoclay. *Journal of Materials Research and Technology*, *15*, 1777–1788. https://
doi.org/10.1016/j.jmrt.2021.08.131

Khan, Y., & Ahmad, A. (2020). A mass approach towards carbon nanotubes in the field of
nanotechnology. In *Intellectual property issues in nanotechnology* (pp. 359–374). CRC
Press.

Kim, H. S., Abdala, A. A., & Macosko, C. W. (2010). Graphene/polymer nanocomposites.
Macromolecules, *43*(16), 6515–6530. doi:10.1021/ma100572e

Kim, H. S., Jin, H.-J., Myung, S. J., Kang, M., & Chin, I.-J. (2006). Carbon nanotube-adsorbed
electrospun nanofibrous membranes of nylon 6. *Macromolecular Rapid Communications*, *27*(2), 146–151. https://doi.org/10.1002/marc.200500617

Kim, J. H., Nam, K.-W., Ma, S. B., & Kim, K. B. (2006). Fabrication and electrochemical
properties of carbon nanotube film electrodes. *Carbon*, *44*(10), 1963–1968. https://doi.
org/10.1016/j.carbon.2006.02.002

King, J. A., Klimek, D. R., Miskioglu, I., & Odegard, G. M. (2013). Mechanical properties of
graphene nanoplatelet/epoxy composites. *Journal of Applied Polymer Science*, *128*(6),
4217–4223. https://doi.org/10.1002/app.38645

Kislenko, V. A., Pavlov, S. V., & Kislenko, S. A. (2020). Influence of defects in graphene on
electron transfer kinetics: The role of the surface electronic structure. *Electrochimica
Acta*, *341*, 136011. https://doi.org/10.1016/j.electacta.2020.136011

Klimm, D. (2014). Electronic materials with a wide band gap: Recent developments. *IUCrJ*,
1(Pt 5), 281–290. doi:10.1107/S2052252514017229

Kotov, N. A. (2006). Carbon sheet solutions. *Nature*, *442*(7100), 254–255. doi:10.1038/442254a

Kuilla, T., Bhadra, S., Yao, D., Kim, N. H., Bose, S., & Lee, J. H. (2010). Recent advances in
graphene based polymer composites. *Progress in Polymer Science*, *35*(11), 1350–1375.
https://doi.org/10.1016/j.progpolymsci.2010.07.005

Kumar, D., & Sharma, R. C. (1998). Advances in conductive polymers. *European polymer
journal*, *34*(8), 1053–1060. https://doi.org/10.1016/S0014-3057(97)00204-8

Lalegül-Ülker, Ö., Elçin, A. E., & Elçin, Y. M. (2018). Intrinsically conductive polymer nanocomposites for cellular applications. In H. J. Chun, C. H. Park, I. K. Kwon, & G. Khang (Eds.),
Cutting-edge enabling technologies for regenerative medicine (pp. 135–153). Springer.

Langer, L., Bayot, V., Grivei, E., Issi, J. P., Heremans, J. P., Olk, C. H., . . . Bruynseraede, Y.
(1996). Quantum transport in a multiwalled carbon nanotube. *Physical Review Letters*,
76(3), 479–482. doi:10.1103/PhysRevLett.76.479

Le, T.-H., Kim, Y., & Yoon, H. (2017). Electrical and electrochemical properties of conducting polymers. *Polymers*, 9(4), 150.

Li, B., Sun, D., Li, B., Tang, W., Ren, P., Yu, J., & Zhang, J. (2020). One-step electrochemically prepared graphene/polyaniline conductive filter membrane for permeation enhancement by fouling mitigation. *Langmuir*, 36(9), 2209–2222.

Li, M., Gao, C., Hu, H., & Zhao, Z. (2013). Electrical conductivity of thermally reduced graphene oxide/polymer composites with a segregated structure. *Carbon*, 65, 371–373. https://doi.org/10.1016/j.carbon.2013.08.016

Li, X., Liu, L., Liu, T., Zhang, D., An, C., & Yang, F. (2020). An active electro-Fenton PVDF/SS/PPy cathode membrane can remove contaminant by filtration and mitigate fouling by pairing with sacrificial iron anode. *Journal of Membrane Science*, 605, 118100.

Li, X., Shao, S., Yang, Y., Mei, Y., Qing, W., Guo, H., . . . Tang, C. Y. (2020). Engineering interface with a one-dimensional RuO2/TiO2 heteronanostructure in an electrocatalytic membrane electrode: Toward highly efficient micropollutant decomposition. *ACS Applied Materials & Interfaces*, 12(19), 21596–21604.

Li, Y., Liu, L., Yang, F., & Ren, N. (2015). Performance of carbon fiber cathode membrane with C–Mn–Fe–O catalyst in MBR–MFC for wastewater treatment. *Journal of Membrane Science*, 484, 27–34.

Li, Y., Yu, H., Zhang, Y., Zhou, N., & Tan, Z. (2022). Kinetics and characterization of preparing conductive nanofibrous membrane by In-situ polymerization of polypyrrole on electrospun nanofibers. *Chemical Engineering Journal*, 433, 133531. https://doi.org/10.1016/j.cej.2021.133531

Li, Z., & Gong, L. (2020). Research progress on applications of polyaniline (PANI) for electrochemical energy storage and conversion. *Materials*, 13(3), 548.

Liang, B., Zhan, W., Qi, G., Lin, S., Nan, Q., Liu, Y., . . . Pan, K. (2015). High performance graphene oxide/polyacrylonitrile composite pervaporation membranes for desalination applications. *Journal of Materials Chemistry A*, 3(9), 5140–5147. doi:10.1039/C4TA06573E

Liang, X., Chang, A. S., Zhang, Y., Harteneck, B. D., Choo, H., Olynick, D. L., & Cabrini, S. (2009). Electrostatic force assisted exfoliation of prepatterned few-layer graphenes into device sites. *Nano Letters*, 9(1), 467–472.

Liu, G., Ye, H., Li, A., Zhu, C., Jiang, H., Liu, Y., . . . Zhou, Y. (2016). Graphene oxide for high-efficiency separation membranes: Role of electrostatic interactions. *Carbon*, 110, 56–61. https://doi.org/10.1016/j.carbon.2016.09.005

Liu, J., Tian, C., Xiong, J., & Wang, L. (2017). Polypyrrole blending modification for PVDF conductive membrane preparing and fouling mitigation. *Journal of Colloid and Interface Science*, 494, 124–129.

Liu, K., Sun, Y., Lin, X., Zhou, R., Wang, J., Fan, S., & Jiang, K. (2010). Scratch-resistant, highly conductive, and high-strength carbon nanotube-based composite yarns. *ACS Nano*, 4(10), 5827–5834. doi:10.1021/nn1017318

Liu, Q., Li, M., Wang, Z., Gu, Y., Li, Y., & Zhang, Z. (2013). Improvement on the tensile performance of buckypaper using a novel dispersant and functionalized carbon nanotubes. *Composites Part A: Applied Science and Manufacturing*, 55, 102–109. https://doi.org/10.1016/j.compositesa.2013.08.011

Liu, Q., Tu, J., Wang, X., Yu, W., Zheng, W., & Zhao, Z. (2012). Electrical conductivity of carbon nanotube/poly(vinylidene fluoride) composites prepared by high-speed mechanical mixing. *Carbon*, 50(1), 339–341. https://doi.org/10.1016/j.carbon.2011.08.051

Liu, S., Wang, Z., Yu, C., Zhao, Z., Fan, X., Ling, Z., & Qiu, J. (2013). Free-standing, hierarchically porous carbon nanotube film as a binder-free electrode for high-energy Li–O2 batteries. *Journal of Materials Chemistry A*, 1(39), 12033–12037. doi:10.1039/C3TA13069J

Liu, X., Gilmore, K. J., Moulton, S. E., & Wallace, G. G. (2009). Electrical stimulation promotes nerve cell differentiation on polypyrrole/poly (2-methoxy-5 aniline sulfonic acid)
composites. *Journal of Neural Engineering*, *6*(6), 065002.

Lovell, M. C., Avery, A. J., & Vernon, M. W. (1976). Electrical properties. In M. C. Lovell,
A. J. Avery, & M. W. Vernon (Eds.), *Physical properties of materials* (pp. 124–152).
Springer Netherlands.

Luo, C., Zuo, X., Wang, L., Wang, E., Song, S., Wang, J., . . . Cao, Y. (2008). Flexible carbon
nanotube–polymer composite films with high conductivity and superhydrophobicity
made by solution process. *Nano Letters*, *8*(12), 4454–4458. doi:10.1021/nl802411d

Luo, X., Yang, G., & Schubert, D. W. (2022). Electrically conductive polymer composite
containing hybrid graphene nanoplatelets and carbon nanotubes: Synergistic effect and
tunable conductivity anisotropy. *Advanced Composites and Hybrid Materials*, *5*(1),
250–262. doi:10.1007/s42114-021-00332-y

Lv, X., Tang, Y., Tian, Q., Wang, Y., & Ding, T. (2020). Ultra-stretchable membrane with high
electrical and thermal conductivity via electrospinning and in-situ nanosilver deposition.
Composites Science and Technology, *200*, 108414.

Ma, J., Meng, Q., Michelmore, A., Kawashima, N., Izzuddin, Z., Bengtsson, C., & Kuan,
H.-C. (2013). Covalently bonded interfaces for polymer/graphene composites. *Journal
of Materials Chemistry A*, *1*(13), 4255–4264.

Ma, P.-C., Siddiqui, N. A., Marom, G., & Kim, J.-K. (2010). Dispersion and functionalization of carbon nanotubes for polymer-based nanocomposites: A review. *Composites Part
A: Applied Science and Manufacturing*, *41*(10), 1345–1367. https://doi.org/10.1016/j.
compositesa.2010.07.003

Ma, T., Ren, W., Zhang, X., Liu, Z., Gao, Y., Yin, L.-C., . . . Cheng, H.-M. (2013). Edge-
controlled growth and kinetics of single-crystal graphene domains by chemical vapor
deposition. *Proceedings of the National Academy of Sciences*, *110*(51), 20386–20391.
doi:10.1073/pnas.1312802110

Madaria, A. R., Kumar, A., Ishikawa, F. N., & Zhou, C. (2010). Uniform, highly conductive, and patterned transparent films of a percolating silver nanowire network on rigid
and flexible substrates using a dry transfer technique. *Nano Research*, *3*(8), 564–573.
doi:10.1007/s12274-010-0017-5

Mahmud, H. N. M. E., Huq, A. O., & Binti Yahya, R. (2016). The removal of heavy metal ions
from wastewater/aqueous solution using polypyrrole-based adsorbents: A review. *RSC
Advances*, *6*(18), 14778–14791.

Mano, N. (2020). Recent advances in high surface area electrodes for bioelectrochemical
applications. *Current Opinion in Electrochemistry*, *19*, 8–13. https://doi.org/10.1016/j.
coelec.2019.09.003

Martínez-Huitle, C. A., Rodrigo, M. A., Sirés, I., & Scialdone, O. (2015). Single and coupled
electrochemical processes and reactors for the abatement of organic water pollutants:
A critical review. *Chemical Reviews*, *115*(24), 13362–13407.

Maruyama, T. (2021). Chapter 6—carbon nanotubes. In S. Thomas, C. Sarathchandran, S. A.
Ilangovan, & J. C. Moreno-Piraján (Eds.), *Handbook of carbon-based nanomaterials*
(pp. 299–319). Elsevier.

McCreery, R. L. (2008). Advanced carbon electrode materials for molecular electrochemistry.
Chemical Reviews, *108*(7), 2646–2687. doi:10.1021/cr068076m

Meng, Q., & Manas-Zloczower, I. (2015). Carbon nanotubes enhanced cellulose nanocrystals
films with tailorable electrical conductivity. *Composites Science and Technology*, *120*,
1–8.

Merkoçi, A., Pumera, M., Llopis, X., Pérez, B., del Valle, M., & Alegret, S. (2005). New
materials for electrochemical sensing VI: Carbon nanotubes. *TrAC Trends in Analytical
Chemistry*, *24*(9), 826–838. https://doi.org/10.1016/j.trac.2005.03.019

Merlini, C., Barra, G. M. D. O., Ramôa, S. D. A. D. S., Contri, G., Almeida, R. D. S., d'Ávila, M. A., & Soares, B. G. (2015). Electrically conductive polyaniline-coated electrospun poly(vinylidene fluoride) mats. *Frontiers in Materials, 2.* doi:10.3389/fmats.2015.00014

Min, C., Shen, X., Shi, Z., Chen, L., & Xu, Z. (2010). The electrical properties and conducting mechanisms of carbon nanotube/polymer nanocomposites: A review. *Polymer-Plastics Technology and Engineering, 49*(12), 1172–1181. doi:10.1080/03602559.2010.496405

Mizutani, U. (2001). *Introduction to the electron theory of metals.* Cambridge University Press.

Mohammed, S., Hegab, H., & Ou, R. (2022). Nanofiltration performance of glutaraldehyde crosslinked graphene oxide-cellulose nanofiber membrane. *Chemical Engineering Research and Design, 183,* 1–12. https://doi.org/10.1016/j.cherd.2022.04.039

Moniruzzaman, M., & Winey, K. I. (2006). Polymer nanocomposites containing carbon nanotubes. *Macromolecules, 39*(16), 5194–5205. doi:10.1021/ma060733p

Neubert, S., Pliszka, D., Thavasi, V., Wintermantel, E., & Ramakrishna, S. (2011). Conductive electrospun PANi-PEO/TiO2 fibrous membrane for photo catalysis. *Materials Science and Engineering: B, 176*(8), 640–646. https://doi.org/10.1016/j.mseb.2011.02.007

Nogales, A., Broza, G., Roslaniec, Z., Schulte, K., Šics, I., Hsiao, B. S., . . . Ezquerra, T. A. (2004). Low percolation threshold in nanocomposites based on oxidized single wall carbon nanotubes and poly(butylene terephthalate). *Macromolecules, 37*(20), 7669–7672. doi:10.1021/ma049440r

Noh, S., An, H., & Song, Y. (2021). Vacuum-filtration fabrication for diverse conductive transparent cellulose electronic devices. *Cellulose, 28*(5), 3081–3096.

Norizan, M. N., Moklis, M. H., Demon, S. Z. N., Halim, N. A., Samsuri, A., Mohamad, I. S., . . . Abdullah, N. (2020). Carbon nanotubes: Functionalisation and their application in chemical sensors. *RSC Advances, 10*(71), 43704–43732.

Omer, S., Forgách, L., Zelkó, R., & Sebe, I. (2021). Scale-up of electrospinning: Market overview of products and devices for pharmaceutical and biomedical purposes. *Pharmaceutics, 13*(2), 286. doi:10.3390/pharmaceutics13020286

Pang, H., Chen, T., Zhang, G., Zeng, B., & Li, Z.-M. (2010). An electrically conducting polymer/graphene composite with a very low percolation threshold. *Materials Letters, 64*(20), 2226–2229. https://doi.org/10.1016/j.matlet.2010.07.001

Park, H. G., & Jung, Y. (2014). Carbon nanofluidics of rapid water transport for energy applications. *Chemical Society Reviews, 43*(2), 565–576.

Park, S. B., Lee, M. S., & Park, M. (2014). Study on lowering the percolation threshold of carbon nanotube-filled conductive polypropylene composites. *Carbon Letters, 15*(2), 117–124.

Pavlov, S. V., Nazmutdinov, R. R., Fedorov, M. V., & Kislenko, S. A. (2019). Role of graphene edges in the electron transfer kinetics: Insight from theory and molecular modeling. *The Journal of Physical Chemistry C, 123*(11), 6627–6634. doi:10.1021/acs.jpcc.8b12531

Perdigão, P., Morais Faustino, B. M., Faria, J., Canejo, J. P., Borges, J. P., Ferreira, I., & Baptista, A. C. (2020). Conductive electrospun polyaniline/polyvinylpyrrolidone nanofibers: Electrical and morphological characterization of new yarns for electronic textiles. *Fibers, 8*(4), 24.

Peres, N. M. R., Rodrigues, J. N. B., Stauber, T., & Lopes dos Santos, J. M. B. (2009). Dirac electrons in graphene-based quantum wires and quantum dots. *Journal of Physics: Condensed Matter, 21*(34), 344202. doi:10.1088/0953-8984/21/34/344202

Persano, L., Camposeo, A., Tekmen, C., & Pisignano, D. (2013). Industrial upscaling of electrospinning and applications of polymer nanofibers: A review. *Macromolecular Materials and Engineering, 298*(5), 504–520. https://doi.org/10.1002/mame.201200290

Poncharal, P., Berger, C., Yi, Y., Wang, Z. L., & de Heer, W. A. (2002). Room temperature ballistic conduction in carbon nanotubes. *The Journal of Physical Chemistry B, 106*(47), 12104–12118. doi:10.1021/jp021271u

Prehn, K., Nunes, S. P., & Schulte, K. (2011). Application of carbon nanotube/polymer composites as electrode for polyelectrolyte membrane fuel cells. *MRS Proceedings, 885*, 0885-A0803–0805. doi:10.1557/PROC-0885-A03-05

Pumera, M. (2007). Electrochemical properties of double wall carbon nanotube electrodes. *Nanoscale Research Letters, 2*(2), 87–93.

Pumera, M. (2009). The electrochemistry of carbon nanotubes: Fundamentals and applications. *Chemistry—A European Journal, 15*(20), 4970–4978. https://doi.org/10.1002/chem.200900421

Pumera, M., & Iwai, H. (2009). Multicomponent metallic impurities and their influence upon the electrochemistry of carbon nanotubes. *The Journal of Physical Chemistry C, 113*(11), 4401–4405. doi:10.1021/jp900069e

Pumera, M., & Miyahara, Y. (2009). What amount of metallic impurities in carbon nanotubes is small enough not to dominate their redox properties? *Nanoscale, 1*(2), 260–265.

Qian, B., Wang, G., Ling, Z., Dong, Q., Wu, T., Zhang, X., & Qiu, J. (2015). Sulfonated graphene as cation-selective coating: A new strategy for high-performance membrane capacitive deionization. *Advanced Materials Interfaces, 2*(16), 1500372. https://doi.org/10.1002/admi.201500372

Quinn, J. J., & Yi, K.-S. (2018). Free electron theory of metals. In J. J. Quinn & K.-S. Yi (Eds.), *Solid state physics: Principles and modern applications* (pp. 83–112). Springer International Publishing.

Randviir, E. P., Brownson, D. A. C., & Banks, C. E. (2014). A decade of graphene research: Production, applications and outlook. *Materials Today, 17*(9), 426–432. https://doi.org/10.1016/j.mattod.2014.06.001

Rashid, M. H.-O., Pham, S. Q. T., Sweetman, L. J., Alcock, L. J., Wise, A., Nghiem, L. D., . . . Ralph, S. F. (2014). Synthesis, properties, water and solute permeability of MWNT buckypapers. *Journal of Membrane Science, 456*, 175–184. https://doi.org/10.1016/j.memsci.2014.01.026

Ravichandran, R., Sundarrajan, S., Venugopal, J. R., Mukherjee, S., & Ramakrishna, S. (2010). Applications of conducting polymers and their issues in biomedical engineering. *Journal of the Royal Society Interface*, rsif20100120.

Roso, M., Lorenzetti, A., Boaretti, C., & Modesti, M. (2016). Electrically conductive membranes obtained by simultaneous electrospinning and electrospraying processes. *Journal of Nanomaterials, 2016*, 8362535. doi:10.1155/2016/8362535

Roy, S. B. (2019). *Electrons in crystalline solids Mott insulators* (pp. 1-1–1-62). IOP Publishing. https://dx.doi.org/10.1088/2053-2563/ab16c9ch1

Rudge, A., Raistrick, I., Gottesfeld, S., & Ferraris, J. P. (1994). A study of the electrochemical properties of conducting polymers for application in electrochemical capacitors. *Electrochimica Acta, 39*(2), 273–287. https://doi.org/10.1016/0013-4686(94)80063-4

Sahu, N., Parija, B., & Panigrahi, S. (2009). Fundamental understanding and modeling of spin coating process: A review. *Indian Journal of Physics, 83*(4), 493–502.

Sakurai, S., Kamada, F., Futaba, D. N., Yumura, M., & Hata, K. (2013). Influence of lengths of millimeter-scale single-walled carbon nanotube on electrical and mechanical properties of buckypaper. *Nanoscale Research Letters, 8*(1), 546. doi:10.1186/1556-276X-8-546

Schlicke, H., Schröder, J. H., Trebbin, M., Petrov, A., Ijeh, M., Weller, H., & Vossmeyer, T. (2011). Freestanding films of crosslinked gold nanoparticles prepared via layer-by-layer spin-coating. *Nanotechnology, 22*(30), 305303.

Schmidt, R. H., Kinloch, I. A., Burgess, A. N., & Windle, A. H. (2007). The effect of aggregation on the electrical conductivity of spin-coated polymer/carbon nanotube composite films. *Langmuir, 23*(10), 5707–5712. doi:10.1021/la062794m

Schoetz, T., Kurniawan, M., Stich, M., Peipmann, R., Efimov, I., Ispas, A., . . . Ueda, M. (2018). Understanding the charge storage mechanism of conductive polymers as hybrid battery-capacitor materials in ionic liquids by in situ atomic force microscopy and electrochemical quartz crystal microbalance studies. *Journal of Materials Chemistry A, 6*(36), 17787–17799.

Schubert, D. W., & Dunkel, T. (2003). Spin coating from a molecular point of view: Its concentration regimes, influence of molar mass and distribution. *Materials Research Innovations*, 7(5), 314–321.

Schubin, S., Wonsowsky, S., & Fowler, R. H. (1934). On the electron theory of metals. *Proceedings of the Royal Society of London. Series A, Containing Papers of a Mathematical and Physical Character*, 145(854), 159–180. doi:10.1098/rspa.1934.0089

Scriven, L. E. (2011). Physics and applications of DIP coating and spin coating. *MRS Proceedings*, 121, 717. doi:10.1557/PROC-121-717

Seshadri, R., Aiyer, H. N., Govindaraj, A., & Rao, C. N. R. (1994). Electron transport properties of carbon nanotubes. *Solid State Communications*, 91(3), 195–199. https://doi.org/10.1016/0038-1098(94)90222-4

Sgobba, V., & Guldi, D. M. (2009). Carbon nanotubes—electronic/electrochemical properties and application for nanoelectronics and photonics. *Chemical Society Reviews*, 38(1), 165–184.

Shante, V. K. S., & Kirkpatrick, S. (1971). An introduction to percolation theory. *Advances in Physics*, 20(85), 325–357. doi:10.1080/00018737100101261

Shao, F., Dong, L., Dong, H., Zhang, Q., Zhao, M., Yu, L., . . . Chen, Y. (2017). Graphene oxide modified polyamide reverse osmosis membranes with enhanced chlorine resistance. *Journal of Membrane Science*, 525, 9–17. https://doi.org/10.1016/j.memsci.2016.12.001

Shariatinia, Z. (2021). Chapter 14—perovskite solar cells as modern nano tools and devices in solar power energy. In S. Devasahayam & C. M. Hussain (Eds.), *Nano tools and devices for enhanced renewable energy* (pp. 377–427). Elsevier.

Shinozuka, Y. (2021). *Electron-lattice interactions in semiconductors*. Jenny Stanford Publishing.

Singleton, J. (2001). *Band theory and electronic properties of solids*. Oxford University Press.

Song, C., Meng, X., Chen, H., Liu, Z., Zhan, Q., Sun, Y., . . . Dai, Y. (2021). Flexible, graphene-based films with three-dimensional conductive network via simple drop-casting toward electromagnetic interference shielding. *Composites Communications*, 24, 100632. https://doi.org/10.1016/j.coco.2021.100632

Song, J. J., Huang, Y., Nam, S.-W., Yu, M., Heo, J., Her, N., . . . Yoon, Y. (2015). Ultrathin graphene oxide membranes for the removal of humic acid. *Separation and Purification Technology*, 144, 162–167. https://doi.org/10.1016/j.seppur.2015.02.032

Song, Y., Cheng, X., Chen, H., Han, M., Chen, X., Huang, J., . . . Zhang, H. (2016). Highly compression-tolerant folded carbon nanotube/paper as solid-state supercapacitor electrode. *Micro & Nano Letters*, 11(10), 586–590.

Spitalsky, Z., Tasis, D., Papagelis, K., & Galiotis, C. (2010). Carbon nanotube–polymer composites: Chemistry, processing, mechanical and electrical properties. *Progress in Polymer Science*, 35(3), 357–401. https://doi.org/10.1016/j.progpolymsci.2009.09.003

Street, G. B., & Clarke, T. C. (1981). Conducting polymers: A review of recent work. *IBM Journal of Research and Development*, 25(1), 51–57. doi:10.1147/rd.251.0051

Sun, M., Wang, X., Winter, L. R., Zhao, Y., Ma, W., Hedtke, T., . . . Elimelech, M. (2021). Electrified membranes for water treatment applications. *ACS ES&T Engineering*, 1(4), 725–752. doi:10.1021/acsestengg.1c00015

Sun, T., Luo, W., Luo, Y., Wang, Y., Zhou, S., Liang, M., . . . Zou, H. (2020). Self-reinforced polypropylene/graphene composite with segregated structures to achieve balanced electrical and mechanical properties. *Industrial & Engineering Chemistry Research*, 59(24), 11206–11218. doi:10.1021/acs.iecr.0c00825

Surana, K., Singh, P. K., Bhattacharya, B., Verma, C. S., & Mehra, R. M. (2015). Synthesis of graphene oxide coated Nafion membrane for actuator application. *Ceramics International*, 41(3, Part B), 5093–5099. https://doi.org/10.1016/j.ceramint.2014.12.080

Syed, J. A., Lu, H., Tang, S., & Meng, X. (2015). Enhanced corrosion protective PANI-PAA/PEI multilayer composite coatings for 316SS by spin coating technique. *Applied Surface Science*, 325, 160–169. https://doi.org/10.1016/j.apsusc.2014.11.021

Szumska, A. A., Maria, I. P., Flagg, L. Q., Savva, A., Surgailis, J., Paulsen, B. D., . . . Nelson, J. (2021). Reversible electrochemical charging of n-type conjugated polymer electrodes in aqueous electrolytes. *Journal of the American Chemical Society*, *143*(36), 14795–14805. doi:10.1021/jacs.1c06713

Taherian, R., & Kausar, A. (2019). *Electrical conductivity in polymer-based composites.* Elsevier.

Tan, X., Hu, C., Zhu, Z., Liu, H., & Qu, J. (2019). Electrically pore-size-tunable polypyrrole membrane for antifouling and selective separation. *Advanced Functional Materials*, *29*(35), 1903081.

Tang, B., Zhang, L., Li, R., Wu, J., Hedhili, M. N., & Wang, P. (2016). Are vacuum-filtrated reduced graphene oxide membranes symmetric? *Nanoscale*, *8*(2), 1108–1116. doi:10.1039/C5NR06797A

Tang, L.-C., Wan, Y.-J., Yan, D., Pei, Y.-B., Zhao, L., Li, Y.-B., . . . Lai, G.-Q. (2013). The effect of graphene dispersion on the mechanical properties of graphene/epoxy composites. *Carbon*, *60*, 16–27. https://doi.org/10.1016/j.carbon.2013.03.050

Tang, S. (2022). Inferring the energy sensitivity and band gap of electronic transport in a network of carbon nanotubes. *Scientific Reports*, *12*(1), 2060. doi:10.1038/s41598-022-06078-x

Terrones, M., Jorio, A., Endo, M., Rao, A. M., Kim, Y. A., Hayashi, T., . . . Dresselhaus, M. S. (2004). New direction in nanotube science. *Materials Today*, *7*(10), 30–45. https://doi.org/10.1016/S1369-7021(04)00447-X

Tosi, M. P. (2005). Electron gas (theory). In F. Bassani, G. L. Liedl, & P. Wyder (Eds.), *Encyclopedia of condensed matter physics* (pp. 52–57). Elsevier.

Tran, B. N., Bhattacharyya, S., Yao, Y., Agarwal, V., & Zetterlund, P. B. (2021). In situ surfactant effects on polymer/reduced graphene oxide nanocomposite films: Implications for coating and biomedical applications. *ACS Applied Nano Materials*, *4*(11), 12461–12471. doi:10.1021/acsanm.1c02950

Urper, O., Çakmak, İ., & Karatepe, N. (2018). Fabrication of carbon nanotube transparent conductive films by vacuum filtration method. *Materials Letters*, *223*, 210–214. https://doi.org/10.1016/j.matlet.2018.03.184

Valentini, L., Cardinali, M., Fortunati, E., Torre, L., & Kenny, J. M. (2013). A novel method to prepare conductive nanocrystalline cellulose/graphene oxide composite films. *Materials Letters*, *105*, 4–7. https://doi.org/10.1016/j.matlet.2013.04.034

van Berkel, S., Klitzke, J. S., Moradi, M.-A., Hendrix, M. M. R. M., Schmit, P., van der Schoot, P., & Schrekker, H. S. (2021). Spin-coated highly aligned silver nanowire networks in conductive latex-based thin layer films. *Thin Solid Films*, *724*, 138599. https://doi.org/10.1016/j.tsf.2021.138599

Van Krevelen, D. W., & Te Nijenhuis, K. (2009). *Properties of polymers: Their correlation with chemical structure; their numerical estimation and prediction from additive group contributions.* Elsevier.

Velický, M., Bradley, D. F., Cooper, A. J., Hill, E. W., Kinloch, I. A., Mishchenko, A., . . . Dryfe, R. A. W. (2014). Electron transfer kinetics on mono- and multilayer graphene. *ACS Nano*, *8*(10), 10089–10100. doi:10.1021/nn504298r

Waite, S. R., & Nazarpour, S. (2016). Graphene technology: The nanomaterials road ahead. In *Graphene technology: From laboratory to fabrication.* Wiley-VCH.

Walatka, V. V., Labes, M. M., & Perlstein, J. H. (1973). Polysulfur nitride—a one-dimensional chain with a metallic ground state. *Physical Review Letters*, *31*(18), 1139–1142. doi:10.1103/PhysRevLett.31.1139

Wallace, G., & Spinks, G. (2007). Conducting polymers—bridging the bionic interface. *Soft Matter*, *3*(6), 665–671. doi:10.1039/B618204F

Wang, J., Jin, X., Liu, Z., Yu, G., Ji, Q., Wei, H., . . . Jiang, K. (2018). Growing highly pure semiconducting carbon nanotubes by electrotwisting the helicity. *Nature Catalysis*, *1*(5), 326–331. doi:10.1038/s41929-018-0057-x

Wang, J., Naguib, H. E., & Bazylak, A. (2012). *Electrospun porous conductive polymer membranes*. Paper presented at the Behavior and Mechanics of Multifunctional Materials and Composites.

Wang, L., Wu, L., Wang, Y., Luo, J., Xue, H., & Gao, J. (2022). Drop casting based superhydrophobic and electrically conductive coating for high performance strain sensing. *Nano Materials Science*. https://doi.org/10.1016/j.nanoms.2021.12.005

Wang, X., Sun, M., Zhao, Y., Wang, C., Ma, W., Wong, M. S., & Elimelech, M. (2020). In situ electrochemical generation of reactive chlorine species for efficient ultrafiltration membrane self-cleaning. *Environmental Science & Technology*, *54*(11), 6997–7007.

Wang, X., Wang, Z., Chen, H., & Wu, Z. (2017). Removal of Cu (II) ions from contaminated waters using a conducting microfiltration membrane. *Journal of Hazardous Materials*, *339*, 182–190.

Wang, Y., Liu, P., Wang, H., Zeng, B., Wang, J., & Chi, F. (2019). Flexible organic light-emitting devices with copper nanowire composite transparent conductive electrode. *Journal of Materials Science*, *54*(3), 2343–2350. doi:10.1007/s10853-018-2986-9

Wang, Y., & Weng, G. J. (2018). Electrical conductivity of carbon nanotube- and graphene-based nanocomposites. In S. A. Meguid & G. J. Weng (Eds.), *Micromechanics and nanomechanics of composite solids* (pp. 123–156). Springer International Publishing.

Watcharotone, S., Dikin, D. A., Stankovich, S., Piner, R., Jung, I., Dommett, G. H. B., ... Ruoff, R. S. (2007). Graphene–silica composite thin films as transparent conductors. *Nano Letters*, *7*(7), 1888–1892. doi:10.1021/nl070477+

Wilson, F. (2012). *Building materials evaluation handbook*. Springer US.

Wilson, M. (2006). Electrons in atomically thin carbon sheets behave like massless particles. *Physics Today*, *59*(1), 21–23. doi:10.1063/1.2180163

Wu, J., Xue, S., Bridges, D., Yu, Y., Zhang, L., Pooran, J., ... Hu, A. (2019). Fe-based ceramic nanocomposite membranes fabricated via e-spinning and vacuum filtration for Cd2+ ions removal. *Chemosphere*, *230*, 527–535. https://doi.org/10.1016/j.chemosphere.2019.05.084

Xi, J., Lou, Y., Jiang, S., Dai, H., Yang, P., Zhou, X., ... Wu, W. (2022). High flux composite membranes based on glass/cellulose fibers for efficient oil-water emulsion separation. *Colloids and Surfaces A: Physicochemical and Engineering Aspects*, *647*, 129016. https://doi.org/10.1016/j.colsurfa.2022.129016

Xie, L., Shu, Y., Hu, Y., Cheng, J., & Chen, Y. (2020). SWNTs-PAN/TPU/PANI composite electrospun nanofiber membrane for point-of-use efficient electrochemical disinfection: New strategy of CNT disinfection. *Chemosphere*, *251*, 126286.

Xie, S., Liu, Y., & Li, J. (2008). Comparison of the effective conductivity between composites reinforced by graphene nanosheets and carbon nanotubes. *Applied Physics Letters*, *92*(24), 243121.

Xu, L., Ma, X., Niu, J., Chen, J., & Zhou, C. (2019). Removal of trace naproxen from aqueous solution using a laboratory-scale reactive flow-through membrane electrode. *Journal of Hazardous Materials*, *379*, 120692.

Xu, W., Ding, Y., Jiang, S., Zhu, J., Ye, W., Shen, Y., & Hou, H. (2014). Mechanical flexible PI/MWCNTs nanocomposites with high dielectric permittivity by electrospinning. *European Polymer Journal*, *59*, 129–135.

Xu, Y., Bai, H., Lu, G., Li, C., & Shi, G. (2008). Flexible graphene films via the filtration of water-soluble noncovalent functionalized graphene sheets. *Journal of the American Chemical Society*, *130*(18), 5856–5857. doi:10.1021/ja800745y

Yang, N., Jiang, X., & Pang, D.-W. (2016). *Carbon nanoparticles and nanostructures*. Springer.

Yang, X., Wang, Y., & Qing, X. (2019). A flexible capacitive sensor based on the electrospun PVDF nanofiber membrane with carbon nanotubes. *Sensors and Actuators A: Physical*, *299*, 111579. https://doi.org/10.1016/j.sna.2019.111579

Yin, R., Yang, S., Li, Q., Zhang, S., Liu, H., Han, J., . . . Shen, C. (2020). Flexible conductive Ag nanowire/cellulose nanofibril hybrid nanopaper for strain and temperature sensing applications. *Science Bulletin*, *65*(11), 899–908.

Yuan, W., Zhou, Y., Li, Y., Li, C., Peng, H., Zhang, J., . . . Shi, G. (2013). The edge- and basal-plane-specific electrochemistry of a single-layer graphene sheet. *Scientific Reports*, *3*(1), 2248. doi:10.1038/srep02248

Zang, X., Zhou, Q., Chang, J., Liu, Y., & Lin, L. (2015). Graphene and carbon nanotube (CNT) in MEMS/NEMS applications. *Microelectronic Engineering*, *132*, 192–206.

Zare, E. N., Lakouraj, M. M., & Ramezani, A. (2015). Effective adsorption of heavy metal cations by superparamagnetic poly (aniline-co-m-phenylenediamine)@ Fe3O4 nano-composite. *Advances in Polymer Technology*, *34*(3).

Zhan, D., Yan, J., Lai, L., Ni, Z., Liu, L., & Shen, Z. (2012). Engineering the electronic structure of graphene. *Advanced Materials*, *24*(30), 4055–4069. https://doi.org/10.1002/adma.201200011

Zhang, P., Qiao, Z.-A., Zhang, Z., Wan, S., & Dai, S. (2014). Mesoporous graphene-like carbon sheet: High-power supercapacitor and outstanding catalyst support. *Journal of Materials Chemistry A*, *2*(31), 12262–12269.

Zhao, C., Xing, L., Xiang, J., Cui, L., Jiao, J., Sai, H., . . . Li, F. (2014). Formation of uniform reduced graphene oxide films on modified PET substrates using drop-casting method. *Particuology*, *17*, 66–73. https://doi.org/10.1016/j.partic.2014.02.005

Zhu, X., Dudchenko, A. V., Khor, C. M., He, X., Ramon, G. Z., & Jassby, D. (2018). Field-induced redistribution of surfactants at the oil/water interface reduces membrane fouling on electrically conducting carbon nanotube UF membranes. *Environmental Science & Technology*, *52*(20), 11591–11600.

Zong, H., Xia, X., Liang, Y., Dai, S., Alsaedi, A., Hayat, T., . . . Pan, J. H. (2018). Designing function-oriented artificial nanomaterials and membranes via electrospinning and electrospraying techniques. *Materials Science and Engineering: C*, *92*, 1075–1091. https://doi.org/10.1016/j.msec.2017.11.007

5 Electrically Conductive Membranes for Fouling Mitigation

5.1 INTRODUCTION

The development of electrically conducting membranes has sought to reduce energy consumption of membrane separation processes by combining the fundamentals of membrane separation with electrochemistry and applying the membrane as an electrode during filtration. Such membrane-electrodes help supply electrical energy directly to the separation interface between membrane and feed. This is dramatically different from placing electrodes on either side of a non-conducting membrane, in which case electrokinetics are slow, and larger amounts of energy are needed.

Electrically conductive membranes offer several advantages over non-conducting membrane materials. There are three main ways in which electrically conductive membranes contribute to performance enhancement in desalination and water treatment processes:

1. Interfacial engineering of conducting fillers to modify matrix membrane properties such as pore size, hydrophobicity, and mechanical behavior, which in turn enhances key performance parameters such as permeate flux and/or permselectivity. In this materials-based approach, the electrical conductivity is not further exploited for electrically or electrochemically assisted filtration (Al Aani et al., 2018; Gu et al., 2020; Yang et al., 2019).
2. Electrically assisted membrane separation in which an electric field is applied to enhance rejection of charged solutes or particles via electrostatic forces. Electrostatic forces may also help detachment and rejection of organic and inorganic foulants from the membrane surface.
3. Electrochemical fouling control involving redox reactions. The membrane acts as an electrode, with controlled evolution of bubbles, whose size and distribution and kinetics result from tuning the membrane structure. As anode, the membrane can cause direct or indirect oxidation of foulants, including inactivation of bacterial species; as cathode, fouling can be prevented or removed through the formation of gas bubbles.

The aforementioned methods ultimately seek to reduce energy costs associated with membrane separation. Although energy analysis is often left out by authors

DOI: 10.1201/9781003144991-5

proposing new materials, development of new membrane materials as well as modified electrofiltration systems are a unique area of study at the intersection of materials engineering, electrochemistry, and membrane separation. As we saw in Chapter 4, electrically conducting membranes can be fabricated as homogeneous self-standing conducting membranes, e.g. buckypaper, by embedding conducting fillers in polymers or ceramics during processing or by coating a conductive layer on prepared membranes. It is also common to employ post-treatment techniques to improve interfacial properties.

In this chapter, we explore developments in electrically conductive membranes for desalination and water treatment, focusing on electrically or electrochemically assisted fouling control, and we also touch on interfacial engineering to incorporate electrically conductive nanofillers in membrane systems. In the first part, we explore mechanisms of de-fouling using conducting membranes. We then review recent literature in the use of electrically conductive membranes to improve filtration for various applications: desalination and wastewater treatment including removal of organic contaminants, microbial decontamination, treatment of oily wastewater, removal of toxic metals, etc.

5.2 MECHANISMS OF FOULING PREVENTION AND CLEANING WITH CONDUCTIVE MEMBRANES

Interactions between membrane surface, feed solution, and foulant(s) ultimately determine the mechanism of fouling prevention and cleaning in electrically conductive membranes. The two main mechanisms of cleaning are (1) electrostatic interactions and (2) redox reactions. Figure 5.1 shows these mechanisms of membrane cleaning schematically. The electrostatic effect refers to foulants being repelled by the membrane surface due to applied electric charge. Electrostatic effects are not limited to fouling mitigation; they are also used to improve rejection of certain charged species, as mentioned above. On the other hand, the membrane may act as anode such that direct or indirect oxidation of foulants takes place at the membrane surface (Figure 5.1c–d) or so that foulants are removed through the physical action of gas bubbles formed at the membrane electrode (anode or cathode). Indirect oxidation is the oxidation of foulants through the production of aqueous oxidants in the electrochemical cell.

The effectiveness of defouling depends on many factors, including surface morphology, structure, and consequent electrochemical properties of the *membrane electrode* (Agarwal et al., 2011; Z. Wu et al., 2008). Table 5.1 summarizes some mechanisms of fouling prevention using conductive membranes.

5.2.1 OXIDATION OF FOULANTS

Inactivation of bacteria using CNT-based membranes was first reported by (Vecitis et al., 2011), who used MWCNTs for microfiltration and established that the inactivation of bacteriophage MS2 was due to direct oxidation at the membrane surface. The extent of virus inactivation is unaffected by the ionic strength of the electrolyte.

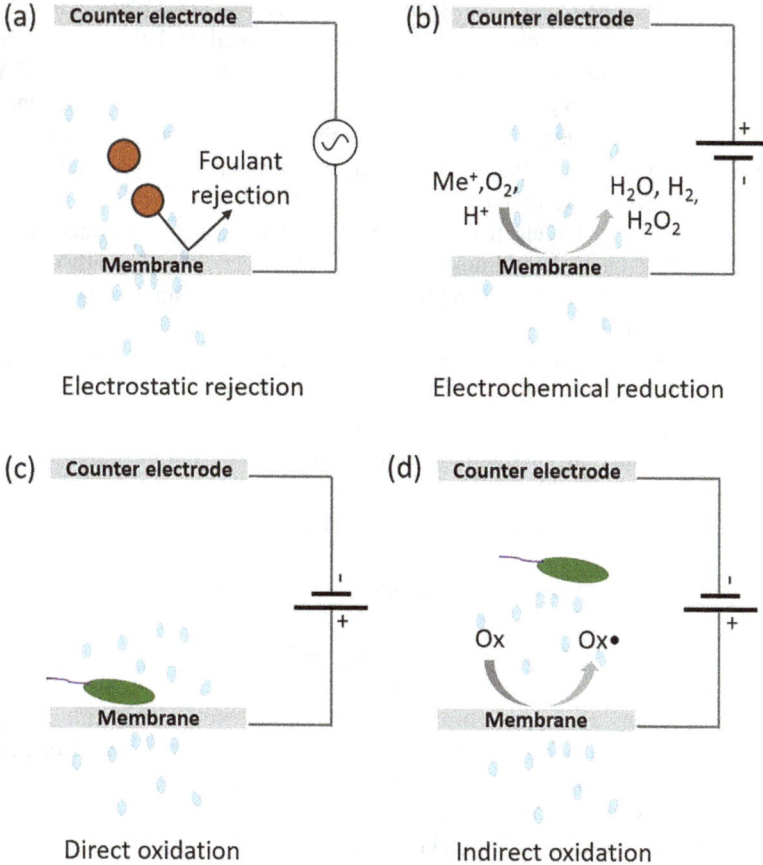

FIGURE 5.1 Mechanisms of electrical/electrochemical de-fouling on electrically conducting membrane surfaces.

TABLE 5.1

Mechanisms of Fouling Prevention with Conductive Membranes (F. Ahmed et al., 2016).

Reference	Mechanism
(S. H. Hong et al., 2008; Sun et al., 2013; Vecitis et al., 2011)	Direct oxidation of virus on membrane surface by applying anodic current
(Z. Wu et al., 2008; Z. Wu et al., 2006)	Generation of nanobubbles to prevent foulant adsorption and remove adsorbed foulant
(Kang et al., 2008; Kang et al., 2007)	Antimicrobial properties of CNTs
(S. H. Hong et al., 2008; Sun et al., 2013)	Bacterial detachment due to cathodic current
(Perez-Roa et al., 2006)	Prevention of formation of biofilm via small electric pulses; application of low voltage pulses (0.5–5 V) to prevent the formation of biofilm

In an electrolytic cell, oxidation, or the loss of electrons, takes place at the positive electrode or anode. Direct loss of electrons disrupts the cell wall of bacterial species, preventing the transport of molecules and thus inactivating the bacteria (T. Y. Wang et al., 2017). Reactive oxygen species (ROS) may also be produced at the anode as intermediates of oxidation of water, such as the hydroxyl radical (•OH) (Zhao & Drlica, 2014), to oxidize foulants. Oxidant agents may also be produced from ions present in the electrolyte or feed solution such as chlorine. Gonzalez-Olmos and coworkers have suggested electro-oxidation as pretreatment to membrane processes to control fouling (Gonzalez-Olmos et al., 2018). However, an electrically conductive membrane eliminates this two-step procedure by enabling a single-membrane electrochemical system.

During the electrolysis of water, the two half-reactions are the OER at the anode and HER at the cathode. OER is a limiting reaction in the process of generating molecular oxygen. OER is the anodic reaction that takes place during water electrolysis. In acidic media, the anodic half-reaction for OER is given by:

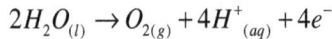

$$2H_2O_{(l)} \rightarrow O_{2(g)} + 4H^+_{(aq)} + 4e^-$$

At the cathode, water is reduced to hydrogen gas and hydroxide ions:

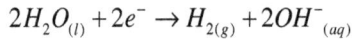

$$2H_2O_{(l)} + 2e^- \rightarrow H_{2(g)} + 2OH^-_{(aq)}$$

When chloride ions are present in the electrolyte, the evolution of chlorine gas at the anode further assists disinfection by cleaning the membrane and decomposing organic foulants (Khalil et al., 2022). On the other hand, the low resistance of polyamide reverse osmosis (RO) membranes requires that chlorine be removed prior to desalination via reduction at the cathode (Khanzada et al., 2022). The electrolytic half reaction for chlorine evolution is given by:

$$2Cl^{-(aq)} \rightleftharpoons Cl_{2\,(g)} + 2e^-$$

Water electrolysis kinetics depend on the efficiency and abundance of electrocatalysts. Both hydrogen evolution and oxygen evolution are important redox reactions that take place at the surface of conductive electrodes. Carbon-based nanomaterials have attracted significant attention in this area. Materials with low overpotentials are particularly sought after. Lu and coworkers showed that after mild surface oxidation, hydrothermal annealing and electrochemical activation, MWCNTs become effective catalysts for water oxidation, initiating oxygen evolution reaction (OER) at small overpotentials of 300 mV in alkaline media (X. Lu et al., 2015). They found that the microscopic structures and conductive inner walls remain preserved and enable efficient electron transport during OER. Figure 5.2 shows the figure of merit for OER.

One must be careful when applying carbon-based materials to electro-oxidation as the membrane itself is prone to oxidation. This is another aspect in which the nanoscale dimensions of CNTs can be advantageous: MWCNTs produced by CVD are more resistant to electrochemical oxidation than carbon black under acidic conditions (Shao et al., 2006). While some attributes may be well studied, the reaction

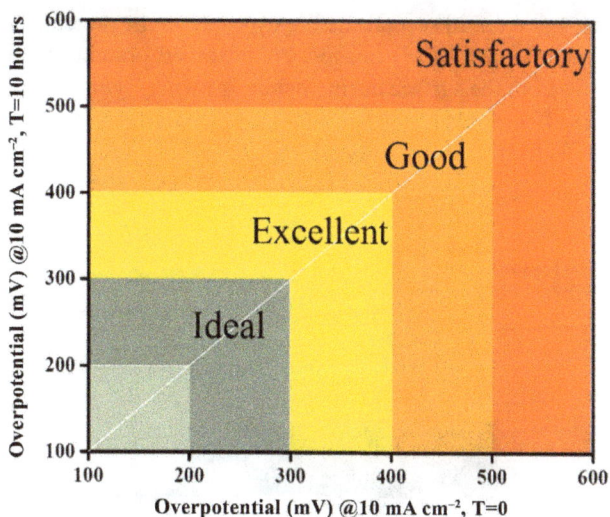

FIGURE 5.2 Performance criteria for performance and robustness for OER electrodes in any pH range.

Source: Tahir, M., Pan, L., Idrees, F., Zhang, X., Wang, L., Zou, J.-J., & Wang, Z. L. (2017). Electrocatalytic oxygen evolution reaction for energy conversion and storage: A comprehensive review. *Nano Energy*, *37*, 136–157. Reprinted with permission.

mechanism and criteria for the ideal catalyst in terms of activity and stability are still ambiguous (Suen et al., 2017). Fabri and Schmidt refer to the oxygen evolution reaction mechanism as "an old but current enigma" (Fabbri & Schmidt, 2018). Currently, OER catalysis is an area of active research (Mittelsteadt et al., 2015). It is hoped that the evolution of materials with enhanced electrocatalytic activity with lower overpotentials (Gonçalves et al., 2019) will play an important role in future membrane development.

5.2.2 ELECTROCHEMICAL BUBBLE GENERATION

The formation of microbubbles or nanobubbles on the conductive membrane surface as a result of redox reactions can prevent fouling and also clean the membrane surface through a physical sweeping mechanism. Microbubbles are bubbles with average diameters of 10–50 μm, whereas nanobubbles have an average diameter of <200 nm (Agarwal et al., 2011). Formation of bubbles creates a physical barrier preventing (organic) foulants to adhere to the membrane surface (Z. Wu et al., 2006). The evolution of gas bubbles also acts to remove foulants already deposited on the membrane surface. Wu et al. (Z. Wu et al., 2008) showed that nanobubbles can remove adsorbed foulants and prevent further adsorption of proteins. In their work, the applied current was controlled to produce nanobubbles of uniform density and size on a highly oriented pyrolytic graphite surface; copper was used as the anode. The adsorption of BSA protein on the graphite surface can be reduced by 26–34% via electrochemical treatment using 1.5 V for 20 seconds before exposure to the protein, as shown using atomic force microscope (AFM) images. Pre-adsorbed protein was also detached and moved away from the surface during bubble growth (Figure 5.3).

Many factors affect bubble evolution kinetics, which in turn determine bubble size and cleaning effectiveness. Many researchers have attempted to link bubble kinetics with electrode surface properties and operating conditions to better understand

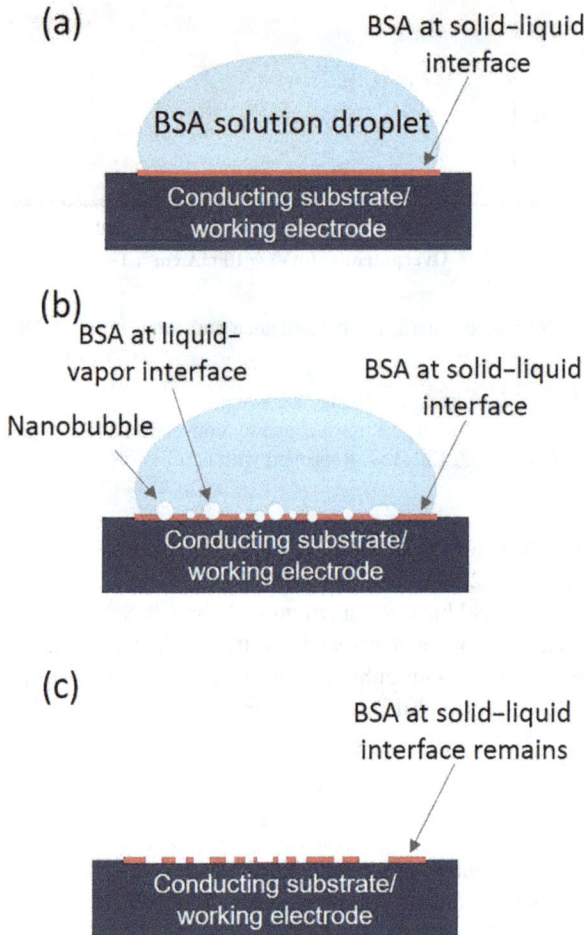

FIGURE 5.3 Schematic mechanism of de-fouling of conductive surface with nanobubbles: (a) protein adsorption to solid–liquid interface; (b) Conductive surface is used as a working electrode to produce nanobubbles on its surface. Foulants molecules are forced to migrate from the solid–liquid interface to the liquid–vapor interface during the growth of the bubbles; (c) the foulant adsorbed at the liquid–vapor interface is labile and readily washed out.

Source: Wu, Z., Chen, H., Dong, Y., Mao, H., Sun, J., Chen, S., & Hu, J. (2008). Cleaning using nanobubbles: Defouling by electrochemical generation of bubbles. *Journal of Colloid and Interface Science*, *328*(1), 10–14. Adapted with permission.

the bubble evolution mechanism. Understanding of bubble evolution kinetics is likely to boost advances in electrochemical de-fouling using bubble generation. Electrochemical measurements of dissolved gas concentration prior to nucleation can be used to estimate the size of the gas bubble nucleus (German et al., 2016). Bubble detachment diameter is directly proportional to current density and inversely proportional to flow velocity (Y. Li, Yang et al., 2019). A lower flow velocity favors coalescence of bubbles (Calabriso et al., 2015). In situ visualization of oxygen micro-bubbles showed that that increasing temperature and current density increases the number, growth rate, and nucleation sites of electrolysis-induced oxygen bubbles (Y. Li et al., 2018). Furthermore, at low current density, there is little effect of electrode surface wettability on bubble dynamics (Y. Li, Kang et al., 2019). Electrochemical de-fouling using bubble evolution will largely improve when the challenges related to understanding bubble evolution on different electrode surfaces and under various fluid conditions are addressed sufficiently. Although the use of electrochemical fouling control is still in its infancy for membrane processes, understanding of bubble evolution mechanisms will be significant especially when upscaling effects are considered and the technology moves to pilot- and industrial-scale studies. Since bubble evolution is fast and at the microscale, tools to capture the process are currently limited, as are mathematic models for bubble dynamics during electrolysis (Y. Li, Yang et al., 2019).

5.2.3 ANTIMICROBIAL ACTIVITY

CNTs also exhibit strong antimicrobial activity (Rita Teixeira-Santos et al., 2021), which has allowed CNT-based materials to be used for biofouling prevention. CNTs may delay bacterial adhesion and reduce biofilm formation (R. Teixeira-Santos et al., 2020). While the mechanism of antimicrobial activity in CNTs remains under scrutiny, some factors that affect antimicrobial activities in CNT-based materials include size and length of tubes, dispersion, number of walls, functionalization, and, in turn, hydrophilicity and biocompatibility (R. Teixeira-Santos et al., 2020). CNT-based composites with polymers and metals show high antimicrobial activity toward a wide range of microorganisms (Saleemi et al., 2021).

Biofouling prevention with CNTs-based membranes relies on their antimicrobial properties. Kang et al. (Kang et al., 2007) find that single-walled CNTs exhibit stronger bacterial toxicity than multiwalled CNTs (Kang et al., 2008), although contradicting results have also been reported in literature. They suggested that bacterial cell death is caused by cell membrane damage from direct contact with CNT aggregates, as shown in Figure 5.4. It is evident that more work needs to be carried out to elucidate the effect of CNT structure, especially number of walls, on their antimicrobial properties.

The application of an electric field also affects the extent of bacterial adhesion on a conductive surface. In 2008, Hong et al. investigated the effect of both anodic and cathodic currents on the adhesion and inactivation of *P. aeruginsa* (PAO1) on indium-tin-oxide-coated glass surfaces (S. H. Hong et al., 2008). While a cathodic current led to successful detachment of the bacteria due to increased mobility, inactivation was not observed. On the other hand, when the conductive surface

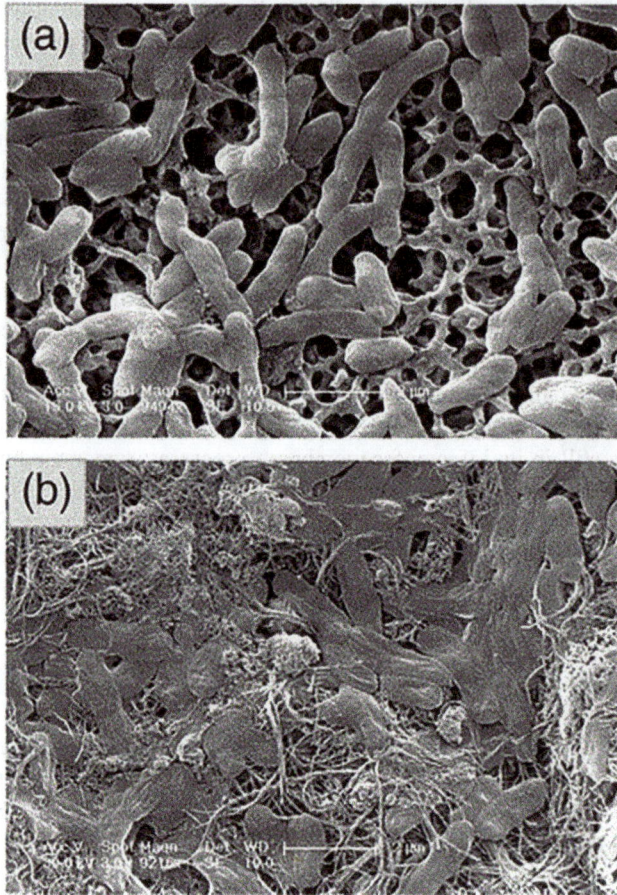

FIGURE 5.4 SEM images of *E. coli* cells incubated for 60 minutes (a) without single SWCNTs maintaining their outer membrane structure and (b) with SWCNTs losing their cellular integrity.

Source: Reprinted with permission from *Langmuir* 2007, 23, 17, 8670–8673. Copyright © 2007 American Chemical Society (Kang et al., 2008).

was used as anode, 85% inactivation of adhered bacteria was observed, along with detachment. This is due to the oxidation effect discussed in Section 9.1.1. In another study, small low voltage pulses (0.5–5 V) on conductive membranes increased bio-fouling resistance, reducing coverage of biofilm by 50% when a pulse of 5 V was applied at 200 Hz (Perez-Roa et al., 2006). It is important to note that both the type of bacterial strain and the electrochemical properties of the conducting surface determine performance in terms of bacterial detachment and inactivation. As discussed in Chapters 2 and 3, fouling mitigation strategy in membrane processes should be targeted toward the specific bacterial strains likely to cause fouling in a given feed solution.

To understand the effect of electrically driven de-fouling on aligned CNT membranes, Sun et al. compared three methods: electro-oxidation, electroreduction, and ionic pumping (Sun et al., 2013). In electro-oxidation and electroreduction, a positive or negative bias was applied to the membrane; they used BSA and naphthalene as foulants. They found that SWCNT membranes were more resistant to biofouling than MWCNT membranes, which they attributed to smaller diameter. This is somewhat counter-intrusive as the number of electrocatalytically active defects increases with the number of walls (Pettes & Shi, 2009). While both electro-oxidation and bubble generation via electroreduction enable de-fouling, electro-oxidation is effective for strongly absorbed foulants but only for a limited number of cycles, as prolonged application of a positive bias damaged the membrane as CNTs are prone to oxidation (Gao et al., 2015). We also verified that electroreduction is more stable for membrane cleaning.

Incorporating CNTs in polyamide membranes has helped reduce roughness and polymer mobility at the molecular scale, which reduces irreversible adsorption by low-molecular-weight humic acid foulants (Cruz-Silva et al., 2019).

5.3 ELECTRICALLY CONDUCTIVE MEMBRANES IN DESALINATION

5.3.1 REVERSE OSMOSIS

David Jassby's group was able to electrochemically prevent and remove $CaSO_4$ and $CaCO_3$ mineral scales on electrically conducting CNT-PA RO membranes (Duan et al., 2014). An intermittent applied anodic potential of 2.5 V caused water oxidation, which led to proton formation and subsequent dissolution of deposited $CaCO_3$ crystals. On the other hand, a continuous application of 1.5 V on the membrane as an anode was effective in delaying $CaSO_4$ scale formation due to a thick layer of counterions that drove $CaSO_4$ crystals away (Duan et al., 2014). Interestingly, they do not touch on the oxidation of CNTs under continuous anodic potential, which is likely to impact the long-term operation of the system.

Our group developed a conductive form of networked cellulose (NC) by incorporating CNTs in the fabrication process (F. E. Ahmed et al., 2019). This conductive NC was blended with PVA to form solution-casted PVA-NC-CNT membranes that exhibited an increase of 93% in permeate flux compared to PVA-NC membranes for RO of 25,000 mg L^{-1} aqueous NaCl. This increase in flux is due to CNTs disrupting the compression of polymer chains under pressure. A cathodic potential of −5 V resulted in removal of foulant (yeast and humic acid) from the membrane surface.

5.3.1.1 Boron Removal

Boron occurs naturally in seawater at an average concentration of 4.6 mg L^{-1}. While it is an important element to plants and animals, the World Health Organization limits boron concentration to <0.5 mg L^{-1} (Taniguchi et al., 2004). The presence of boron in water beyond these low limits is toxic to plants, animals, and humans (Kot, 2015). In plants, high concentrations cause damage to leaves and premature ripening, which reduces crop yields (Nable et al., 1997). In humans, exposure to boron causes reproductive and development problems (Bolt et al., 2020).

Boron rejection remains a challenge in seawater reverse osmosis (SWRO) plants producing universal drinking water. Although RO membranes can remove borate ions under alkaline solutions, the existence of boron as undissociated boron molecules under neutral and acidic conditions makes removal via RO boron removal difficult (Tang et al., 2017). Boron rejection is also affected by operating temperature and pressure. To bring boron concentrations to the allowed levels, about 90% of the boric acid present in seawater must be removed, which existing single-pass RO processes cannot do. Often, the target is reached by adopting a multipass process, which increases operating costs by 10–20% (Sagiv & Semiat, 2004). It is therefore important to gear single-pass processes toward higher removal of boron.

Much of the work in this area has involved membrane modification to improve boron rejection. Very recently, electrically conductive membranes have been applied specifically to enhance boron rejection in RO membranes. Jung et al. demonstrated that applying cathodic potentials to CNT-PVA-coated RO membranes increases localized pH near the membrane surface, which increases boron rejection (Jung et al., 2020). A cell potential of 5 V resulted in >90% boron removal using 35 g L^{-1} NaCl + boric acid mixture. However, the increased pH also has undesirable side effects in the form of active generation of OH- on the CNT-PVA membrane, which causes membrane scaling with $Mg(OH)_2$ and subsequent flux decline. They suggested electrically enhanced boron rejection as a chemical-free low-cost alternative to chemical pH adjustment (Jung et al., 2020). Instead of modifying membranes with nanomaterials, Bao and coworkers placed a porous carbon cloth on a commercial RO membrane and applied a cathodic potential of 4 V to increase boron rejection from 75 to 93.8% (Bao et al., 2021), also by elevating pH near the membrane surface. At high pH, boron changes from boric acid to a negatively charged and larger borate ion, which is more easily rejected by the RO membrane. Application of electrically conducting membranes to improve the boron rejection in RO is a novel concept, with much more investigation needed to optimize material and applied potential frequency and amplitude. With the other advantages offered by electrically conducting materials, they may be the answer to the challenges associated with fouling as well as low boron rejection.

5.3.2 NANOFILTRATION

With high multivalent ion rejection, nanofiltration (NF) is a promising technology for softening water or for pretreatment of RO feed. It also finds many uses in the food and dairy industries where only larger ions need to be removed. Citing the low water permeability of interfacial polymerization (IP) nanofiltration (NF) membranes, Professor Gai's group in China reported a high-performance CNT-PA NF membrane with covalent coupling of the CNT interlayer and the PA functional layer (Chen et al., 2021). The interpenetrating network structure leads to a high rejection of 99.47% of Congo red as well as high selectivity of NaCl to dye (150.32). Furthermore, the membrane shows a pH-responsive molecular sieving as the rejection switches rapidly between low (2%) and high (>99%) between alkaline and neutral conditions (Chen et al., 2021).

There has been an interest in electrically tuning membrane surface properties to control ion rejection of conducting NF membranes. Zhang and coworkers applied electrically assisted enhancement of surface charge density to control rejection rates for Na_2SO_4 and NaCl (H. Zhang et al., 2019). Increasing voltage from 0 to 2.5 V increases the surface charge density of the membrane by a factor of 6.1 from 11.9 to 73.0 mC m^{-2}. This increase in surface charge density improves Na_2SO_4 rejection from 81.6 to 93% and also improves NaCl rejection from 53.8 to 82.4%, while retaining high permeability (H. Zhang et al., 2019). Under applied electrical potential, the Donnan potential difference between the membrane and bulk solution increases, which leads to increased resistance to ion transfer.

Hilal's group fabricated electrically conducting NF membranes from NC and CNTs via vacuum filtration of a sonicated suspension, followed by drying at 40 °C (F. E. Ahmed et al., 2017). The improved surface hydrophilicity of CNS-NC membrane enables regeneration of the electrode surface via electrolysis. NC serves to compact the pore structure, which enables a rejection of 60% for $MgSO_4$ and 47% for $CaCl_2$ while retaining a high flux of 100 L m^{-2} h^{-1}.

Rohani and Yusoff used chemically oxidative IP to synthesize conductive polyaniline (PANI)-coated PVDF membranes for NF with tunable separation selectivity (Rohani & Yusoff, 2019). A higher electrical conductivity of 6.7 S cm^{-1} was achieved when the PANI layer was synthesized using diffusion cell polymerization as opposed to solution polymerization (7.6×10^{-2} S cm^{-1}). An external electric potential field reduced water contact angle by 30–40% (Rohani & Yusoff, 2019). The effect of the electric field on the contact angle is not well reported in the literature. Membranes with complete PANI coverage demonstrated electrical tuning of permeability for neutral polyethylene glycols. Molecular weight cut-off (MWCO) values obtained for 7 V applied potential are higher than those obtained in the absence of applied potential.

5.3.3 Membrane Distillation

Dumée et al. used CVD to grow CNTs that were then prepared as self-supporting CNT membranes through vacuum filtration (L. F. Dumée et al., 2010). The prepared CNTs were randomly oriented and had extremely high aspect ratios, with diameters of 10–15 nm and lengths of 150–300 µm (Figure 5.5). Sonication is used to disperse CNTs and prevent agglomeration. The membranes prepared in this study had a water contact angle of 113° and a porosity of 90%, which made them suitable for desalination by membrane distillation (MD).

The same group later sought to improve the MD performance of CNT membranes through chemical modification using a two-step alkoxysilane treatment before vacuum filtration (Dumée et al., 2011). To prepare the CNTs reaction with alkoxysilane-based groups, the CNTs were first functionalized using UV/ozone to create hydroxyl groups. A silicon content of up to 2.5% in the CNTs resulted in a contact angle of 140°. Increased hydrophobicity led to a greater liquid entry pressure and increased mass flux during DCMD, as shown in Figure 5.6.

To prepare polymer-CNT composite membranes for MD, the same group employed three techniques: hot pressing a CNT membrane between PP layers; vacuum filtration

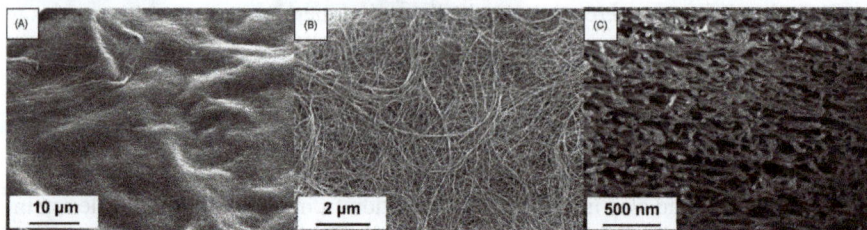

FIGURE 5.5 SEM images of (a,b) surface and (c) cross section of self-supporting CNT membrane.

Source: Dumée, L. F., Sears, K., Schütz, J., Finn, N., Huynh, C., Hawkins, S., . . . Gray, S. (2010). Characterization and evaluation of carbon nanotube Bucky-Paper membranes for direct contact membrane distillation. *Journal of Membrane Science, 351*(1), 36–43. Reprinted with permission.

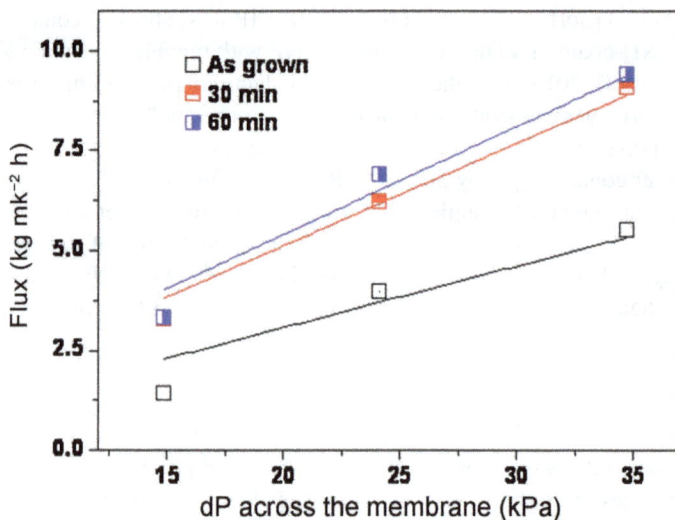

FIGURE 5.6 Effect of prior silane treatment on the DCMD flux of CNT membranes using 3.5 wt% NaCl.

Source: Dumée, L., Germain, V., Sears, K., Schütz, J., Finn, N., Duke, M., . . . Gray, S. (2011). Enhanced durability and hydrophobicity of carbon nanotube bucky paper membranes in membrane distillation. *Journal of Membrane Science, 376*(1), 241–246. Reprinted with permission.

of CNTs on PP, followed by hot pressing of another PP layer; and vacuum filtration of a 5 wt% polymer solution through a CNT membrane (L. Dumée et al., 2010). While all three composites performed better than CNT-only membranes, the composite formed by sandwiching a CNT layer between PP layers had a salt rejection of 98.5% compared to 90% for CNT membranes during 40 hours of DCMD operation.

Electrically conducting CNT membranes were also developed by Roy et al., who incorporated polyvinylidene difluoride (PVDF) as a surface binder (Roy et al., 2014). They sonicated CNTs in acetone for 3 hours and dissolved PVDF in acetone before mixing it with the CNT suspension. The mixture was sonicated further and vacuum filtered on a porous polypropylene (PP) support. When the feed temperature was kept at 70 °C, a high mass flux of 36.8 L m^{-2} h^{-1} was obtained for the CNT-COOH-PP membrane with a salt rejection of 99.9%. The polar carboxylated CNTs acted as sorbent sites for moisture transport and repelled liquid water, resulting in a mass transfer coefficient 1.5 times greater than that of unmodified PP. The functionalization of –COOH groups on the CNTs positively affected the interaction of water molecules with the membrane surface, also allowing better vapor transport (Roy et al., 2014). These studies are important in relating the functionalization of CNTs with desalination performance for MD. MD is a promising technique that combines membrane separation with thermal processes and can be used for hypersaline feed solutions, unlike RO, the use of which is limited to moderate-/high-salinity solutions ≤35,000 mg L^{-1}.

Recently, Kim et al. modified PVDF membranes by applying a thin multilayered SWCNT coating using vacuum filtration to enhance the electrical performance of the membranes (Kim et al., 2022). Their findings reveal that electrical repulsion using SWCNT/PVDF membranes effectively prevented membrane fouling and wetting, when they used synthetic seawater as feed and BSA and HA as model foulants at varying concentrations. With electrical repulsion at a low potential of 0.5 V, contaminants were prevented from reaching the membrane surface. They also did not notice a drop in current throughout the operation, indicating a stable CNT coating with no signs of detachment under low applied voltage.

5.3.4 OTHER

Other emerging desalination to which electrically conducting membranes have been applied include forward osmosis (FO) and capacitive deionization. In an attempt to prepare high performing ceramic membranes for FO, Zhang et al. introduced a nanocomposite interlayer of TiO$_2$ and CNTs between a defect-free PA layer and a ceramic substrate (M. Zhang et al., 2020). The compatibility of the PA layer with high cross-linking was favored by low-roughness nanocomposite interlayers with reduced pore size. The resulting membrane had a higher water permeability (2 L m^{-2} h^{-1} bar) and NaCl rejection (98%), compared to the membrane with no interlayer. In this case, the interlayer acted to provide more active sites for the formation of a defect-free PA layer while simultaneously providing a 3D network structure for fast water transport (M. Zhang et al., 2020).

CDI is a desalination process in which an electric potential difference is applied across an aqueous solution using porous electrodes with alternating cation exchange and anion exchange membranes positioned between them (Biesheuvel & van der Wal, 2010). Selective transport of ions through the ion exchange membranes and eventual adsorption on the electrical double layer of the electrodes cause salt ions to be removed from the feed (Porada et al., 2013). The salt removal efficiency of CDI systems depends on electrode material and geometry; carbon-based materials are most commonly used (Angeles & Lee, 2021).

Gleb Yushin's group incorporated electrically conductive polyaniline onto non-woven CNT fabric via electrodeposition (Benson et al., 2013) and found that these high-strength conductive membranes showed fast ion adsorption and high specific capacitances, making them suitable for low-energy desalination via capacitive deionization (CDI). In the same year as Yushin's group, Ma et al. used the high electrical conductivity of CNTs with the high ion sorption of zeolite to fabricate CNT/Ca-selective zeolite composite electrodes for CDI (Youhai Liu et al., 2013). These electrodes were prepared by electrophoretic deposition and tested for the removal of Ca^{2+} ions. With CNT content of 25 wt%, they achieved an electrosorption capacity of 25 mg g^{-1} for Ca^{2+} ions at an electrosorption voltage of 2 V. The composite membrane showed higher selectivity for Ca^{2+} ions as compared to Mg^{2+} and Na^+. Since CDI relies on suitable electrode materials, there is ample literature on the use of electrically conductive surfaces developed for CDI. However, since our focus is on the mitigation of fouling and related phenomena, we only provide limited examples of CNT-based composites for CDI desalination.

Pervaporation membranes based on polyvinylamine incorporated with CNTs were also recently developed by Hu et al. (S. Y. Hu et al., 2012). Adding CNTs increased the permeation flux as well as rejection rate for the dehydration of ethylene glycol, especially at low feed concentrations. At 2 wt% CNTs, the membrane showed a permeation flux of 146 g m^{-2} h^{-1}. Pervaporation uses a non-porous membrane to separate liquid mixtures using chemical potential gradient as a driving force (Low & Murugaiyan, 2021). While it has traditionally been used to separate liquid mixtures with close boiling points, e.g. dehydration of organic solvents, it has been proposed as a promising technique for desalination as well (Q. Wang et al., 2016).

Interestingly, there are no reported studies to date on the use of electrically conducting membranes for highly saline feed solutions, such as the treatment of desalination brine. This is an area that needs more attention as growing desalination capacity is generating large volumes of brine. Additionally, the high salt content of brine makes this type of feed a suitable candidate as electrolyte.

5.4 ELECTRICALLY CONDUCTIVE MEMBRANES IN WATER TREATMENT

Highly electrical conductive membranes were prepared by de Lannoy et al. by vacuum filtering a thin layer of polyvinyl alcohol (PVA) cross-linked with carboxylated CNTs and succinic acid on a cellulose nitrate membrane (C. F. de Lannoy et al., 2012). After curing the composite membrane at 100 °C, the CNT-PVA active layer had a conductivity >20 orders of magnitude greater than neat PVA. Electrical conductivity of the composite increases with curing time and CNT content, as shown in Figure 5.7. Curing increased cross-linking of the CNTs with PVA as well as bonding with the support membrane.

The highest rejection (>90%) was achieved with low CNT content of 2 and 5 wt% for rejection of PEO under 550 kPa (Figure 5.10).

There are two drawbacks of using individual CNTs in membranes: (1) Membranes with individual CNTs that are not in any way bonded to each other have an irregular pore size (Shah et al., 2013) that will change with compaction under pressure-driven

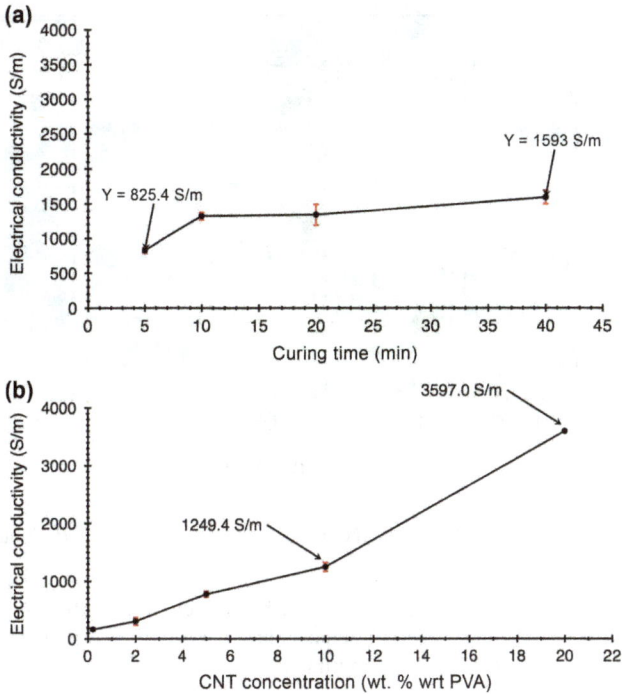

FIGURE 5.7 Conductivity of the PVA–CNT membranes vs. (a) curing time and (b) CNT concentration.

Source: de Lannoy, C. F., Jassby, D., Davis, D. D., & Wiesner, M. R. (2012). A highly electrically conductive polymer–multiwalled carbon nanotube nanocomposite membrane. *Journal of Membrane Science, 415–416,* 718–724. Reprinted with permission.

operation; (2) there is a greater risk of leakage, which raises health and safety concerns (C. F. de Lannoy et al., 2012). The latter is discussed in more detail in subsequent sections.

Zhang and Vecitis developed a conducting CNT-polyvinylidene (PVDF) porous electrode on a PES UF membrane and applied a cathodic current to negatively charge its surface (Q. Zhang & Vecitis, 2014). This method of inducing capacitive charging resulted in reduced organic fouling. The CNT-PVDF membrane was fabricated via phase inversion; a carbon cloth was used as a counter-electrode. Figure 5.9 shows the effect of a negative charge on the rejection of Suwannee River fulvic acid (SRFA) and sodium alginate (ALG).

They observed a decrease in fouling rate constant when capacitive charging was used. Capacitive organic fouling reduction also translates into less frequent cleaning as compared to the control PES membrane. The authors suggested that potential-induced cathodic negative surface charges increase the energy barrier and reduce the collision of negatively charged organic substances with the membrane surface. As a result, given a payback time of 1.5–5 years, operational energy costs can be reduced by 37–84% (Q. Zhang & Vecitis, 2014).

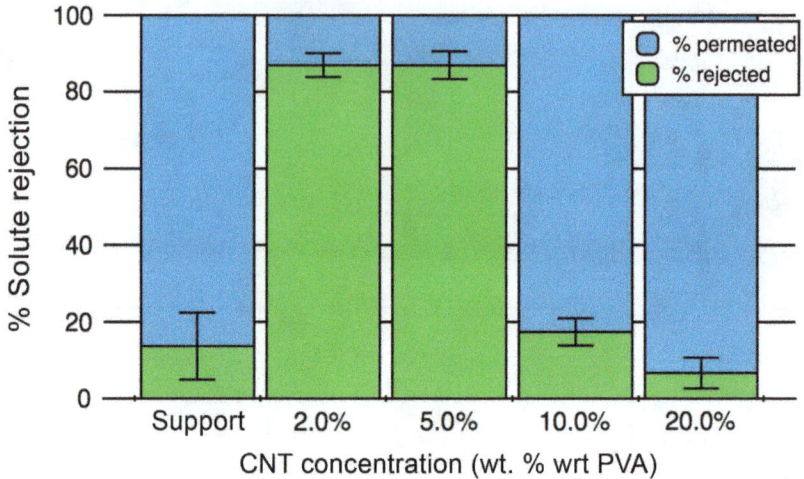

FIGURE 5.8 PEO rejection as a function of CNT concentration in CNT-PVA nanocomposite membranes.

Source: de Lannoy, C. F., Jassby, D., Davis, D. D., & Wiesner, M. R. (2012). A highly electrically conductive polymer–multiwalled carbon nanotube nanocomposite membrane. *Journal of Membrane Science*, *415–416*, 718–724. Reprinted with permission.

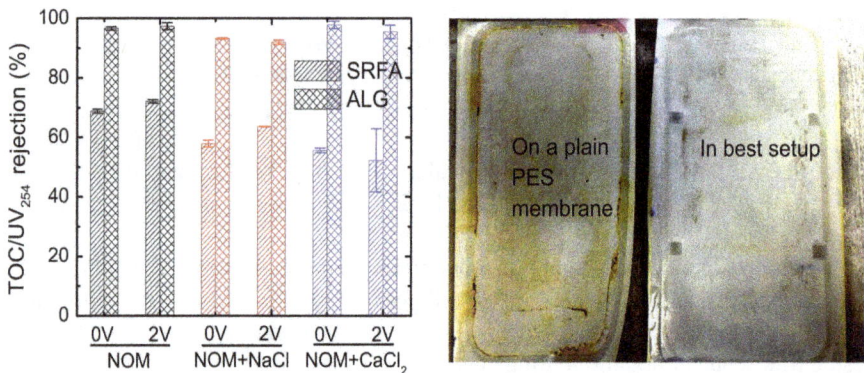

FIGURE 5.9 Organic matter removal percentage under different solution conditions in the cross-flow filtration (left) and photos of the membrane after a 3 hour test in [SRFA] = 10 ppm and [NaCl] = 1 mM with and without 2 V cell voltage. SRFA refers to Suwannee River fulvic acid (SRFA) and sodium alginate (SA).

Source: Zhang, Q., & Vecitis, C. D. (2014). Conductive CNT-PVDF membrane for capacitive organic fouling reduction. *Journal of Membrane Science*, *459*, 143–156. Reprinted with permission.

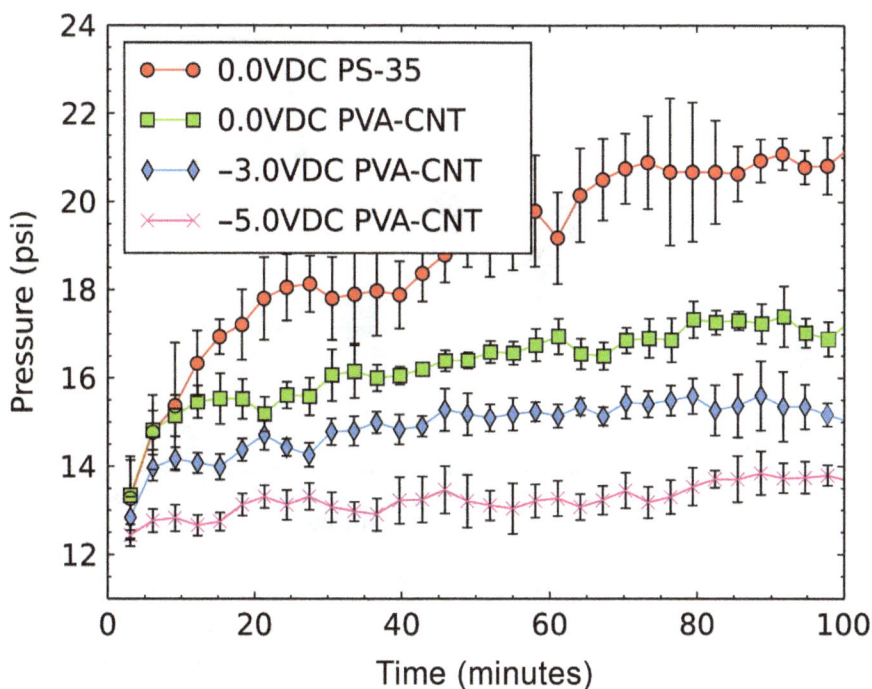

FIGURE 5.10 Fouling of 5 g/L AA on PS-35 and PVA–CNT membranes at different applied membrane cell potentials.

Source: Dudchenko, A. V., Rolf, J., Russell, K., Duan, W., & Jassby, D. (2014). Organic fouling inhibition on electrically conducting carbon nanotube–polyvinyl alcohol composite ultrafiltration membranes. *Journal of Membrane Science, 468,* 1–10. Reprinted with permission.

Using carboxylated CNTs, Dudchenko et al. prepared PVA-CNT-COOH layers on polysulfone (PSf) membranes via sequential deposition and cross-linking (Dudchenko et al., 2014). They studied the effect of applied potential on alginic acid (AA) and polyethylene oxide (PEO) fouling. The conductive film had an electric conductivity of 25 S cm^{-1}. Fouling was observed through monitoring operating pressure at constant flux mode with 250 mg L^{-1} of 200 kDa PEO. The operating pressure was 45% higher in the case of PS alone as compared to the composite membrane after 100 minutes, indicating slower fouling in the composite membrane due to greater hydrophilicity. An applied potential of −3 V and −5 V helped reduce fouling due to 5 g L^{-1} AA but had no effect on PEO fouling as PEO has a relatively neutral charge (ξ = −7.5 mV) as compared to that of AA (ξ = −68.1 mV). Fouling reductions of 33.7 and 51.1% were observed when −3 V and −5 V were applied, respectively (Figure 5.10).

Our group carried out periodic electrolysis for in situ cleaning of CNT-based membranes via bubble generation by coating MWCNTs on a commercial microfiltration

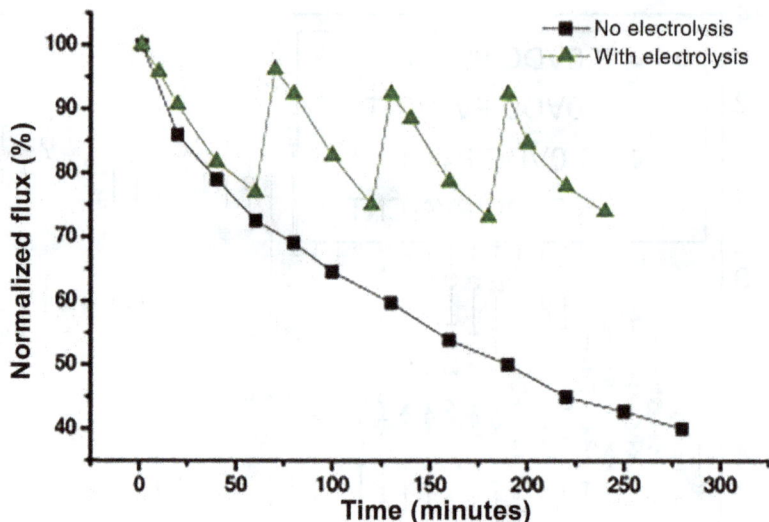

FIGURE 5.11   Effect of periodic electrolysis on flux recovery for yeast filtration.

Source: Lalia, B. S., Ahmed, F. E., Shah, T., Hilal, N., & Hashaikeh, R. (2015). Electrically conductive membranes based on carbon nanostructures for self-cleaning of biofouling. *Desalination*, *360*, 8–12. Reprinted with permission.

membrane (Hashaikeh et al., 2014; Lalia et al., 2015). The resulting membrane had a surface electrical conductivity of 10 S cm^{-1}, as measured using the four- probe method, which was enhanced by printing silver electrodes to enable electrolysis. MWCNT was coated onto a commercial membrane by vacuum filtration.

In this work, we applied the membrane as cathode in an electrofiltration cell with a stainless steel electrode and saltwater as electrolyte. Filtration was stopped at intervals to carry out electrolysis at 2 V for 2–3 minutes, allowing the evolution of hydrogen bubbles to lift off the foulant layer, which in turn led to significant flux recovery. Periodic electrolytic cleaning was effective for both organic and inorganic foulants. In another work, we developed self-standing buckypaper membranes using PVDF as binder in a networked CNT structure (Lalia et al., 2015). These CNTs are fabricated via a scalable CVD method, and the resulting tubes contain a large number of defects but offer ease of film formation. After four filtration cycles using yeast as foulant, the flux was still at >70% of initial flux, whereas it had declined to 40% when no voltage was applied (see Figure 5.11).

Fan et al. prepared CNT/ceramic hollow fiber membranes for fouling mitigation via electropolarization (Fan et al., 2016). The application of square-wave potentials between +1 V and −1 V resulted in a permeate flux >8 times greater than with no electropolarization for bacterial feedwater and 1.5 times higher than without electropolarization for NOM, which confirms its role in the mitigation of both organic and biological fouling. According to the authors, fouling mitigation was due to a combination of electrostatic repulsion, electrochemical oxidation, and electrokinetic behavior.

Research in membrane technology has recently extended to ceramic materials due to the high chemical resistance, greater temperature range, high chemical stability, and long lifespan compared to their polymeric counterparts (Asif & Zhang, 2021; Kayvani Fard et al., 2018). Existing methods for ceramic membrane fabrication suffer from operational challenges that make it difficult to make dense membranes (C. Li et al., 2020), rendering ceramic membranes suitable for wastewater treatment and for pretreatment to RO desalination. Today, porous ceramic MF and UF membranes are commercially available, and there is an increasing interest in developing functional ceramic membranes for emerging desalination processes such as NF and MD (Arumugham et al., 2021). Through rapid development of inorganic membranes as well as electrically conductive membranes, a few researchers have recently combined these two exciting areas to realize electrically conductive ceramic membranes for fouling mitigation.

Anis et al. developed electrically conducting ceramic MF membranes for biofouling control from nanozeolite and carbon nanotubes (Anis et al., 2021). The composite membrane recovered 95 and 90% of the original flux during filtration with yeast and SA, respectively, when −1.5 V were applied for 4–5 minutes after the first cycle of electrolytic cleaning.

It was also shown that higher negative ionic charge for PPy-coated polysulfone membranes results in superior electrostatic repulsion of ions from the surface and high rejection (Bhattacharya et al., 2008).

Alvarez and coworkers developed polycarbonate (PC)/CNT hybrid MF membranes using drawable nanotubes for biofouling control (Alvarez et al., 2017). A drawable CNT array is a type of CNT forest in which CNTs are aligned vertically during CVD growth (Miao, 2015). Resistive heating of the membrane at a voltage of 20–25 V and power consumption of 2.6–3 W elevated effective temperatures to >100 °C, which killed *E. coli* in the feed suspension. In about 60 seconds, the heating effect helped eliminate 99% of the *E. coli* from the membrane surface (Alvarez et al., 2017).

Demirel's group embedded silica-coated MWCNTs (SiO_2-CNT) into a pristine PVDF polymer matrix to improve membrane permeability by 25% from 303 to 377 L m^{-2} h^{-1} due to incorporation of hydrophilic nanoparticles and improved morphological properties resulting from nanoparticle dispersion (Demirel & Dadashov, 2021). Embedding silica-coated CNTs also increased rejection of sodium alginate (SA) and flux recovery from 74.2 to 94.7% due to superior antifouling (Demirel & Dadashov, 2021).

5.4.1 Removal and/or Degradation of Organic Contaminants

Organic pollutants in aqueous streams pose health concerns for the end user. As such, many studies focus on applying membranes to the removal of such contaminants. Mavukkandy et al. developed PVDF membranes with CNT/polyvinylpyrrolidone (PVP) immobilized on the top layer to increase permeability and improve removal of methylene blue (MB) (Mavukkandy et al., 2018). Mónica Silva and coworkers mixed different amounts of pristine MWCNTs with cellulose acetate non-solvent-induced phase separation membranes for dye removal, but percolation of CNTs was not confirmed at 1 wt% CNT loading due to aggregation (Silva et al., 2020). In their work, CNTs imparted increased surface hydrophilicity and improved the removal of MB.

As mentioned previously, self-supporting CNT membranes with a paper-like structure are commonly used as membrane materials. While defects may adversely affect mechanical properties such as tensile strength by up to 85% (Carraher, 2013), the same defects contribute to stable porous CNT membranes (Shah et al., 2013). Additionally, defects provide more active sites for electrocatalytic activity and are useful in applications where electrochemical properties are combined with filtration. Buckypaper membranes prepared with fast, scalable processes have thus been an area of interest (L. F. Dumée et al., 2010; Rashid et al., 2014).

Rashid et al. (Rashid et al., 2014) fabricated freestanding buckypaper membranes with multiwall carbon nanotubes (MWCNTs) dispersed with different surfactants, including Triton X-100 (Trix), and macrocyclic ligands. MWCNTs were either as grown, amine functionalized or carboxylic acid functionalized. Interestingly, all the membranes were hydrophilic with contact angles ranging from $28°$ to $55°$. Electrical conductivities for all MWCNT membranes fell within the range of 24–58 S cm^{-1}. The average pore size of the membranes ranged from 20 to 26 nm. The electrical conductivity of MWCNTs was found to be lower than that of SWCNT membranes prepared with the same surfactant. The membrane was applied to filtration of bisphenol A (BPA) solution and a mixture of 12 trace organic contaminants including pharmaceuticals, personal care products, and pesticides, all with molecular weights under 400 g mol^{-1}. BPA was removed via adsorption with a constant rejection of 90% for all membranes tested. The authors showed that choice of surfactant had an effect on specific surface area and hence rejection for specific organic contaminants. MWCNT-Trix was not suitable for removal of primidone, a hydrophilic neutral pharmaceutical.

Salehi and Madaeni (Salehi & Madaeni, 2010) investigated the impact of membrane conductivity on BSA adsorption on its surface. They prepared membranes by polymerizing pyrrole on polysulfone UF membranes. Their findings suggest that the adsorptive capacity of BSA can be increased by increasing membrane conductivity: The adsorptive capacity of BSA on the conductive membrane is 19.2 g m^{-2}, compared to 140 g m^{-2} for the non-conducting PS membrane. The negatively charged conductive membrane shows lower protein adsorption as the surface repels the negatively charged BSA molecules. Additionally, since the membrane is covered with PPy, there may be fewer active adsorptive sites on the surface for protein adsorption. Karimi et al. developed conducting polymer membranes using polypyrrole/polyacrylonitrile (PPy-PAN) membranes for electroseparation of dyes from synthetic wastewaters (Karimi et al., 2014). They synthesized PPy layers on either side of the PAN substrate films from an ethanol solution containing 5 wt% pyrrole and FeCl$_3$ as oxidant. They investigated the effect of applying an electric current for the separation of various dyes in an attempt to determine optimal electric current for dye molecules transport. Optimum currents of 2 and 1.5 mA were found for the Basacryl Red GL and Basacryl Blue GL dyes (Karimi et al., 2014). Ma and coworkers developed a carbon paper–polyether sulfone (Car-PES) UF membrane via phase inversion with carbon paper as the substrate and PES as the active layer (C. Ma et al., 2020). Under application of –3 V, electrostatic repulsion prevented transport of negatively charged foulants (BSA, SA, yeast, and emulsified oil) and improved TOC removal efficiency by up to 97%. Wang et al. used non-solvent-induced phase separation (NIPS) to prepare PANI membranes with a pure water flux of 15.5 L m^{-2} h^{-1} bar^{-1} (K. Wang et al., 2019). The application of an external voltage alleviated BSA fouling, as indicated by decreased flux decline.

Anis et al. reported the fabrication of UF membranes based on nanozeolite and CNTs for the simultaneous removal and degradation of crystal violet dye from a solution of NaCl and Na_2SO_4. The salt solution acts as electrolyte to assist electrolysis (Anis et al., 2022). At –3 V, they achieved dye rejection of 100% with a flux of 210 L m^{-2} h^{-1}. They used mass spectroscopy to confirm that dye was not only removed but also degraded by applying a potential to the conducting inorganic membrane. They also observed that Donnan steric repulsion at higher potentials improved cation permeation and increased resistance for anion transport. In their later work, Anis and coworkers coated commercial Al_2O_3 membranes with titanium via an e-beam deposition process (Anis et al., 2022). The resulting conductive membrane was superrhydrophilic and displayed decent electrocatalytic activity for hydrogen evolution, with overpotentials of 0.45 and 0.4 V (vs. the reference hydrogen electrode) in acid and base solutions, respectively. They applied the titanium-coated membranes for fouling control using yeast and SA, obtaining high flux recovery after each subsequent electrolysis cycle, via hydrogen bubbles physically sweeping foulants away from the surface.

Enhanced hydrophilicity has several benefits in pressure-driven filtration applications such as ultrafiltration (UF). Functionalization of MWCNTs with carboxyl or hydroxyl groups can increase hydrophilicity, which in turn can increase water flux and reduce fouling. Arockiasamy et al. (Lawrence Arockiasamy et al., 2013) incorporated carboxylated MWCNTs in poly(phenylene sulfone) PPSU membranes as UF membranes for protein separation and water treatment. They found that the increased hydrophilicity of the blend membranes caused slower fouling and higher protein adsorption. Majeed et al. incorporated hydroxylated MWCNTs with polyacrylonitrile (PAN) and prepared phase inversion UF membranes (Majeed et al., 2012). The water flux increased drastically by 63% when filler content was only 0.5 wt%; the authors attributed this improved flux to an increase in hydrophilicity. Tensile strength was almost double that of near PAN membranes when a filler loading of 2 wt% was used, and the composite membranes were more resistant to compaction. Often, more than tensile strength membrane, resistance to compaction plays an important role in pressure-driven processes where the membrane is compacted by the applied pressure.

Mantel and Ernst coated a 40 nm thick layer of gold on both sides of a hydrophilic flat-sheet PA membrane with permeability between MF and UF (Mantel et al., 2021). The membrane was coated on both sides to eliminate the need for a separate counterelectrode. Applying +2.5 V to the active side induced electrosorption of negatively charged substances onto the membrane. The removal rate for NOM, humic acid, and Brilliant Blue ionic dye increased from 1–5% at 0 V to 75, 93, and 99% when a potential of +2.5 V was applied, indicating that electrosorptive UF reaches an NOM removal rate comparable to NF membranes.

Lee et al. developed CNT membranes for simultaneous chlorine removal and oxidation of organic compounds during filtration (Lee et al., 2021). They were able to remove up to 80% of chlorine in the feed, although chlorine removal depended on the mass of CNTs within the membrane and applied pressure. A cathodic current regenerates the membranes by providing electrons for the reductive decomposition of chlorine. They suggested that MF/UF pretreatment should be replaced by electroconductive membranes that are able to remove chlorine in addition to organic contaminants without the use of any chemicals.

Du et al. fabricated composite hollow fiber membranes based on CNTs and electrospun PAN/PVDF nanofibers (Du et al., 2020). Electrically assisted filtration at a cathodic potential of 2 V enhanced removal of turbidity and TOC to 93 and 43%, respectively, compared to 73 and 31% without applied potential. Their findings suggest that both positive and negative potentials enhance removal of turbidity and TOC, but the enhancement effect is more pronounced with negative potentials. This enhanced removal is attributed to the electrostatic interaction between the hollow fiber membrane and negatively charged organic matter and colloids.

5.4.2 Microbial Decontamination

Due to the antimicrobial activity of CNTs just described, there has been much interest in polymer membranes embedded with CNTs for bacterial inactivation and biofouling prevention under the effect of an electric bias (C.-F. de Lannoy et al., 2013; Dumée et al., 2011). De Lannoy et al. fabricated polymer nanocomposite membranes using carboxylated MWCNTs with a diameter <8 nm and length between 10–30 μm embedded in interfacial polymerization polyamide films (C.-F. de Lannoy et al., 2013). The CNTs were first dispersed in water with sodium dodecylbenzene as surfactant and then vacuum filtered through a PES membrane. The CNT-coated PES membrane was immersed in a 2% (w/w) aqueous solution of *m*-phenylenediamine for 30 seconds, slightly dried, and then immersed into a solution of trimesoyl chloride (TMC) dissolved in hexane. This method was used to form thin film composite polyamide membranes on the CNT-coated PES membrane, as shown in Figure 5.12; FTIR analysis showed that the CNTs were embedded with the polyamide and formed ester bonds with TMC.

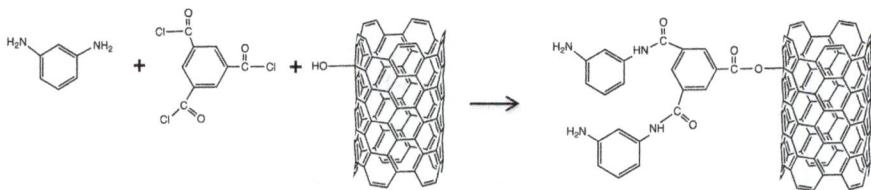

FIGURE 5.12 Condensation reaction between trimesoyl chloride, *m*-phenylenediamine, and hydroxyl group on CNT sidewall leading to the incorporation of the CNTs into the membrane matrix.

Source: Reprinted with permission from *Environ. Sci. Technol.* 2013, 47, 6, 2760–2768. Copyright © 2013 American Chemical Society. https://doi.org/10.1021/es3045168

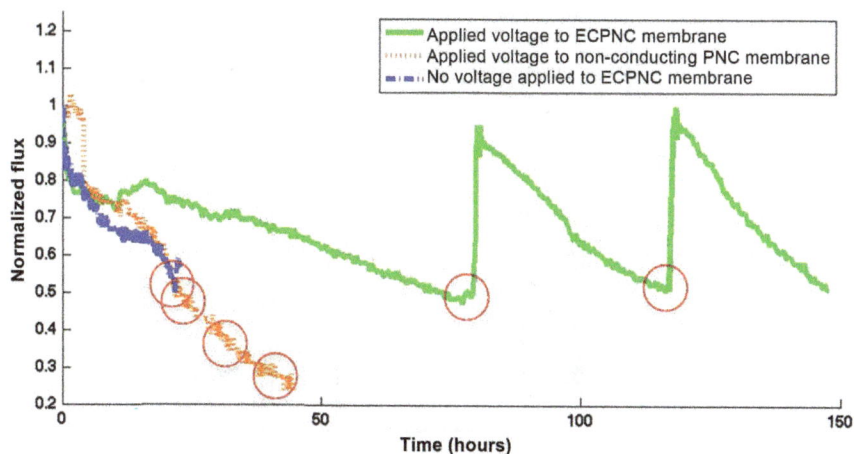

FIGURE 5.13　Membrane flux for control and electrically conductive polymer nanocomposite (ECPNC) membranes without applied voltage and with applied voltage during filtration of *P. aeruginosa* suspension (10^8 colonies forming units mL^{-1} of *P. aeruginosa* in 10% lysogeny broth). Red circles represents membrane flushing points.

Source: Reprinted with permission from *Environ. Sci. Technol.* 2013, 47(6), 2760–2768. Copyright © 2013 American Chemical Society. https://doi.org/10.1021/es3045168

With an electrical conductivity of 400 S m^{-1} and NaCl rejection >95%, the membranes were further applied to biofouling control in a cross-flow filtration cell (C.-F. de Lannoy et al., 2013). First, highly bacterially contaminated feed solution was used to induce fouling, and filtration was carried out until flux declined by 45%, indicating a biofilm-induced non-reversible fouling. When an electric potential in the form of a 1.5 V square wave at 16.7 mHz was applied to the membrane, bacteria were still deposited on the membrane, but there was no attachment, allowing full recovery of flux following a short rinse and without the use of cleaning chemicals. Figure 5.13 shows flux behavior for control and electrically conductive membranes with and without applied voltage during filtration of *P. aeruginosa* bacterial suspension. Red circles represent membrane flushing points (C.-F. de Lannoy et al., 2013).

Vecitis et al. applied a small voltage of 2 and 3 V to MWCNT microfilters for 30 seconds post-filtration for the removal and inactivation of viruses and bacteria. The applied voltage inactivated >75% of the bacteria and removed more than 99.6% of the adsorbed viruses from the membrane, as shown in Figure 5.14 (Rahaman et al., 2012; Vecitis et al., 2011).

Another group sprayed CNT dispersions on porous substrates to produce composite membranes with the CNT layers as the active layer (D. Ma et al., 2021). One block of the copolymer adsorbs on the CNT surface via π–π interaction, while the other blocks are solvated, allowing homogeneous and stable CNT dispersions. The resulting membrane demonstrates 100% rejection to *E. coli* and has a pure water flux of ~3300 L m^{-2} h^{-1} bar^{-1} (D. Ma et al., 2021). The thickness of the CNT layer could be controlled via spray coating volume, although a thinner selective layer favors higher

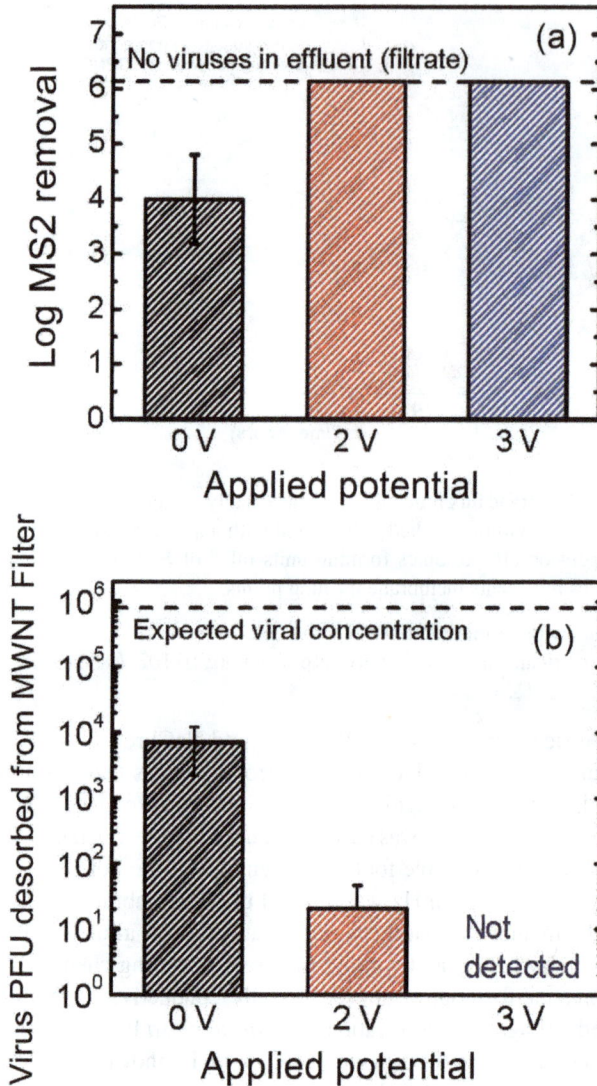

FIGURE 5.14 Electrochemical removal and/or inactivation for different potentials applied (a) during filtration and (b) post-filtration for 30 seconds with MWCNT microfilters (Vecitis et al., 2011).

Source: Reprinted with permission from *Environ. Sci. Technol.* 2011, *45*(8), 3672–3659. Copyright © 2011 American Chemical Society. https://doi.org/10.1021/es2000062.

water flux according to the Hagen–Poiseuille equation (Epstein, 1989). Liu et al. showed that applying a voltage of 100 mV to PVDF membrane blended with PPy as a conductive additive and sodium dodecyl benzene sulfonate (SDBS) as dopant increased retention of yeast (J. Liu et al., 2017).

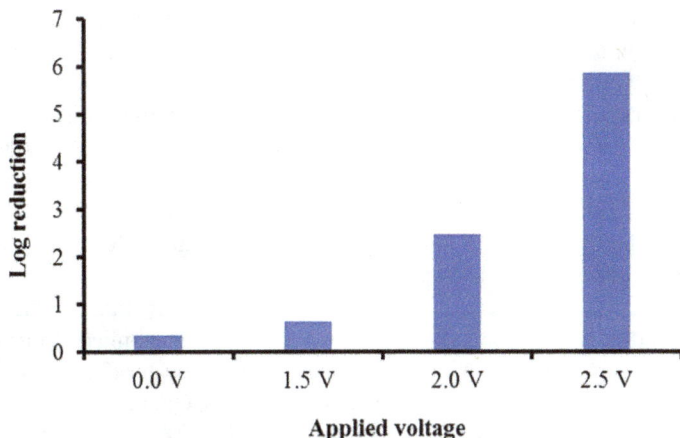

FIGURE 5.15 Effect of applied potential on bacterial cell viability with two LIG-PVA membranes stacked in dead-end filtration mode; feed: mixed bacterial cultural suspension ($\sim10^6$ CFU mL^{-1} in 0.9% NaCl solution); flow rate: $\sim$500 L m^{-2} h^{-1}.

Source: Reprinted with permission from *ACS Appl. Mater. Interfaces* 2019, 11, 11, 10914–10921. Copyright © 2019 American Chemical Society. https://doi.org/10.1021/acsami.9b00510.

Another group in China built on their work to control biofouling by in situ electrochemical cleaning using 2.5 V for 30 minutes to restore 85% of the initial water flux (Y.-J. Wang et al., 2022). Electrogenerated oxidative species removed deposited bacterial cells and EPS from the membrane surface.

Thakur et al. developed laser-induced graphene (LIG)-PVA composite membranes (LIG-PVA), which, when used as electrodes, showed complete elimination of mixed bacterial culture viability (Thakur et al., 2019). Complete disinfection was observed at 2.5 V as compared to 39% bacterial removal when no potential was applied. Figure 5.15 shows the effect of applied voltage on bacterial cell viability with two LIG-PVA membranes stacked in dead-end filtration mode. They attributed voltage-dependent antimicrobial action to both electrical and chemical effects. As bacteria contact the electrode surface, rapid physical destruction or direct oxidation may occur. Additionally, electrochemical generation of toxic species such as hydrogen peroxide may exist in high concentrations near the electrode. The same group made conductive composite UF membranes using LIG and GO and found that increasing the amount of cross-linked GO on the LIG surface increased both BSA and bacterial rejection. The films exhibited 83% less biofilm growth compared to a typical polymer UF membrane under non-filtration conditions. When an anodic potential of 3 V was applied, flux improved by 11%.

5.4.3 Oily Wastewater Treatment

Oily wastewater is generated as a by-product across several industries including oil and gas, food and beverage, shipping, tanning, textile, and metal manufacturing

(Cheryan & Rajagopalan, 1998; Tanudjaja et al., 2019). Oil-water separation is an essential step in treating oily wastewater as the effects of direct disposal can be detrimental to the environment. In particular, treatment of produced water from oil and gas extraction could help recover oil from waste that is otherwise injected back into the wells, while simultaneously providing freshwater. Produced water is the largest waste stream generated during oil and gas extraction (Fakhru'l-Razi et al., 2009). The efficient treatment of produced water is a contemporary challenge as we learn of the harmful effects of direct discharge on the environment. Removing organics is considered an essential step in treating produced water.

Geng and Chen modified the inner layer of an Al_2O_3 MF membrane with Magnéli Ti_4O_7 and studied the antifouling of the membrane under an applied electric field for three different feed solutions: peanut-oil-based oily wastewater (200 mg L^{-1}), humic acid (6 mg L^{-1}), and BSA (1 g L^{-1}) (Geng & Chen, 2016). Operating the membrane at critical electric potentials (30–40 V) increased specific flux by over a factor of three as compared to the uncoated membrane. The critical electrical potential is the potential that offsets the migration of charged species toward the membrane by convection (Qiu et al., 2018; Ratnaningsih et al., 2021). During fouling control experiments, the Ti_4O_7/Al_2O_3 composite membrane was used as anode. The initial flux of the composite membrane is >25% greater than that of the uncoated Al_2O_3 membrane; when a potential of 40 V is applied, the membrane sustains 91.8% of the initial specific flux during 1 hour of operation, whereas specific permeate flux of the uncoated membrane drops to 15.6% of its initial value in the same period. Applied electric potential causes a strong electrophoretic effect inside the cell, and oil particles are dragged away from the anode in the presence of the cationic surfactant cetrimonium bromide (CTAB), which also results in an energy consumption reduction of 58%. Even without the application of applied potential, the composite membrane exhibits stronger antifouling due to greater hydrophilicity.

Professor Liu's group in China prepared a reduced graphene oxide/PVDF membrane by phase inversion of PVDF and graphene oxide on carbon fiber and then heat treating the resulting membrane in hydroiodic acid (Y. Zhang et al., 2016). GO results in smaller pore size, and the thermal treatment helped partially reduce GO to rGO, which rendered the membrane more conductive but less hydrophilic. The membrane was used to remove polyacrylamide (PAM), a macromolecule pollutant in oil field wastewater; filtration was assisted with an external electric field of 0.6 V cm^{-1} (Y. Zhang et al., 2016).

Li et al. coated an α-alumina membrane support with dopamine and CNTs and applied this electrically conductive layers for the fouling control of flat-sheet ceramic membranes for the filtration of oily wastewater (P. Li et al., 2021). A negative charge induced by applying 2 V reduced membrane fouling by 50%, and the energy consumption of the system was 22.2 × 10^{-3} kWh m^{-3} when filtration was paused for 15 seconds for cleaning. Electrostatic force prevented foulants from attaching to the surface.

Due to several desirable characteristics such as high permeabilities, open pore structures, high porosity, and tunable morphologies, electrospun membranes have been the focus of several reports addressing the treatment of oily wastewater (Ejaz Ahmed et al., 2014; Ge et al., 2018; S. K. Hong et al., 2018; W. Ma et al., 2019; J.

Wang et al., 2018; M. Zhang et al., 2019; Zhu et al., 2019). To date, only one group has used electrospinning to develop electrically conductive membranes to treat oily water. They fabricated superhydrophobic membranes by decorating sonicated CNTs on polyurethane nanofibers, followed by methyltrichlorosilane (MTS) modification (Huang et al., 2019). CNTs were added to improve conductivity, thermal stability, and mechanical properties of the nanofibrous membrane. Cyclic stretching did not negatively impact electrical conductivity and superhydrophobicity of the hierarchical membrane structure. With a water contact angle of 152 and k_{el} of 7.6 S m^{-1}. The membrane modified for 30 minutes was used to separate oil from pure water as well as corrosive solutions including salt, acid, and alkali; membrane flux was maintained above 1200 L m^{-2} h^{-1} with a separation efficiency >96% even after 30 cycles.

In another study, electrically conductive TFC membranes embedded with CNTs were applied to fouling control in forward osmosis during oil-water separation (Fan et al., 2018). With draw solution with 2 M NaCl and pure water as feed, the electrically conductive membrane had a flux 3.5 times greater than a commercial cellulose triacetate FO membrane. Furthermore, the electrically conductive membrane exhibited high organic and microbial fouling resistance with electrochemical assistance. Compared to open circuit conditions, a flux improvement of 50% was observed for oil-water separation under 2 V. Conductive fillers thus not only improve permeability and mechanical stability, but their ability to be used as an electrode also assists antifouling and mitigation strategies in the filtration of oily wastewater. To fabricate conductive membranes, they dispersed carboxylated CNTs using SDS as surfactant and loaded appropriate amounts of CNT dispersion onto an MF PES supporting layer via vacuum filtration. The PES-CNT membrane acted as support for interfacial polymerization of a TFC PA active layer. Shakeri and Rastgar fabricated graphene/PANI membranes through laminating via a pressure-assisted technique onto a polyamide-imide (PAI) support layer (Shakeri et al., 2019). Cross-linking of PANI with graphene nanosheets and thermal treatment helped improve mechanical stability. Flux recovery rate and fouling resistance both improved when 2 V anodic potential was applied to the conductive membranes using SA as a model organic foulant in the feed solution.

Yi et al. developed CNT-PVA composite membranes and used them as cathode during filtration of n-hexadecane-in-water emulsion (n-emulsion) and cutting fluid emulsion to enhance oil removal efficiency and reduce membrane fouling (Yi et al., 2018). The application of -1.5V to the conductive membrane increases the oil removal rate from 85 to 98% for n-emulsion and from 57 to 83% for c-emulsion. They also observed an increase in flux by a factor of >2 with applied potential. The membranes showed stability with five runs, although each run was only carried out for 60 minutes.

MD has emerged as a promising technology for the treatment of oily wastewater (Kalla, 2021). Han et al. developed an electrically conductive membrane cathode by coating CNTs on a commercial MD membrane for the treatment of concentrated hexadecane-in-saline water emulsions (2000 mg L^{-1}) using DCMD (Han et al., 2021). At a cell potential of 3 V, the modified membrane suffered a flux decline of less than 5%, and the anti-oil fouling behavior was confirmed over three cycles. The antifouling performance is due to an energy barrier caused by electrostatic repulsion.

Ho and coworkers fabricated membranes for the treatment of palm oil mill efflu-ent (POME) by blending GO and MWCNTs with PVDF membrane matrix via phase inversion (Ho, Teoh et al., 2021; Ho et al., 2019). Applying response surface meth-odology (RSM), they determined an optimum nanomaterials concentration of 4.22 wt% and an electric field of 221 V cm^{-1} applied for 6 minutes (at 32 minute intervals) in intermittent mode can achieve high normalized flux in an electrically enhanced filtration system. RSM is a statistical tool used to optimize systems, in this case either high flux or minimum environmental impact. Later, they found that the optimum weight ratio of GO:MWCNTs is 1:9 compared to using pristine MWCNTs, in terms of environmental impacts.

Studies on self-cleaning inorganic membranes is limited but crucial to improving the fouling resistance of ceramic membranes. Anis et al. applied ceramic nanozeo-lite/CNT membranes to oil-water separation with periodic electrolysis for flux recov-ery (Anis et al., 2021). Periodic electrolysis was carried out at 7 V for 30 seconds between each filtration cycle with the ceramic membrane as cathode and a graphite counter-electrode. They attributed the rapid permeation of water with flux > 400 L m^{-2} h^{-1} to the hierarchical structure of the membrane, which has hydrophilic zeolite nanoparticles embedded with CNTs. The CNTs acted as binder for zeolite nanoparti-cles and imparted flexibility and electrical conductivity. Periodic electrolysis allowed flux recovery to be as high as 80% at the end of 10 cycles through the generation of hydrogen gas bubbles. Figure 5.16 shows flux decline and rejection with and without periodic electrolytic cleaning.

Treatment of oily wastewater remains an exciting avenue for future water treat-ment research. Electrically active materials that serve as electrochemical sites may be at the forefront of improving oily wastewater treatment systems: electrochemi-cal cleaning, electrochemically enhanced fouling, and electrochemical degradation of oily contaminants are all ways in which electrically conductive materials can advance oil-water separation.

5.4.4 REMOVAL OF TOXIC METALS

Heavy metal contamination in aqueous streams poses a serious health and environ-ment hazard (Vilela et al., 2016). Heavy metals are present in industrial wastewater in unsafe amounts and may end up affecting aquatic ecosystems (Izah et al., 2022). Current research in wastewater treatment demands membrane materials for fast and efficient removal of pollutants and heavy metals from water.

The adsorption capacity for metal ions from aqueous solutions increases when the surface of CNTs is oxidized, as compared to pristine CNTs (Kandah & Meunier, 2007; Rao et al., 2007). Rao et al. attribute sorption to chemical inter-actions between metal ions and surface functional groups of CNTs (C. Lu & Liu, 2006).

Modification of CNTs is often carried out with the goal of increasing sorption capacity of resulting electrodes for the removal of metal ions. Zhan et al. incor-porated chitosan to CNTs and found that chitosan-modified CNT electrodes have a 25% higher removal ratio than pristine CNTs for Cu^{2+} ions (Zhan et al., 2011). This improved removal results from a greater number of binding sites for metal ions,

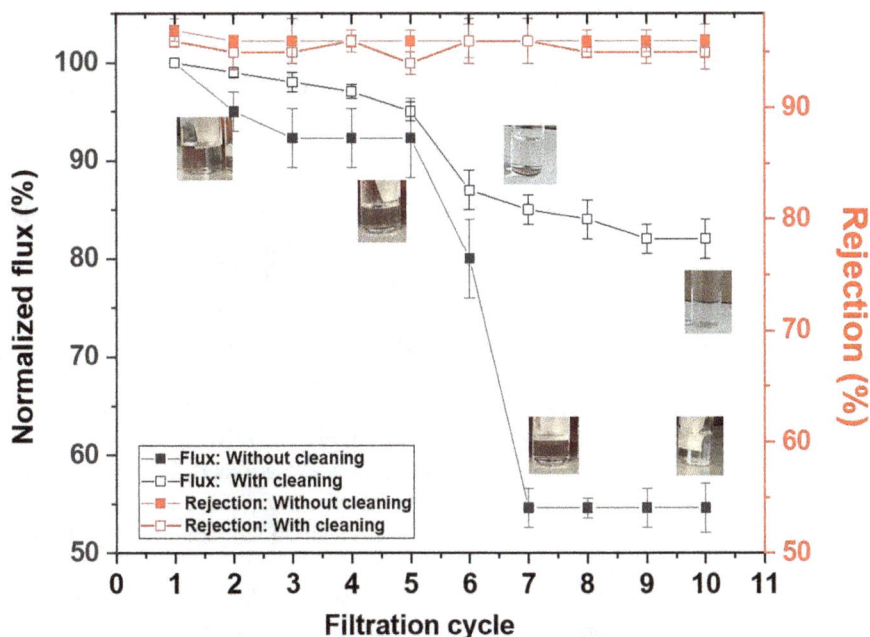

FIGURE 5.16 Flux and rejection during crude oil filtration for nanozeolite/CNT composite membranes with 60 wt% zeolite for 10 filtration cycles with and without electrolysis (feed: 600 ppm crude oil and 2500 ppm NaCl; pressure applied: 0.7 bar).

Source: Anis, S. F., Lalia, B. S., Hashaikeh, R., & Hilal, N. (2021). Hierarchical underwater oleophobic electroceramic/carbon nanostructure membranes for highly efficient oil-in-water separation. *Separation and Purification Technology*, *275*, 119241. Reprinted with permission.

lower zeta potential, higher surface area, and increased hydrophilicity (Zhan et al., 2011). Interestingly, research in electrically conductive membranes for removal of toxic metals has not progressed significantly. Dr. Hesham Hamed's group developed electrically conductive hybrid composite membranes based on electrospun Fe_3O_4/MWCNTs/polyamide 6 nanofibers for the removal of Pb(II) from aqueous solution, favored in weak acid (pH 6) media (Bassyouni et al., 2019). They reported that the membranes were easily regenerated for four adsorption-desorption cycles, making the composite promising as an advanced adsorbent for industrial-scale heavy metal removal. It will be interesting to see how this membrane performs for removal of toxic metals from more complex systems and whether the electrical conductivity can be exploited to improve separation.

5.4.5 Other

Duan and coworkers applied an electrically conducting UF and NF membrane system, along with an anaerobic sequencing batch reactor, for the treatment of high strength industrial wastewater containing large concentrations of paint stripper (benzyl alcohol) (Duan et al., 2016). UF membranes were prepared by vacuum filtering CNT-COOH and then PVA on a PSF support such that the PVA:CNT ratio remained 3:1. The PVA layer was cross-linked after deposition using glutar-aldehyde and hydrochloric acid in water. The NF membranes were made by first depositing CNT-COOH on a PSF support and then carrying out IP to fabricate a PA layer on top. When a cathodic potential of 5 V is applied to the UF membranes, fouling is significantly reduced as compared to 0 V (Duan et al., 2016). On the other hand, application of a negative electric potential did not impact NF membrane fouling, although positively charged membranes led to polymerization of fouling molecules on the NF surface, which in turn increased membrane fouling and prevented simple cleaning and flux recovery. The same conducting UF membranes were previously applied to limit microbial attachment using low voltages (Ronen et al., 2015). A potential of 1.5 V applied to the electrochemical cell using titanium as counter-electrode was considerably more effective in inactivating *E. coli* bacteria than 1 V. At low potentials, small concentrations of hydrogen peroxide were generated through electroreduction of oxygen; exposure to electrochemically generated hydrogen peroxide prevented bacterial attachment during filtration (Ronen et al., 2015).

Hou et al. developed electrically conductive membrane electrodes and used them in conjunction with microbial electrolysis for removal of ammonia from wastewater (Hou et al., 2018). Membrane fabrication consisted of electrodepositing a thin-film nickel layer on a hydrophobic PP support. They showed that the integrated membrane electrode had 40% greater NH_3-N recovery rate and 11% higher current density. Applying a negative potential helped the membrane repel negatively charged organics and microbes and reduced fouling.

5.4.5.1 Electrically Aligned CNTs

Very recently, electric fields have been used to manipulate the orientation of CNTs during the preparation of CNT membranes. Aligned CNTs offer excellent fluid transport properties, as reported in the last decade (Yingchun Liu et al., 2005; Nicholls et al., 2012). However, CNT-based materials suffer from agglomeration, and the control of CNT alignment within the membrane matrix remains a challenge in the development of CNT-based membranes (Sears et al., 2010; Street et al., 2014). Wu et al. applied an alternating electric field to improve MWCNT dispersion and align MWCNTs in a polystyrene membrane (B. Wu et al., 2014). Figure 5.17 shows a schematic of the electric field apparatus used for the preparation of their MWCNT-polystyrene membranes, prepared using 3 wt% MWCNTs with varying AC frequencies (B. Wu et al., 2014). "Electro-casting" reduced aggregation considerably; in addition, increasing frequency from 1 to 100 Hz led to improved dispersion. An electric field is considered a reliable method of aligning CNTs as the conductive tubes align themselves in the direction of the applied field. The highly anisotropic electric polarizability allows CNTs to orient in a polymer matrix under electric field, as reported by Zhang et al. (D. Zhang et al., 2021). They studied the kinetics of CNT migration in a CNT/epoxy mixture under the effect of a direct current field and found that CNT migration toward electrodes depends on electric field

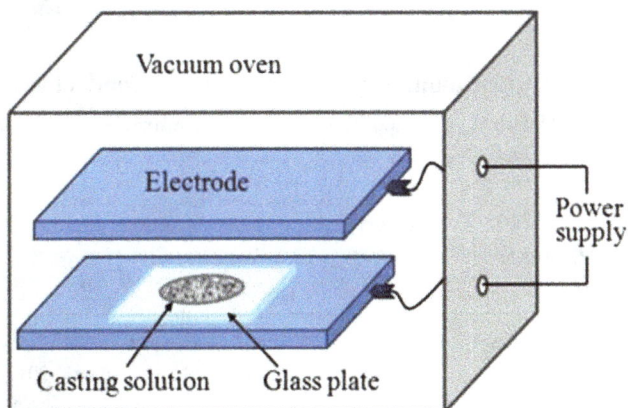

FIGURE 5.17 Schematic presentation of apparatus used for electrical alignment of MWCNTs in polymer matrix.

Source: Wu, B., Li, X., An, D., Zhao, S., & Wang, Y. (2014). Electro-casting aligned MW-CNTs/polystyrene composite membranes for enhanced gas separation performance. *Journal of Membrane Science*, *462*, 62–68. Reprinted with permission.

strength, whereas CNT concentration and length distribution do not significantly affect migration velocity. Such a fundamental understanding of the behavior of CNTs enables control over spatial distribution of CNTs to manipulate the mechanical, thermal, and electrical properties of resulting CNT/polymer composites (D. Zhang, She et al., 2021). Castellano and coworkers combined an AC+DC electric field with a solvent-deposition approach to fabricate vertically aligned CNT membranes (Castellano et al., 2020). Application of an electric field increases CNT alignment from 18 to 72%, and the high alignment is retained during polymer infusion and curing. They also confirmed flow through CNT pores using non-destructive He-N$_2$ flowrate-ratio testing as well as filtration with Au nanoparticles and dye exclusion. Compared to CVD-grown aligned CNT forests, these field-aligned solution-fabricated vertically aligned CNT membranes exhibited a 100–300 times flow enhancement. Swain and coworkers also included MWCNTs and reduced graphene oxide (rGO) in polysulfone membranes and applied an AC electric field to align the nanofillers for selective separation of O$_2$/N$_2$ gas. The introduction of nanofillers increased permeability of the membranes due to enhanced permeation from CNTs and created a long, tortuous path from rGO nanosheets, which allowed enhanced selectivity.

Liu and coworkers used a DC electric field to align CNTs in the fabrication of PES membrane as the substrate for TFC NF membranes (C. Liu et al., 2020). The membrane with aligned CNTs demonstrated improved chlorine resistance and antifouling ability.

Table 5.2 shows selected studies of electrically conductive polymer membranes.

TABLE 5.2

Characteristics and Applications of Selected Electrically Conductive Membranes.

Material	Electrical Conductivity (S m^{-1})	Application	Performance	Reference
Nanozeolite/CNT	4.9×10^3	Oil-water separation	80% flux recovery after 10 cycles of periodic electrolysis at 7V; flux > 400 L m^{-2} h^{-1}	(Anis et al., 2021)
PANI	2.2×10^{-4}	Ultrafiltration	Application of 1V improved BSA fouling resistance	(K. Wang et al., 2019)
Au coated on PES	3×10^7	Electrosorptive ultrafiltration	Positive charge of 2.5 V leads to electrosorption of dyes and NOMs; NOM removal rate reaches that of NF membranes.	(Mantel et al., 2021)
Ni/PP	1×10^3	Ammonia recovery	Membrane electrode showed 40% higher NH$_3$–N recovery rate and 11% higher current density; negative potential reduced fouling.	(Hou et al., 2018)
CNT-PVA	2.5×10^2	Oily wastewater treatment	Permeation flux after an operation of 60 min with −1.5 V is over 2 times of that without; oil removal also enhanced under applied potential.	(Yi et al., 2018)
PANI/bacterial cellulose (BC)	5×10^{-2}	—	—	(W. Hu et al., 2011)
PAN/MWCNTs	39	Turbidity and TOC removal	Applied voltage mitigates membrane fouling and enhances efficiency.	(Du et al., 2020)
Graphene/PES	1.7×10^3	Removal of organic and inorganic pollutants	At 3 V, 90% of Cr(VI) was removed after 6 hours.	(Thamaraiselvan et al., 2021)
PPy-PS		Protein rejection and protein adsorption	BSA adsorption decreases from 139.9 to 19.17 g m^{-1}.	(Madaeni & Molaeipour, 2010; Salehi & Madaeni, 2010)
PPy/PAN	2×10^{-1}	Electroseparation of dyes	High rejection of azo dyes	(Karimi et al., 2014)

TABLE 5.2 *(Continued)*

Characteristics and Applications of Selected Electrically Conductive Membranes.

Material	Electrical Conductivity (S m^{-1})	Application	Performance	Reference
CNT	5.8×10^3	Removal of trace organic contaminants	BPA removal of 90% was achieved, whereas other trace contaminants were removed with efficiency greater than 80% by MWCNT-Trix.	(Rashid et al., 2014)
CNT	—	DCMD	99% salt rejection and flux of ~12 kg m^{-2} h at a vapor partial pressure difference of 22.7 kPa	(Dumée et al., 2011; L. F. Dumée et al., 2010)
CNT	—	Electrochemical inactivation and removal of bacteria and viruses	Inactivation of more than 75% of the bacteria and more than 99.6% of adsorbed viruses on membrane was achieved by applying 2–3 V for 30 seconds after filtration.	(Rahaman et al., 2011; Vecitis et al., 2011)
CNT– PVDF	3.2×10^{-4}	Capacitive organic fouling reduction	Reduction in fouling rate constant from 0.022 h^{-1} to 0.0015 h^{-1} for NaCl feed and from 0.030 h^{-1} to 0.016 h^{-1} for CaCl$_2$, compared to control membrane	(Q. Zhang & Vecitis, 2014)
PS/PVA-CNT-COOH	Dry: 2.4×10^3 Soaked in water: 1.7×10^3	UF membranes for fouling inhibition	Applying voltage of 3 V and 5 V after 100 minutes of operation decreased operating pressure by 33 and 51%, respectively, compared to application of no voltage in a constant flux mode of 20 LMH.	(Dudchenko et al., 2014)
MWCNT on PVDF support	1×10^3	Self-cleaning via periodic electrolysis using CaCO$_3$ and yeast foulants	Voltage of 2 V applied every 30 minutes during 2 hours filtration of CaCO$_3$, flux was maintained at 63%, compared to 26% without cleaning.	(Hashaikeh et al., 2014)

(Continued)

TABLE 5.2 *(Continued)*

Characteristics and Applications of Selected Electrically Conductive Membranes.

Material	Electrical Conductivity (S m^{-1})	Application	Performance	Reference
CNS-PVDF	5×10^3	Periodic electrolysis using yeast foulant	Flux maintained at 70% of original value after 4 filtration cycles of 60 minutes each with periodic electrolysis of 2–3 V.	(Lalia et al., 2015)
Polyamide-CNT		NF	Conductivity of 400 S m^{-1} with NaCl rejection > 95%	(C.-F. de Lannoy et al., 2013)
CNT-COOH-PP	—	DCMD	Flux increased by 51.5% compared to unmodified PP membrane	(Roy et al., 2014)
MWCNT/ poly(phenylene sulfone) (PPSU) blend membranes	—	UF	Addition of MWCNTs resulted in better separation and improved water flux for protein rejection.	(Lawrence Arockiasamy et al., 2013)
MWCNT-PAN blend membranes	—	UF	Water flux increased by 63% at 0.5 wt% CNT loading. Tensile strength increased by over 97% at 2 wt% CNT loading.	(Majeed et al., 2012)

5.5 CONCLUSION

The concept of membrane electrodes using multifunctional materials enables merging electrochemical cells with membrane filtration units to enhance performance and mitigate fouling. Some mechanisms employed by electrically conductive membranes for fouling control are electrostatic interactions, bubble generation via redox reactions, and/or oxidation through redox reactions. This chapter showed much of the innovation in materials development and process optimization in exploiting electric fields for the mitigation of fouling and related phenomena. It also described the various applications of electrically conductive membranes, including heavy metal removal, microbial decontamination, treatment of oily wastewater, and seawater and brackish water desalination. We also addressed the shortcomings of common electrically conductive materials applied to membrane separation. A major challenge surrounding electrically conductive membranes is the seeming trade-off between electrical conductivity and separation properties. Additionally, commonly used materials such as CNTs are prone to oxidation under anodic voltages, which affects membrane integrity, especially if electrolysis is carried out for long periods or at high voltages. The lack of long-term studies, however, makes it difficult to predict membrane stability after being subjected to electrolysis. More work is needed to ensure robust materials with long-term operational stability and scalable configurations.

BIBLIOGRAPHY

Agarwal, A., Ng, W. J., & Liu, Y. (2011). Principle and applications of microbubble and nano-bubble technology for water treatment. *Chemosphere, 84*(9), 1175–1180. https://doi.org/10.1016/j.chemosphere.2011.05.054

Ahmed, F. E., Hashaikeh, R., & Hilal, N. (2019). Fouling control in reverse osmosis membranes through modification with conductive carbon nanostructures. *Desalination, 470*, 114118. https://doi.org/10.1016/j.desal.2019.114118

Ahmed, F. E., Lalia, B. S., Hilal, N., & Hashaikeh, R. (2017). Electrically conducting nanofiltration membranes based on networked cellulose and carbon nanostructures. *Desalination, 406*, 60–66. https://doi.org/10.1016/j.desal.2016.09.005

Ahmed, F. E., Lalia, B. S., Kochkodan, V., Hilal, N., & Hashaikeh, R. (2016). Electrically conductive polymeric membranes for fouling prevention and detection: A review. *Desalination, 391*, 1–15. https://doi.org/10.1016/j.desal.2016.01.030

Al Aani, S., Haroutounian, A., Wright, C. J., & Hilal, N. (2018). Thin film nanocomposite (TFN) membranes modified with polydopamine coated metals/carbon-nanostructures for desalination applications. *Desalination, 427*, 60–74. https://doi.org/10.1016/j.desal.2017.10.011

Alvarez, N. T., Noga, R., Chae, S.-R., Sorial, G. A., Ryu, H., & Shanov, V. (2017). Heatable carbon nanotube composite membranes for sustainable recovery from biofouling. *Biofouling, 33*(10), 847–854. doi:10.1080/08927014.2017.1376322

Angeles, A. T., & Lee, J. (2021). Carbon-based capacitive deionization electrodes: Development techniques and its influence on electrode properties. *The Chemical Record, 21*(4), 820–840. https://doi.org/10.1002/tcr.202000182

Anis, S. F., Lalia, B. S., Hashaikeh, R., & Hilal, N. (2021). Hierarchical underwater oleophobic electro-ceramic/carbon nanostructure membranes for highly efficient oil-in-water separation. *Separation and Purification Technology, 275*, 119241. https://doi.org/10.1016/j.seppur.2021.119241

Anis, S. F., Lalia, B. S., Hashaikeh, R., & Hilal, N. (2022). Titanium coating on ultrafiltration inorganic membranes for fouling control. *Separation and Purification Technology, 282*, 119997. https://doi.org/10.1016/j.seppur.2021.119997

Anis, S. F., Lalia, B. S., Khair, M., Hashaikeh, R., & Hilal, N. (2021). Electro-ceramic self-cleaning membranes for biofouling control and prevention in water treatment. *Chemical Engineering Journal, 415*, 128395. https://doi.org/10.1016/j.cej.2020.128395

Anis, S. F., Lalia, B. S., Lesimple, A., Hashaikeh, R., & Hilal, N. (2022). Electrically conductive membranes for contemporaneous dye rejection and degradation. *Chemical Engineering Journal, 428*, 131184. https://doi.org/10.1016/j.cej.2021.131184

Arumugham, T., Kaleekkal, N. J., Gopal, S., Nambikkattu, J., K, R., Aboulella, A. M., . . . Banat, F. (2021). Recent developments in porous ceramic membranes for wastewater treatment and desalination: A review. *Journal of Environmental Management, 293*, 112925. https://doi.org/10.1016/j.jenvman.2021.112925

Asif, M. B., & Zhang, Z. (2021). Ceramic membrane technology for water and wastewater treatment: A critical review of performance, full-scale applications, membrane fouling and prospects. *Chemical Engineering Journal, 418*, 129481. https://doi.org/10.1016/j.cej.2021.129481

Bao, X., Long, W., Liu, H., & She, Q. (2021). Boron and salt ion transport in electrically assisted reverse osmosis. *Journal of Membrane Science, 637*, 119639. https://doi.org/10.1016/j.memsci.2021.119639

Bassyouni, D., Mohamed, M., El-Ashtoukhy, E.-S., El-Latif, M. A., Zaatout, A., & Hamad, H. (2019). Fabrication and characterization of electrospun Fe3O4/o-MWCNTs/polyamide 6 hybrid nanofibrous membrane composite as an efficient and recoverable adsorbent for removal of Pb (II). *Microchemical Journal, 149*, 103998. https://doi.org/10.1016/j. microc.2019.103998

Benson, J., Kovalenko, I., Boukhalfa, S., Lashmore, D., Sanghadasa, M., & Yushin, G. (2013). Multifunctional CNT-polymer composites for ultra-tough structural supercapacitors and desalination devices. *Advanced Materials, 25*(45), 6625–6632. doi:10.1002/adma.201301317

Bhattacharya, A., Mukherjee, D. C., Gohil, J. M., Kumar, Y., & Kundu, S. (2008). Preparation, characterization and performance of conducting polypyrrole composites based on polysulfone. *Desalination, 225*(1–3), 366–372. http://dx.doi.org/10.1016/j.desal.2006.07.018

Biesheuvel, P. M., & van der Wal, A. (2010). Membrane capacitive deionization. *Journal of Membrane Science, 346*(2), 256–262. https://doi.org/10.1016/j.memsci.2009.09.043

Bolt, H. M., Başaran, N., & Duydu, Y. (2020). Effects of boron compounds on human reproduction. *Archives of Toxicology, 94*(3), 717–724. doi:10.1007/s00204-020-02700-x

Calabriso, A., Borello, D., Cedola, L., Del Zotto, L., & Santori, S. G. (2015). Assessment of CO2 bubble generation influence on direct methanol fuel cell performance. *Energy Procedia, 75*, 1996–2002.

Carraher, C. E. (2013). *Carraher's polymer chemistry* (9th ed.). CRC Press.

Castellano, R. J., Praino, R. F., Meshot, E. R., Chen, C., Fornasiero, F., & Shan, J. W. (2020). Scalable electric-field-assisted fabrication of vertically aligned carbon nanotube membranes with flow enhancement. *Carbon, 157*, 208–216.

Chen, L.-Y., Jiang, M.-Y., Zou, Q., Xiong, S.-W., Wang, Z.-G., Cui, L.-S., . . . Gai, J.-G. (2021). Highly permeable carbon nanotubes/polyamide layered membranes for molecular sieving. *Chemical Engineering Journal, 425*, 130684. https://doi.org/10.1016/j.cej.2021.130684

Cheryan, M., & Rajagopalan, N. (1998). Membrane processing of oily streams. Wastewater treatment and waste reduction. *Journal of Membrane Science, 151*(1), 13–28. https://doi. org/10.1016/S0376-7388(98)00190-2

Cruz-Silva, R., Takizawa, Y., Nakaruk, A., Katouda, M., Yamanaka, A., Ortiz-Medina, J., . . . Endo, M. (2019). New insights in the natural organic matter fouling mechanism of polyamide and nanocomposite multiwalled carbon nanotubes-polyamide membranes. *Environmental Science & Technology, 53*(11), 6255–6263. doi:10.1021/acs.est.8b07203

de Lannoy, C. F., Jassby, D., Davis, D. D., & Wiesner, M. R. (2012). A highly electrically conductive polymer–multiwalled carbon nanotube nanocomposite membrane. *Journal of Membrane Science, 415–416*, 718–724. https://doi.org/10.1016/j.memsci.2012.05.061

de Lannoy, C.-F., Jassby, D., Gloe, K., Gordon, A. D., & Wiesner, M. R. (2013). Aquatic biofouling prevention by electrically charged nanocomposite polymer thin film membranes. *Environmental Science & Technology, 47*(6), 2760–2768. doi:10.1021/es3045168

Demirel, E., & Dadashov, S. (2021). Fabrication of a novel PVDF based silica coated multiwalled carbon nanotube embedded membrane with improved filtration performance. *Chemical Engineering Communications*, 1–26. doi:10.1080/00986445.2021.1935253

Du, L., Quan, X., Fan, X., Wei, G., & Chen, S. (2020). Conductive CNT/nanofiber composite hollow fiber membranes with electrospun support layer for water purification. *Journal of Membrane Science, 596*, 117613. https://doi.org/10.1016/j.memsci.2019.117613

Duan, W., Dudchenko, A., Mende, E., Flyer, C., Zhu, X., & Jassby, D. (2014). Electrochemical mineral scale prevention and removal on electrically conducting carbon nanotube—polyamide reverse osmosis membranes. *Environmental Science: Processes & Impacts, 16*(6), 1300–1308. doi:10.1039/C3EM00635B

Duan, W., Ronen, A., de Leon, J. V., Dudchenko, A., Yao, S., Corbala-Delgado, J., . . . Jassby, D. (2016). Treating anaerobic sequencing batch reactor effluent with electrically conducting ultrafiltration and nanofiltration membranes for fouling control. *Journal of Membrane Science, 504*, 104–112. https://doi.org/10.1016/j.memsci.2016.01.011

Dudchenko, A. V., Rolf, J., Russell, K., Duan, W., & Jassby, D. (2014). Organic fouling inhibition on electrically conducting carbon nanotube–polyvinyl alcohol composite ultrafiltration membranes. *Journal of Membrane Science, 468*, 1–10. https://doi.org/10.1016/j.
memsci.2014.05.041

Dumée, L. F., Germain, V., Sears, K., Schütz, J., Finn, N., Duke, M., . . . Gray, S. (2011).
Enhanced durability and hydrophobicity of carbon nanotube bucky paper membranes
in membrane distillation. *Journal of Membrane Science, 376*(1), 241–246. https://doi.
org/10.1016/j.memsci.2011.04.024

Dumée, L. F., Sears, K., Schütz, J. r., Finn, N., Duke, M., & Gray, S. (2010). Carbon nanotube
based composite membranes for water desalination by membrane distillation. *Desalination and Water Treatment, 17*(1–3), 72–79. doi:10.5004/dwt.2010.1701

Dumée, L. F., Sears, K., Schütz, J., Finn, N., Huynh, C., Hawkins, S., . . . Gray, S. (2010).
Characterization and evaluation of carbon nanotube Bucky-Paper membranes for direct
contact membrane distillation. *Journal of Membrane Science, 351*(1), 36–43. https://doi.
org/10.1016/j.memsci.2010.01.025

Ejaz Ahmed, F., Lalia, B. S., Hilal, N., & Hashaikeh, R. (2014). Underwater superoleophobic
cellulose/electrospun PVDF–HFP membranes for efficient oil/water separation. *Desalination, 344*, 48–54. https://doi.org/10.1016/j.desal.2014.03.010

Epstein, N. (1989). On tortuosity and the tortuosity factor in flow and diffusion through porous
media. *Chemical Engineering Science, 44*(3), 777–779.

Fabbri, E., & Schmidt, T. J. (2018). Oxygen evolution reaction—the enigma in water electrolysis. *ACS Catalysis, 8*(10), 9765–9774. doi:10.1021/acscatal.8b02712

Fakhru'l-Razi, A., Pendashteh, A., Abdullah, L. C., Biak, D. R. A., Madaeni, S. S., & Abidin,
Z. Z. (2009). Review of technologies for oil and gas produced water treatment. *Journal of Hazardous Materials, 170*(2), 530–551. https://doi.org/10.1016/j.jhazmat.2009.
05.044

Fan, X., Liu, Y., Quan, X., & Chen, S. (2018). Highly permeable thin-film composite forward
osmosis membrane based on carbon nanotube hollow fiber scaffold with electrically
enhanced fouling resistance. *Environmental Science & Technology, 52*(3), 1444–1452.
doi:10.1021/acs.est.7b05341

Fan, X., Zhao, H., Quan, X., Liu, Y., & Chen, S. (2016). Nanocarbon-based membrane filtration integrated with electric field driving for effective membrane fouling mitigation.
Water Research, 88, 285–292. https://doi.org/10.1016/j.watres.2015.10.043

Gao, G., Pan, M., & Vecitis, C. D. (2015). Effect of the oxidation approach on carbon nanotube
surface functional groups and electrooxidative filtration performance. *Journal of Materials Chemistry A, 3*(14), 7575–7582. doi:10.1039/C4TA07191C

Ge, J., Zong, D., Jin, Q., Yu, J., & Ding, B. (2018). Biomimetic and superwettable nanofibrous
skins for highly efficient separation of oil-in-water emulsions. *Advanced Functional
Materials, 28*(10), 1705051. https://doi.org/10.1002/adfm.201705051

Geng, P., & Chen, G. (2016). Magnéli Ti4O7 modified ceramic membrane for electrically-
assisted filtration with antifouling property. *Journal of Membrane Science, 498*, 302–314.
https://doi.org/10.1016/j.memsci.2015.07.055

German, S. R., Edwards, M. A., Chen, Q., Liu, Y., Luo, L., & White, H. S. (2016). Electrochemistry of single nanobubbles. Estimating the critical size of bubble-forming nuclei for
gas-evolving electrode reactions. *Faraday Discussions, 193*(0), 223–240. doi:10.1039/
C6FD00099A

Gonçalves, J. M., Matias, T. A., Toledo, K. C. F., & Araki, K. (2019). Chapter six—electrocatalytic
materials design for oxygen evolution reaction. In R. van Eldik & C. D. Hubbard (Eds.),
Advances in inorganic chemistry (Vol. 74, pp. 241–303). Academic Press.

Gonzalez-Olmos, R., Penadés, A., & Garcia, G. (2018). Electro-oxidation as efficient pretreatment to minimize the membrane fouling in water reuse processes. *Journal of Membrane
Science, 552*, 124–131. https://doi.org/10.1016/j.memsci.2018.01.041

Gu, Q.-A., Li, K., Li, S., Cui, R., Liu, L., Yu, C., . . . Xiao, G. (2020). In silico study of structure and water dynamics in CNT/polyamide nanocomposite reverse osmosis membranes. *Physical Chemistry Chemical Physics*, *22*(39), 22324–22331. doi:10.1039/D0CP03864D

Han, M., Wang, Y., Yao, J., Liu, C., Chew, J. W., Wang, Y., . . . Han, L. (2021). Electrically conductive hydrophobic membrane cathode for membrane distillation with super anti-oil-fouling capability: Performance and mechanism. *Desalination*, *516*, 115199. https://doi.org/10.1016/j.desal.2021.115199

Hashaikeh, R., Lalia, B. S., Kochkodan, V., & Hilal, N. (2014). A novel in situ membrane cleaning method using periodic electrolysis. *Journal of Membrane Science*, *471*(0), 149–154. http://dx.doi.org/10.1016/j.memsci.2014.08.017

Ho, K. C., Teoh, Y. X., Teow, Y. H., & Mohammad, A. W. (2021). Life cycle assessment (LCA) of electrically-enhanced POME filtration: Environmental impacts of conductive-membrane formulation and process operating parameters. *Journal of Environmental Management*, *277*, 111434. https://doi.org/10.1016/j.jenvman.2020.111434

Ho, K. C., Teow, Y. H., & Mohammad, A. W. (2019). Optimization of nanocomposite conductive membrane formulation and operating parameters for electrically-enhanced palm oil mill effluent filtration using response surface methodology. *Process Safety and Environmental Protection*, *126*, 297–308. https://doi.org/10.1016/j.psep.2019.03.019

Hong, S. H., Jeong, J., Shim, S., Kang, H., Kwon, S., Ahn, K. H., & Yoon, J. (2008). Effect of electric currents on bacterial detachment and inactivation. *Biotechnology and Bioengineering*, *100*(2), 379–386.

Hong, S. K., Bae, S., Jeon, H., Kim, M., Cho, S. J., & Lim, G. (2018). An underwater superoleophobic nanofibrous cellulosic membrane for oil/water separation with high separation flux and high chemical stability. *Nanoscale*, *10*(6), 3037–3045. doi:10.1039/C7NR08199E

Hou, D., Iddya, A., Chen, X., Wang, M., Zhang, W., Ding, Y., . . . Ren, Z. J. (2018). Nickel-based membrane electrodes enable high-rate electrochemical ammonia recovery. *Environmental Science & Technology*, *52*(15), 8930–8938. doi:10.1021/acs.est.8b01349

Hu, S. Y., Zhang, Y., Lawless, D., & Feng, X. (2012). Composite membranes comprising of polyvinylamine-poly(vinyl alcohol) incorporated with carbon nanotubes for dehydration of ethylene glycol by pervaporation. *Journal of Membrane Science*, *417–418*(0), 34–44. http://dx.doi.org/10.1016/j.memsci.2012.06.010

Hu, W., Chen, S., Yang, Z., Liu, L., & Wang, H. (2011). Flexible electrically conductive nanocomposite membrane based on bacterial cellulose and polyaniline. *The Journal of Physical Chemistry B*, *115*(26), 8453–8457. doi:10.1021/jp204422v

Huang, X., Li, B., Song, X., Wang, L., Shi, Y., Hu, M., . . . Xue, H. (2019). Stretchable, electrically conductive and superhydrophobic/superoleophilic nanofibrous membrane with a hierarchical structure for efficient oil/water separation. *Journal of Industrial and Engineering Chemistry*, *70*, 243–252. https://doi.org/10.1016/j.jiec.2018.10.021

Izah, S. C., Aigberua, A. O., & Srivastav, A. L. (2022). Chapter 4—factors influencing the alteration of microbial and heavy metal characteristics of river systems in the Niger Delta region of Nigeria. In S. Madhav, S. Kanhaiya, A. Srivastav, V. Singh, & P. Singh (Eds.), *Ecological significance of river ecosystems* (pp. 51–78). Elsevier.

Jung, B., Kim, C. Y., Jiao, S., Rao, U., Dudchenko, A. V., Tester, J., & Jassby, D. (2020). Enhancing boron rejection on electrically conducting reverse osmosis membranes through local electrochemical pH modification. *Desalination*, *476*, 114212. https://doi.org/10.1016/j.desal.2019.114212

Kalla, S. (2021). Use of membrane distillation for oily wastewater treatment—A review. *Journal of Environmental Chemical Engineering*, *9*(1), 104641. https://doi.org/10.1016/j.jece.2020.104641

Kandah, M. I., & Meunier, J.-L. (2007). Removal of nickel ions from water by multi-walled carbon nanotubes. *Journal of Hazardous Materials*, *146*(1), 283–288. https://doi.org/10.1016/j.jhazmat.2006.12.019

Kang, S., Herzberg, M., Rodrigues, D. F., & Elimelech, M. (2008). Antibacterial effects of carbon nanotubes: Size does matter! *Langmuir, 24*(13), 6409–6413. doi:10.1021/la800951v

Kang, S., Pinault, M., Pfefferle, L. D., & Elimelech, M. (2007). Single-walled carbon nanotubes exhibit strong antimicrobial activity. *Langmuir, 23*(17), 8670–8673. doi:10.1021/la701067r

Karimi, M., Mohsen-Nia, M., & Akbari, A. (2014). Electro-separation of synthetic azo dyes from a simulated wastewater using polypyrrole/polyacrylonitrile conductive membranes. *Journal of Water Process Engineering, 4*, 6–11. http://dx.doi.org/10.1016/j.jwpe.2014.08.008

Kayvani Fard, A., McKay, G., Buekenhoudt, A., Al Sulaiti, H., Motmans, F., Khraisheh, M., & Atieh, M. (2018). Inorganic membranes: Preparation and application for water treatment and desalination. *Materials, 11*(1). doi:10.3390/ma11010074

Khalil, A., Ahmed, F. E., Hashaikeh, R., & Hilal, N. (2022). 3D printed electrically conductive interdigitated spacer on ultrafiltration membrane for electrolytic cleaning and chlorination. *Journal of Applied Polymer Science, n/a*(n/a), 52292. https://doi.org/10.1002/app.52292

Khanzada, N. K., Jassby, D., & An, A. K. (2022). Conductive reverse osmosis membrane for electrochemical chlorine reduction and sustainable brackish water treatment. *Chemical Engineering Journal, 435*, 134858. https://doi.org/10.1016/j.cej.2022.134858

Kim, J., Yun, E.-T., Tijing, L., Shon, H. K., & Hong, S. (2022). Mitigation of fouling and wetting in membrane distillation by electrical repulsion using a multi-layered single-wall carbon nanotube/polyvinylidene fluoride membrane. *Journal of Membrane Science*, 120519. https://doi.org/10.1016/j.memsci.2022.120519

Kot, F. S. (2015). Boron in the environment. *Boron Separation Processes, 1*, 33.

Lalia, B. S., Ahmed, F. E., Shah, T., Hilal, N., & Hashaikeh, R. (2015). Electrically conductive membranes based on carbon nanostructures for self-cleaning of biofouling. *Desalination, 360*, 8–12. https://doi.org/10.1016/j.desal.2015.01.006

Lawrence Arockiasamy, D., Alam, J., & Alhoshan, M. (2013). Carbon nanotubes-blended poly(phenylene sulfone) membranes for ultrafiltration applications. *Applied Water Science, 3*(1), 93–103. doi:10.1007/s13201-012-0063-0

Lee, H.-J., Zhang, N., Ganzoury, M. A., Wu, Y., & de Lannoy, C.-F. (2021). Simultaneous dechlorination and advanced oxidation using electrically conductive carbon nanotube membranes. *ACS Applied Materials & Interfaces, 13*(29), 34084–34092. doi:10.1021/acsami.1c06137

Li, C., Sun, W., Lu, Z., Ao, X., & Li, S. (2020). Ceramic nanocomposite membranes and membrane fouling: A review. *Water Research, 175*, 115674. https://doi.org/10.1016/j.watres.2020.115674

Li, P., Yang, C., Sun, F., & Li, X.-y. (2021). Fabrication of conductive ceramic membranes for electrically assisted fouling control during membrane filtration for wastewater treatment. *Chemosphere, 280*, 130794. https://doi.org/10.1016/j.chemosphere.2021.130794

Li, Y., Kang, Z., Deng, X., Yang, G., Yu, S., Mo, J., . . . Zhang, F.-Y. (2019). Wettability effects of thin titanium liquid/gas diffusion layers in proton exchange membrane electrolyzer cells. *Electrochimica Acta, 298*, 704–708.

Li, Y., Kang, Z., Mo, J., Yang, G., Yu, S., Talley, D. A., . . . Zhang, F.-Y. (2018). In-situ investigation of bubble dynamics and two-phase flow in proton exchange membrane electrolyzer cells. *International Journal of Hydrogen Energy, 43*(24), 11223–11233. https://doi.org/10.1016/j.ijhydene.2018.05.006

Li, Y., Yang, G., Yu, S., Kang, Z., Mo, J., Han, B., . . . Zhang, F.-Y. (2019). In-situ investigation and modeling of electrochemical reactions with simultaneous oxygen and hydrogen microbubble evolutions in water electrolysis. *International Journal of Hydrogen Energy, 44*(52), 28283–28293. https://doi.org/10.1016/j.ijhydene.2019.09.044

Liu, C., Wang, W., Zhu, L., Cui, F., Xie, C., Chen, X., & Li, N. (2020). High-performance nanofiltration membrane with structurally controlled PES substrate containing electrically aligned CNTs. *Journal of Membrane Science, 605*, 118104. https://doi.org/10.1016/j.memsci.2020.118104

Liu, J., Tian, C., Xiong, J., & Wang, L. (2017). Polypyrrole blending modification for PVDF conductive membrane preparing and fouling mitigation. *Journal of Colloid and Interface Science, 494*, 124–129.

Liu, Y., Ma, W., Cheng, Z., Xu, J., Wang, R., & Gang, X. (2013). Preparing CNTs/Ca-Selective zeolite composite electrode to remove calcium ions by capacitive deionization. *Desalination, 326*, 109–114. http://dx.doi.org/10.1016/j.desal.2013.07.022

Liu, Y., Wang, Q., Wu, T., & Zhang, L. (2005). Fluid structure and transport properties of water inside carbon nanotubes. *The Journal of Chemical Physics, 123*(23). http://dx.doi.org/10.1063/1.2131070

Low, S. C., & Murugaiyan, S. V. (2021) Thermoplastic polymers in membrane separation. In Hashmi, S. (Ed.), *Reference module in materials science and materials engineering*. Elsevier Inc. https://doi.org/10.1016/B978-0-12-820352-1.00083-3

Lu, C., & Liu, C. (2006). Removal of nickel (II) from aqueous solution by carbon nanotubes. *Journal of Chemical Technology & Biotechnology: International Research in Process, Environmental & Clean Technology, 81*(12), 1932–1940.

Lu, X., Yim, W.-L., Suryanto, B. H. R., & Zhao, C. (2015). Electrocatalytic oxygen evolution at surface-oxidized multiwall carbon nanotubes. *Journal of the American Chemical Society, 137*(8), 2901–2907. doi:10.1021/ja509879r

Ma, C., Yi, C., Li, F., Shen, C., Wang, Z., Sand, W., & Liu, Y. (2020). Mitigation of membrane fouling using an electroactive polyether sulfone membrane. *Membranes, 10*(2), 21.

Ma, D., Li, H., Meng, Z., Zhang, C., Zhou, J., Xia, J., & Wang, Y. (2021). Absolute and fast removal of viruses and bacteria from water by spraying-assembled carbon-nanotube membranes. *Environmental Science & Technology, 55*(22), 15206–15214. doi:10.1021/acs.est.1c04644

Ma, W., Zhang, M., Liu, Z., Kang, M., Huang, C., & Fu, G. (2019). Fabrication of highly durable and robust superhydrophobic-superoleophilic nanofibrous membranes based on a fluorine-free system for efficient oil/water separation. *Journal of Membrane Science, 570–571*, 303–313. https://doi.org/10.1016/j.memsci.2018.10.035

Madaeni, S., & Molaeipour, S. (2010). Investigation of filtration capability of conductive composite membrane in separation of protein from water. *Ionics, 16*(1), 75–80. doi:10.1007/s11581-009-0372-y

Majeed, S., Fierro, D., Buhr, K., Wind, J., Du, B., Boschetti-de-Fierro, A., & Abetz, V. (2012). Multi-walled carbon nanotubes (MWCNTs) mixed polyacrylonitrile (PAN) ultrafiltration membranes. *Journal of Membrane Science, 403–404*(0), 101–109. http://dx.doi.org/10.1016/j.memsci.2012.02.029

Mantel, T., Jacki, E., & Ernst, M. (2021). Electrosorptive removal of organic water constituents by positively charged electrically conductive UF membranes. *Water Research, 201*, 117318. https://doi.org/10.1016/j.watres.2021.117318

Mavukkandy, M. O., Zaib, Q., & Arafat, H. A. (2018). CNT/PVP blend PVDF membranes for the removal of organic pollutants from simulated treated wastewater effluent. *Journal of Environmental Chemical Engineering, 6*(5), 6733–6740. https://doi.org/10.1016/j.jece.2018.10.029

Miao, M. (2015). Carbon nanotube yarns for electronic textiles. *Electronic Textiles*, 55–72.

Mittelsteadt, C., Norman, T., Rich, M., & Willey, J. (2015). Chapter 11—PEM electrolyzers and PEM regenerative fuel cells industrial view. In P. T. Moseley & J. Garche (Eds.), *Electrochemical energy storage for renewable sources and grid balancing* (pp. 159–181). Elsevier.

Nable, R. O., Bañuelos, G. S., & Paull, J. G. (1997). Boron toxicity. *Plant and Soil, 193*(1), 181–198. doi:10.1023/A:1004272227886

Nicholls, W., Borg, M., Lockerby, D., & Reese, J. (2012). Water transport through (7,7) carbon nanotubes of different lengths using molecular dynamics. *Microfluidics and Nanofluidics, 12*(1–4), 257–264. doi:10.1007/s10404-011-0869-3

Perez-Roa, R. E., Tompkins, D. T., Paulose, M., Grimes, C. A., Anderson, M. A., & Noguera, D. R. (2006). Effects of localised, low-voltage pulsed electric fields on the development and inhibition of Pseudomonas aeruginosa biofilms. *Biofouling*, *22*(6), 383–390. doi:10.1080/08927010601053541

Pettes, M. T., & Shi, L. (2009). Thermal and structural characterizations of individual single-, double-, and multi-walled carbon nanotubes. *Advanced Functional Materials*, *19*(24), 3918–3925. https://doi.org/10.1002/adfm.200900932

Porada, S., Zhao, R., van der Wal, A., Presser, V., & Biesheuvel, P. M. (2013). Review on the science and technology of water desalination by capacitive deionization. *Progress in Materials Science*, *58*(8), 1388–1442. https://doi.org/10.1016/j.pmatsci.2013.03.005

Qiu, P., Agne, M. T., Liu, Y., Zhu, Y., Chen, H., Mao, T., . . . Snyder, G. J. (2018). Suppression of atom motion and metal deposition in mixed ionic electronic conductors. *Nature Communications*, *9*(1), 2910. doi:10.1038/s41467-018-05248-8

Rahaman, M. S., Vecitis, C. D., & Elimelech, M. (2011). Electrochemical carbon-nanotube filter performance toward virus removal and inactivation in the presence of natural organic matter. *Environmental Science & Technology*, *46*(3), 1556–1564. doi:10.1021/es203607d

Rahaman, M. S., Vecitis, C. D., & Elimelech, M. (2012). Electrochemical carbon-nanotube filter performance toward virus removal and inactivation in the presence of natural organic matter. *Environmental Science & Technology*, *46*(3), 1556–1564. doi:10.1021/es203607d

Rao, G. P., Lu, C., & Su, F. (2007). Sorption of divalent metal ions from aqueous solution by carbon nanotubes: A review. *Separation and Purification Technology*, *58*(1), 224–231. https://doi.org/10.1016/j.seppur.2006.12.006

Rashid, M. H.-O., Pham, S. Q. T., Sweetman, L. J., Alcock, L. J., Wise, A., Nghiem, L. D., . . . Ralph, S. F. (2014). Synthesis, properties, water and solute permeability of MWNT buckypapers. *Journal of Membrane Science*, *456*, 175–184. https://doi.org/10.1016/j.memsci.2014.01.026

Ratnaningsih, E., Reynard, R., Khoiruddin, K., Wenten, I. G., & Boopathy, R. (2021). Recent advancements of UF-based separation for selective enrichment of proteins and bioactive peptides—a review. *Applied Sciences*, *11*(3), 1078.

Rohani, R., & Yusoff, I. I. (2019). Towards electrically tunable nanofiltration membranes: Polyaniline-coated polyvinylidene fluoride membranes with tunable permeation properties. *Iranian Polymer Journal*, *28*(9), 789–800. doi:10.1007/s13726-019-00744-0

Ronen, A., Duan, W., Wheeldon, I., Walker, S., & Jassby, D. (2015). Microbial attachment inhibition through low-voltage electrochemical reactions on electrically conducting membranes. *Environmental Science & Technology*, *49*(21), 12741–12750. doi:10.1021/acs.est.5b01281

Roy, S., Bhadra, M., & Mitra, S. (2014). Enhanced desalination via functionalized carbon nanotube immobilized membrane in direct contact membrane distillation. *Separation and Purification Technology*, *136*(0), 58–65. http://dx.doi.org/10.1016/j.seppur.2014.08.009

Sagiv, A., & Semiat, R. (2004). Analysis of parameters affecting boron permeation through reverse osmosis membranes. *Journal of Membrane Science*, *243*(1), 79–87. https://doi.org/10.1016/j.memsci.2004.05.029

Saleemi, M. A., Kong, Y. L., Yong, P. V. C., & Wong, E. H. (2021). An overview of antimicrobial properties of carbon nanotubes-based nanocomposites. *Advanced Pharmaceutical Bulletin*, *12*(3), 449–465. https://doi.org/10.34172/apb.2022.049

Salehi, E., & Madaeni, S. S. (2010). Influence of conductive surface on adsorption behavior of ultrafiltration membrane. *Applied Surface Science*, *256*(10), 3010–3017. http://dx.doi.org/10.1016/j.apsusc.2009.11.065

Sears, K., Dumée, L., Schütz, J., She, M., Huynh, C., Hawkins, S., . . . Gray, S. (2010). Recent developments in carbon nanotube membranes for water purification and gas separation. *Materials*, *3*(1), 127–149.

Shah, T. K., Malecki, H. C., Basantkumar, R. R., Liu, H., Fleischer, C. A., Sedlak, J. J., . . . Goldfinger, J. M. (2013). *Carbon nanostructures and methods of making the same.* Google Patents.

Shakeri, A., Salehi, H., & Rastgar, M. (2019). Antifouling electrically conductive membrane for forward osmosis prepared by polyaniline/graphene nanocomposite. *Journal of Water Process Engineering, 32,* 100932. https://doi.org/10.1016/j.jwpe.2019.100932

Shao, Y., Yin, G., Zhang, J., & Gao, Y. (2006). Comparative investigation of the resistance to electrochemical oxidation of carbon black and carbon nanotubes in aqueous sulfuric acid solution. *Electrochimica Acta, 51*(26), 5853–5857. https://doi.org/10.1016/j.electacta.2006.03.021

Silva, M. A., Hilliou, L., & de Amorim, M. T. P. (2020). Fabrication of pristine-multiwalled carbon nanotubes/cellulose acetate composites for removal of methylene blue. *Polymer Bulletin, 77*(2), 623–653. doi:10.1007/s00289-019-02769-0

Street, A., Sustich, R., Duncan, J., & Savage, N. (2014). *Nanotechnology applications for clean water: Solutions for improving water quality.* Elsevier Science.

Suen, N.-T., Hung, S.-F., Quan, Q., Zhang, N., Xu, Y.-J., & Chen, H. M. (2017). Electrocatalysis for the oxygen evolution reaction: Recent development and future perspectives. *Chemical Society Reviews, 46*(2), 337–365. doi:10.1039/C6CS00328A

Sun, X., Wu, J., Chen, Z., Su, X., & Hinds, B. J. (2013). Fouling characteristics and electrochemical recovery of carbon nanotube membranes. *Advanced Functional Materials, 23*(12), 1500–1506. doi:10.1002/adfm.201201265

Tang, Y. P., Luo, L., Thong, Z., & Chung, T. S. (2017). Recent advances in membrane materials and technologies for boron removal. *Journal of Membrane Science, 541,* 434–446. https://doi.org/10.1016/j.memsci.2017.07.015

Taniguchi, M., Fusaoka, Y., Nishikawa, T., & Kurihara, M. (2004). Boron removal in RO seawater desalination. *Desalination, 167,* 419–426. https://doi.org/10.1016/j.desal.2004.06.157

Tanudjaja, H. J., Hejase, C. A., Tarabara, V. V., Fane, A. G., & Chew, J. W. (2019). Membrane-based separation for oily wastewater: A practical perspective. *Water Research, 156,* 347–365. https://doi.org/10.1016/j.watres.2019.03.021

Teixeira-Santos, R., Gomes, M., Gomes, L. C., & Mergulhão, F. J. (2021). Antimicrobial and anti-adhesive properties of carbon nanotube-based surfaces for medical applications: A systematic review. *Iscience, 24*(1), 102001.

Teixeira-Santos, R., Gomes, M., & Mergulhão, F. J. (2020). Carbon nanotube-based antimicrobial and antifouling surfaces. In S. Snigdha, S. Thomas, E. K. Radhakrishnan, & N. Kalarikkal (Eds.), *Engineered antimicrobial surfaces* (pp. 65–93). Springer Singapore.

Thakur, A. K., Singh, S. P., Kleinberg, M. N., Gupta, A., & Arnusch, C. J. (2019). Laser-induced graphene–PVA composites as robust electrically conductive water treatment membranes. *ACS Applied Materials & Interfaces, 11*(11), 10914–10921. doi:10.1021/acsami.9b00510

Thamaraiselvan, C., Thakur, A. K., Gupta, A., & Arnusch, C. J. (2021). Electrochemical removal of organic and inorganic pollutants using robust laser-induced graphene membranes. *ACS Applied Materials & Interfaces, 13*(1), 1452–1462. doi:10.1021/acsami.0c18358

Vecitis, C. D., Schnoor, M. H., Rahaman, M. S., Schiffman, J. D., & Elimelech, M. (2011). Electrochemical multiwalled carbon nanotube filter for viral and bacterial removal and inactivation. *Environmental Science & Technology, 45*(8), 3672–3679. doi:10.1021/es2000062

Vilela, D., Parmar, J., Zeng, Y., Zhao, Y., & Sánchez, S. (2016). Graphene-based microbots for toxic heavy metal removal and recovery from water. *Nano Letters, 16*(4), 2860–2866. doi:10.1021/acs.nanolett.6b00768

Wang, J., Hou, L. a., Yan, K., Zhang, L., & Yu, Q. J. (2018). Polydopamine nanocluster deco-rated electrospun nanofibrous membrane for separation of oil/water emulsions. *Journal of Membrane Science, 547,* 156–162. https://doi.org/10.1016/j.memsci.2017.10.028

Wang, K., Xu, L., Li, K., Liu, L., Zhang, Y., & Wang, J. (2019). Development of polyaniline conductive membrane for electrically enhanced membrane fouling mitigation. *Journal of Membrane Science, 570–571,* 371–379. https://doi.org/10.1016/j.memsci.2018.10.050

Wang, Q., Li, N., Bolto, B., Hoang, M., & Xie, Z. (2016). Desalination by pervaporation: A review. *Desalination, 387,* 46–60. https://doi.org/10.1016/j.desal.2016.02.036

Wang, T. Y., Libardo, M. D. J., Angeles-Boza, A. M., & Pellois, J. P. (2017). Membrane oxida-tion in cell delivery and cell killing applications. *ACS Chemical Biology, 12*(5), 1170–1182. doi:10.1021/acschembio.7b00237

Wang, Y.-J., Huang, L., Fang, Z., Wang, X.-M., Gao, M., Liu, H.-Q., . . . Huang, T.-Y. (2022). Electrochemically self-cleanable carbon nanotube interlayered membrane for enhanced forward osmosis in wastewater treatment. *Journal of Environmental Chemical Engineer-ing, 10*(3), 107399. https://doi.org/10.1016/j.jece.2022.107399

Wu, B., Li, X., An, D., Zhao, S., & Wang, Y. (2014). Electro-casting aligned MWCNTs/pol-ystyrene composite membranes for enhanced gas separation performance. *Journal of Membrane Science, 462,* 62–68. https://doi.org/10.1016/j.memsci.2014.03.015

Wu, Z., Chen, H., Dong, Y., Mao, H., Sun, J., Chen, S., . . . Hu, J. (2008). Cleaning using nano-bubbles: Defouling by electrochemical generation of bubbles. *Journal of Colloid and Interface Science, 328*(1), 10–14. https://doi.org/10.1016/j.jcis.2008.08.064

Wu, Z., Zhang, X., Zhang, X., Li, G., Sun, J., Zhang, Y., . . . Hu, J. (2006). Nanobubbles influence on BSA adsorption on mica surface. *Surface and Interface Analysis, 38*(6), 990–995.

Yang, G., Xie, Z., Cran, M., Ng, D., & Gray, S. (2019). Enhanced desalination performance of poly (vinyl alcohol)/carbon nanotube composite pervaporation membranes via inter-facial engineering. *Journal of Membrane Science, 579,* 40–51. https://doi.org/10.1016/j.memsci.2019.02.034

Yi, G., Chen, S., Quan, X., Wei, G., Fan, X., & Yu, H. (2018). Enhanced separation perfor-mance of carbon nanotube–polyvinyl alcohol composite membranes for emulsified oily wastewater treatment under electrical assistance. *Separation and Purification Technol-ogy, 197,* 107–115. https://doi.org/10.1016/j.seppur.2017.12.058

Zhan, Y., Pan, L., Nie, C., Li, H., & Sun, Z. (2011). Carbon nanotube–chitosan composite elec-trodes for electrochemical removal of Cu(II) ions. *Journal of Alloys and Compounds, 509*(18), 5667–5671. http://dx.doi.org/10.1016/j.jallcom.2011.02.118

Zhang, D., He, Y., Wang, R., & Taub, A. I. (2021). Kinetics of single-walled carbon nano-tube migration in epoxy resin under DC electric field. *Applied Physics A, 128*(1), 10. doi:10.1007/s00339-021-05170-9

Zhang, D., She, X., He, Y., Chapkin, W. A., Bregman, A. T., Wang, R., & Taub, A. (2021). *Bridging of carbon fibers in CF/epoxy composites using electrostatically induced CNT alignment.* Paper presented at the Proceedings of the American Society for Composites—Thirty-Sixth Technical Conference on Composite Materials.

Zhang, H., Quan, X., Fan, X., Yi, G., Chen, S., Yu, H., & Chen, Y. (2019). Improving ion rejection of conductive nanofiltration membrane through electrically enhanced surface charge den-sity. *Environmental Science & Technology, 53*(2), 868–877. doi:10.1021/acs.est.8b04268

Zhang, M., Jin, W., Yang, F., Duke, M., Dong, Y., & Tang, C. Y. (2020). Engineering a nanocom-posite interlayer for a novel ceramic-based forward osmosis membrane with enhanced performance. *Environmental Science & Technology, 54*(12), 7715–7724. doi:10.1021/acs.est.0c02809

Zhang, M., Ma, W., Wu, S., Tang, G., Cui, J., Zhang, Q., . . . Huang, C. (2019). Electrospun frogspawn structured membrane for gravity-driven oil-water separation. *Journal of Col-loid and Interface Science, 547,* 136–144. https://doi.org/10.1016/j.jcis.2019.03.099

Zhang, Q., & Vecitis, C. D. (2014). Conductive CNT-PVDF membrane for capacitive organic fouling reduction. *Journal of Membrane Science, 459*, 143–156. https://doi.org/10.1016/j.memsci.2014.02.017

Zhang, Y., Liu, L., & Yang, F. (2016). A novel conductive membrane with RGO/PVDF coated on carbon fiber cloth for fouling reduction with electric field in separating polyacrylamide. *Journal of Applied Polymer Science, 133*(26). https://doi.org/10.1002/app.43597

Zhao, X., & Drlica, K. (2014). Reactive oxygen species and the bacterial response to lethal stress. *Current Opinion in Microbiology, 21*, 1–6. doi:10.1016/j.mib.2014.06.008

Zhu, Z., Li, Z., Zhong, L., Zhang, R., Cui, F., & Wang, W. (2019). Dual-biomimetic superwetting silica nanofibrous membrane for oily water purification. *Journal of Membrane Science, 572*, 73–81. https://doi.org/10.1016/j.memsci.2018.10.071

6 Electrically Conductive Spacers for Fouling Mitigation in Desalination and Water Treatment

6.1 INTRODUCTION

Feed spacers play an important role in improving mass transfer in all membrane separation processes. Currently, ample research activity focuses on exploiting modified feed spacer performance. Little has been done to mitigate membrane fouling from the perspective of spacer design. However, since spacers are typically the first to foul, there is growing interest in developing antifouling spacers as well as electrically conducting spacers for fouling mitigation. This is also accompanied by the effect of spacer parameters on fouling dynamics, into which in situ microscopic monitoring techniques such as OCT can give a deeper insight. Understanding fouling on membrane spacers will eventually lead to design improvements and more efficient membrane modules. Today, as membrane development has reached new heights, considering module improvements in other components such as feed spacers is more important than ever in enhancing performance. Electrically conducting spacers may have certain advantages over electrically conducting membranes in that often electrically conducting modification of membrane materials compromises key membrane properties such as selectivity; it is still a challenge to tailor electrically conducting membranes with specific permeate flux and solute rejection. Similar to conducting membranes, electrically conducting spacers can be used for electrolytic self-cleaning for control of fouling and flux recovery. Instead, an increased interest in developing electrically conducting spacers in order to mitigate fouling is thus not surprising and will help enhance the overall process for membrane technologies in desalination and water treatment. Other strategies that have been adopted to optimize engineered feed spacers for fouling mitigation involve 3D-printed spacers, surface coatings, and modified geometry. Modification of novel feed spacers has emerged as a promising area, especially as new research in nanomaterials continues to take form on a daily basis.

A major challenge faced by membrane separation processes is that of fouling. Although the membrane itself is at the heart of any membrane process, the

DOI: 10.1201/9781003144991-6

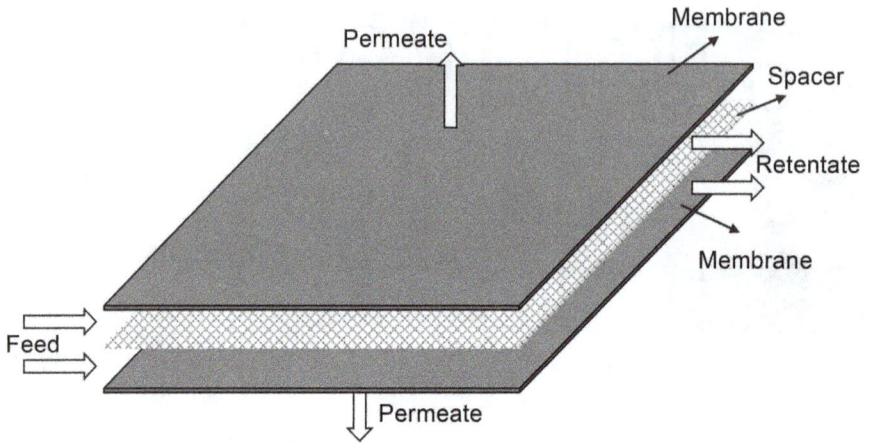

FIGURE 6.1   Schematic of feed channel spacer mesh between two membrane elements.

successful implementation of membrane technology relies on optimizing several other components, operating conditions, and feed constituents. Feed spacers are one such major component. Spacers serve as turbulence promoters to strengthen flow mixing and enhance mass transfer in the membrane feed channel (Winograd et al., 1973) and are crucial to industrial-scale membrane systems (Bucs et al., 2014). Figure 6.1 shows a schematic of a feed channel spacer between two membrane elements as in a module. In improving membrane module performance, researchers focus on one (or more) of three areas: the membrane itself, the feed channel spacer design (Fimbres-Weihs & Wiley, 2010), and module configuration. When no temperature gradient exists, as is the case with microfiltration (MF) (Hartinger et al., 2020), ultrafiltration (UF), nanofiltration (NF), forward osmosis (FO) (Zhang et al., 2014), and reverse osmosis (RO), the feed spacer acts to minimize the effect of concentration polarization (CP). In thermally driven processes like membrane distillation, the feed spacer takes on the dual role of reducing CP and temperature polarization (TP).

CP and TP are boundary layer effects arising from the difference in concentration or temperature between the bulk feed and at the membrane surface, respectively. Tony Fane has been widely credited for his contributions in analyzing boundary layer mass transfer (Field & Wu, 2018), which includes aspects of both these phenomena and helps us better understand membrane separation. CP is a primary cause for flux decline during the initial period of any membrane separation process (Sablani et al., 2001). It is the phenomenon in which retained solutes accumulate near the membrane surface due to membrane selectivity, increasing the concentration at the membrane surface, which also increases resistance to permeate flow (Ang & Mohammad, 2015). CP is affected by filtration mode, cross-flow velocity, temperature, and module geometry (Guo et al., 2012). There is some ambiguity regarding the classification of CP; some consider it a form of fouling (Guo et al., 2012), while others maintain that the two are

(a)

(b)

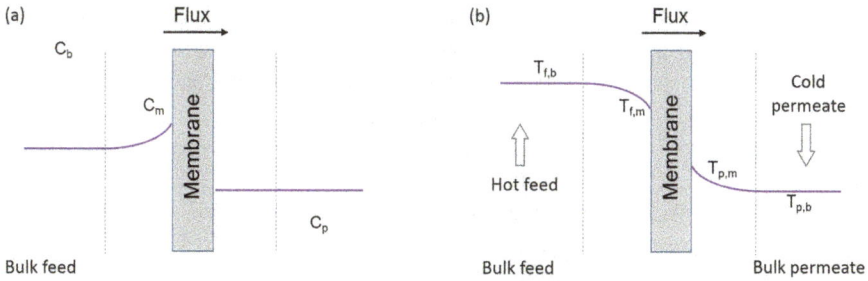

FIGURE 6.2   Schematic of (a) concentration polarization effect in all membrane processes and (b) temperature polarization effect in membrane distillation.

closely related phenomena with coupled effects (F. Wang & Tarabara, 2007). In general, the greater the CP, the more severe the fouling because at higher concentrations close to the membrane, constituents will interact and deposit more intensely on its surface (Shirazi et al., 2010). CP increases fouling potential and reduces permeate water quality and permeate flux (Matsuura, 2020). External fouling (i.e. on the membrane surface) also exacerbates CP as solutes retained at the surface find it difficult to diffuse back into the bulk solution in the presence of fouling (Chong & Fane, 2009; Hoek & Elimelech, 2003); this is known as cake-enhanced concentration polarization (CECP). Figure 6.2 shows both CP and TP schematically.

Similarly, the formation of a thermal boundary layer, referred to as TP, is a limiting factor in heat transfer efficiency in MD. The lower temperature at the membrane–water interface results in a smaller temperature gradient across the membrane, which decreases the driving force for mass transfer and lowers mass flux. The temperature polarization coefficient (*TPC*) is an indicator of the extent of TP and can be calculated from:

$$TPC = \frac{T_{f,m} - T_{p,m}}{T_f - T_p}$$

Equation 6.1

where $T_{f,m}$ and $T_{p,m}$ are the feed and permeate temperature at the membrane surface, respectively, and T_f and T_p are the bulk feed and permeate temperatures.

To overcome these challenges, considerable effort has been made on spacer material and geometry design, as well as modification of feed spacers, in order to improve membrane performance (Al Ashhab et al., 2014; Baker & Dudley, 1998; Ben-Sasson et al., 2014; Schwinge et al., 2004; Linares et al., 2014). The advantage of working on the spacer rather than the membrane is that you can improve performance without compromising on membrane permeability. There is also the added benefit of using nanomaterials on the spacer rather than the membrane, as the toxic effects of these nanomaterials are less understood as they may leach into the permeate. While the role of spacers in the mitigation of CP and, where relevant, TP has already been established, their role in fouling control is considerably less explored. Only 31 research articles focus on spacer engineering for fouling mitigation, with most of the interest taking place in 2017 and beyond. Rahmawati et al. recently reviewed progress in the development of spacers for membrane modules (Rahmawati et al., 2021).

Spacer-focused fouling control strategies are mostly aimed at arguably the most problematic fouling type: biofouling. Biofilm growth in the spacer-filled channel degrades overall performance by inducing feed channel pressure drop, reducing permeate flux and enhancing biofilm-related CP. The first step in biofouling control through spacer design is the online monitoring of this type of fouling through non-invasive tools that help characterize the spatiotemporal development of biofilm growth on the spacer or membrane surface. Next, the effect of hydrodynamic conditions is often studied using CFD tools. The last phase of biofouling mitigation in membrane feed spacer channels involves fabrication techniques as well as spacer modification to induce self-cleaning, e.g. electrically conducting surfaces for fouling control. The role of optimized spacer design in slowing down the buildup of foulants is often undermined (Kerdi et al., 2019).

Spacer design can be optimized to slow down the deposition of foulants on the membrane surface (Koo et al., 2021). While numerical modeling via computational fluid dynamics is considered an excellent tool to predict membrane performance and optimize spacer design (Koutsou et al., 2007; F. Li et al., 2002), it is usually helpful in predicting performance only in the absence of membrane fouling (Toh et al., 2020); numerical studies correlating fouling and spacer design are rare (Picioreanu et al., 2009). The challenge in feed spacer development stems partly from unconventional methods needed to fabricate complex structures and from the lack of sufficient experimental studies on advanced feed spacer designs. Several distinct approaches have been adopted to modify feed spacers, including coating (Ronen et al., 2015), altering geometry (Siddiqui et al., 2017b), applying electrically conductive spacers (Baek et al., 2014), and fabricating spacers via 3D printing (Tan et al., 2017).

In this chapter, we address the role of spacers in fouling mitigation, including the effect of spacer material, fabrication method, geometry, and surface modification. We then extend our focus to electrically conducting spacers that have been employed in desalination and water treatment and how the electrical and electrochemical properties of conducting spacers can be exploited for the mitigation of fouling and related phenomena. We discuss the main issues and challenges with the aim of guiding future investigations in this rapidly evolving and exciting area.

Membrane spacers are employed in almost all membrane separation processes including MF, UF, NF, RO, FO, ED, and MD. As industrial modules often contain multiple membrane elements, spacers also separate one membrane from another to act as stabilizers and form feed channels. The other more prominent role of spacers is to encourage mixing and minimize boundary layer effects just discussed. Compared to systems without feed spacers, mass transfer improvements in membrane systems with spacers can be linked to increased economic viability (A. Da Costa et al., 1991). Yet certain spacer properties and geometries favor increased mass transfer more than others, while limiting energy losses (A. R. Da Costa & Fane, 1994).

6.2 EFFECT OF SPACER GEOMETRY

The choice of spacer configuration not only reduces CP and TP but also lowers accumulation of foulants (Kavianipour et al., 2017; Gu et al., 2017). Two types of feed spacers are currently in use commercially: woven and non-woven (Li et al., 2002; Gu et al., 2017). Figure 6.3 shows the key parameters of feed spacer geometries.

Optimization of feed spacer geometry has mostly focused on fluid dynamics and mass transfer of the membrane system through both experimental and computational modeling studies. Feed spacer design may also determine the force needed to overcome foulant-related resistance in fluid flow. Many studies have elucidated the effect of spacer thickness, mesh size, and shape on permeate flux (Schwinge et al., 2004; Siddiqui et al., 2017b). Filament mesh spacing also affects fluid flow patterns, mass transfer coefficient, wall shear stress distribution, and pressure drop (Saeed et al., 2015a). Geometric modifications aimed at minimizing the impact of fouling (Fritzmann et al., 2013) provide insights into mass transfer through SWM (Siddiqui et al., 2016), as well as the potential for alternative designs to control fouling, with emphasis on biofouling (Ronen et al., 2015; Siddiqui et al., 2017a). These studies demonstrate how feed spacer geometry affects feed channel hydrodynamics and consequently flow-rate factors. Design factors that affect system performance include dissipation of water and shear stress distribution at the feed spacer filament as well as membrane (A. R. Da Costa & Fane, 1994). The impact of feed spacer geometry should therefore be investigated from a fouling perspective (Haaksman et al., 2017).

Computational fluid dynamics (CFD) studies are most commonly employed to study the influence of design parameters on filtration performance, with a wide range of operating parameters under investigation in SWMs. CFD has been used to investigate solute transport using different geometries. For instance, Ranade and Kumar showed that water flow differed only slightly when a curved feed spacer was used as compared to a flat spacer due to negligible secondary flows stemming from the curvature (Ranade & Kumar, 2006). Previous experimental studies by Schock and Miquel indicated similar results experimentally (Haaksman et al., 2017; F. Li et al., 2002). It can be concluded that feed spacers in SWMs can be described using flat surfaces. The various thicknesses and geometries available for spacers ultimately affect flow hydrodynamics and turbulence at the membrane surface, which affects not only concentration polarization but also foulants that are accumulated near the surface (Haidari et al., 2018).

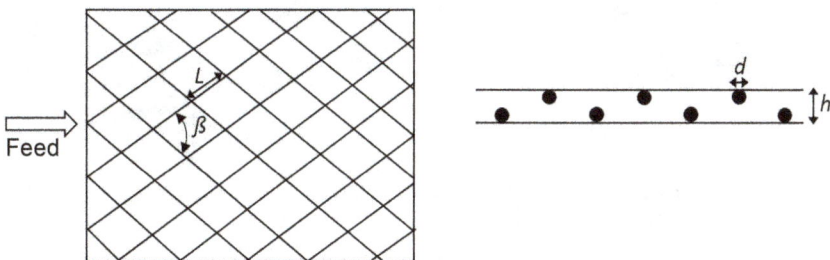

FIGURE 6.3   Feed spacer geometry features.

Two-dimensional CFD studies of membrane systems remain more common than 3D models due to the complexity and large time consumption of the latter. Radu et al. investigated the impact of spacer strand size and alignment on inorganic and biological fouling in an RO system (Radu et al., 2014). Apart from cylindrical stands, unconventional cross-sectional designs such as triangular, sawtooth (J. Liu et al., 2015), elliptical (Sousa et al., 2014), and square (Ahmad et al., 2005) have all been explored for feed spacers. Some studies have used 3D models to design feed spacers for RO and UF systems (Fimbres-Weihs & Wiley, 2010). Among cylindrical feed spacers, the internal strand angle β and flow attack angle α affects channel height and solute transport (Fimbres-Weihs & Wiley, 2007; Ranade & Kumar, 2006; Shakaib, 2009). Gu et al. applied 3D CFD simulations to study 20 feed spacer configurations differing in mesh geometry and operating factors (Gu et al., 2017). They found that a fully woven feed spacer performed better in terms of reducing CP and increasing water flow. Although the fully woven spacer undergoes a slightly greater pressure drop ΔP than in non-woven spacers, the higher angle between axial filaments helps reduce CP. Figure 6.4 shows the four filament configurations of feed spacers used in their study. Pressure drop is also sensitive to the attack angle, which is the angle between axial filaments and direction of inflow. Another group also applied 3D modeling to compare four feed spacer shapes, namely ladder type, triple, wavy, and submerged using the Spacer Configuration Efficacy (SCE). The concept of SCE was first introduced by Saeed et al. to compare spacer geometries and flow distributions (Saeed et al., 2015b). The

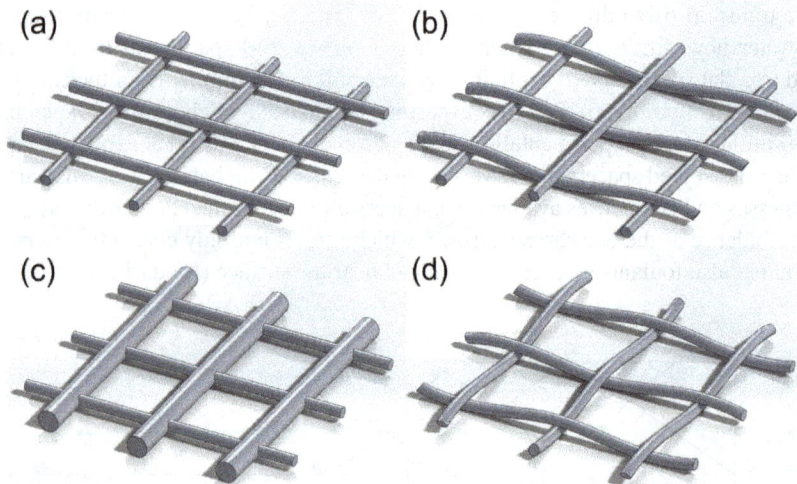

FIGURE 6.4   Feed spacer filament configurations: (a) non-woven, (b) partially woven, (c) middle layer, (d) fully woven spacers.

Source: Gu, B., Adjiman, C. S., & Xu, X. Y. (2017). The effect of feed spacer geometry on membrane performance and concentration polarisation based on 3D CFD simulations. *Journal of Membrane Science, 527,* 78–91. Reprinted with permission.

SCE is a dimensionless number that captures mass transfer and energy required and therefore quantifies mixing quality on the feed side of the membrane for a given spacer geometry (Kavianipour et al., 2017).

Siddiqui and coworkers compared two commercial feed spacers with different spacer thicknesses (0.86 and 0.79 mm), as well as four modified feed spacers: thinner strands, larger mesh size, etc.) (Siddiqui et al., 2017a). The modified spacers exhibited a lower pressure drop, i.e. greater biofouling resistance as compared to the unmodified commercial spacers, with similar results for short-term and long-term biofouling studies. Thicker feed spacers have been shown to reduce pressure drop and minimize clogging in pressure-driven membrane processes for water treatment; however, thicker spacers may negatively impact the active membrane area (Bartels et al., 2008). Previously, Park et al. studied the effect of feed spacer thickness on membrane fouling mitigation with long-term RO pilot tests (659 hours) (Park et al., 2016). Their work revealed that a thicker membrane spacer can reduce membrane fouling and therefore decrease the frequency of membrane cleaning.

Feed spacer channel porosity has also been shown to affect membrane biofouling in spiral wound elements (Lin et al., 2022), although fouling resistance is more sensitive to filament diameter and spacer thickness. High porosity feed spacer channels (with porosity exceeding 85%) showed a slower increase of feed channel pressure drop in the presence of biofouling but with more accumulated biomass in the middle due to lower shear stress. On the other hand, a block effect dominates the development of biofilm in the interspace region between spacer filaments and membrane surface, leading to severe biofouling. Lin et al. suggested a channel porosity of 85% as optimal for biofouling resistance.

Kerdi et al. proposed a modified spacer design for fouling mitigation by intrinsically modifying cylindrical spacer filaments through perforations and investigating the effect of perforations on permeate flux, feed channel pressure drop, and membrane fouling (Kerdi et al., 2018). They used 3D printing to fabricate four symmetric spacers: standard with no perforation (0-Hole), spacer with one hole at the filament intersection (1-Hole), spacer with one hole at the filament intersection and one hole at the center of filament (2-Hole), and spacer with one hole at the filament intersection and two holes on the filament (3-Hole), as shown in Figure 6.5. Filament diameter was kept constant for all printed spacers. Their study revealed that perforations lowered net pressure drop across the spacer filled channel. Performance enhancement compared to the 0-Hole is attributed to the formation of microjets in the spacer cell, which add to turbulence and clean the membrane surface. The membrane of the 1-Hole spacer was the least fouled and also exhibited the highest permeate flux due to greater turbulence at the outlet of the microjets and shear stress fluctuations created inside the cells. The 1-Hole spacer experienced a 75% increase in permeate flux at constant pressure. Their group also investigated helical structures with one, two, or three helices per filament to generate additional vortices and mitigate fouling (Kerdi et al., 2020). They found that the presence of microhelices helps improve performance and that the structure with three helices per filament showed the least biofouling, as monitored by OCT.

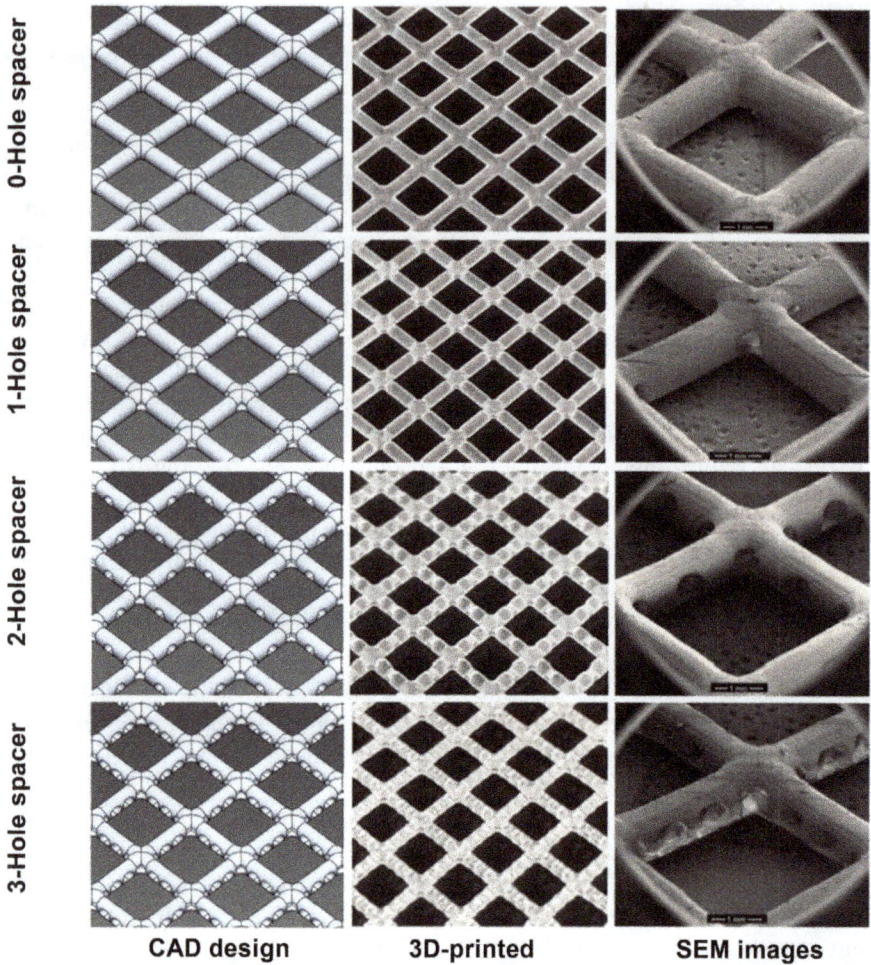

FIGURE 6.5 CAD design models, photographs, and SEM images of 3D-printed feed spacers with perforations.

Source: Kerdi, S., Qamar, A., Vrouwenvelder, J. S., & Ghaffour, N. (2018). Fouling resilient perforated feed spacers for membrane filtration. *Water Research*, *140*, 211–219. Reprinted with permission.

Other unique approaches have been adopted by researchers to control fouling and related phenomena. A study by Tan et al. demonstrated the use of vibrating 3D spacers to control biofouling by implementing wave-like protrusions in spacers, which improves turbulence and shear rate (Y. Z. Tan et al., 2019). Previously, the same group demonstrated that 3D sinusoidal spacer geometry was 48 and 25% more effective in achieving fouling reduction than 1D and 2D spacers, respectively (Wu et al., 2019). Even though optimization of membrane and spacer material is given much attention, one must consider the role of spacer geometry in fouling control (Siddiqui et al., 2017a).

6.3 ROLE OF SPACERS IN FOULING MITIGATION

Monitoring of full-scale and pilot-scale RO modules has revealed that biofouling is a feed spacer problem more than it is a membrane problem. This finding is of utmost importance as so much emphasis is placed on antifouling membranes, yet the role of spacers is often underestimated. In both spiral wound NF and RO systems, feed channel pressure drop caused by biomass accumulation was greater when a feed spacer was present (Vrouwenvelder et al., 2009). The findings of Vrouwenvelder et al. are based on in situ visual observations of fouling accumulation, differences in pressure drop and biomass developments, and non-destructive observation of fouling using magnetic resonance imaging (MRI). Since then, researchers have dedicated more attention, albeit not enough, to low fouling feed spacers and control of hydrodynamic conditions in addition to fouling resistant membrane materials. An image of biofouling localized near feed spacer strands is shown in Figure 6.6.

Due to the widespread use of spiral wound modules (SWM) for reverse osmosis desalination, feed spacers have long been implemented in SWM to improve flow dynamics. This is because of the benefits offered by SWM in the form of high density, large surface area, and consequently high productivity per unit volume. Faisal Ibney Hai's group was among the first to acknowledge the role of spacers in increasing the fouling resistance in hollow fiber modules in a submerged membrane reactor

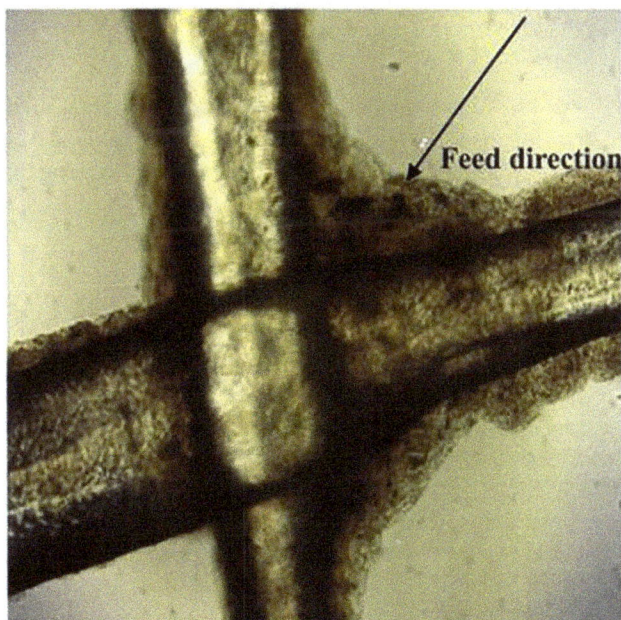

FIGURE 6.6   Biofouling localized near feed spacer strands.

Source: Reid, K., Dixon, M., Pelekani, C., Jarvis, K., Willis, M., & Yu, Y. (2014). Biofouling control by hydrophilic surface modification of polypropylene feed spacers by plasma polymerisation. *Desalination*, *335*(1), 108–118. Reprinted with permission.

treating high strength industrial wastewater (Hai et al., 2008). They reported that the use of a spacer prevented fouling for one month as compared to no spacer, in which case severe fouling was observed in one day.

Several cleaning techniques used to recover the membrane surface have been extended to spacer regeneration as well. For example, gas sparging via injection of air reduces fouling deposition and/or intermittently removes foulants from the membrane surface. Ngene and coworkers extended gas sparging to spacer cleaning and developed a method based on CO_2 nucleation in membrane spacer channels (Ngene et al., 2010). When water saturated with dissolved CO_2 flows at higher pressures and the spacer channel is rinsed at a lower pressure, depressurization causes CO_2 bubbles to nucleate at spacer filaments and clean their surface. When comparing the three methods, namely water rinsing, water/N_2 sparging, and water/CO_2 (dissolved) nucleation, foulant removal of 40, 85, and 100% were achieved, respectively. Cleaning efficiency was measured as a percentage of clean channel hydraulic resistance that recovered after cleaning. Yang et al. used bubble generation from direct H_2O_2 dosing to the feedwater to clean fouled CuO spacers (W. Yang et al., 2019). Periodic dosing of H_2O_2 reduced biofouling by 69% via physically shearing off biofilms through O_2 generation as well as chemically killing bacteria. Reduction in biofouling was indicated by feed channel pressure drop. Similar to cleaning methods devised for membrane surfaces, these techniques boast chemical-free effective cleaning of the spacer surface.

6.3.1 SURFACE MODIFICATION

Ever since the role of spacers in fouling aggravation was established, an interest in developing fouling resistant spacer materials has naturally developed. Some researchers have adopted surface modification approaches by coating the feed spacer surface to reduce the effect of fouling.

6.3.1.1 Antimicrobial Coatings

Antimicrobial or antibacterial surface coatings have the ability to kill or prevent the reproduction of bacteria. Antimicrobial agents can be incorporated into the bulk material or as a surface coating to prevent unwanted biological growth on surfaces. Antibacterial coatings use three main mechanisms to neutralize bacteria and other biological agents: (1) photocatalytic reactions that generate aggressive reactive species, (2) biological attack to bacteria functions by interrupting electron/ion transport and disrupting cell respiration, and (3) direct mechanical interaction to kill bacteria by puncturing cellular walls (Casini, 2022).

Hausman et al. reported the fabrication of a functionalized polypropylene feed spacer containing a spacer arm with metal chelating ligands, iminodiacetic acid (IDA) (Hausman et al., 2009). The increased hydrophilicity and antimicrobial property of the Cu-charged feed spacer prevented adhesion of microbial cells and consequent biofilm formation by an order of magnitude compared to unmodified feed spacers (Hausman & Escobar, 2011) over 168 hours as copper is believed to by cytotoxic. Due to its antimicrobial properties, silver was also applied by the authors as a chelated, antimicrobial metal to modify PP spacers, but leaching of silver resulted in poor long-term performance compared to Cu-charged spacers (Hausman & Escobar, 2013). FTIR studies also confirmed that the silver-charged spacer was more susceptible to colloidal fouling than traditional EPS-controlled biofouling. In another study, the group concluded that spacer thickness, orientation, coating with either silver, copper, or gold, and using a biostatic feed spacer also did not prevent biofouling (Araújo et al., 2012). Hydrodynamic conditions such as lower linear feed flow velocity played a more important role in reducing biofouling and feed channel drop.

Yang et al. applied a chemical reduction technique to coat both membrane and feed spacer with silver nanoparticles for biofouling control in RO systems (H.-L. Yang et al., 2009). Silver has long been used as an antimicrobial compound; the antimicrobial action of silver or silver compounds is related to the release of bioactive Ag^+ and their availability to interact with bacterial cell membranes (Lansdown, 2006).

They compared permeate flux decline between two systems—modified membrane with unmodified feed spacer and modified feed spacer with unmodified membrane—in a cross-flow flat-sheet membrane cell as part of a pilot plant at the Wukan desalination plant. They found that both systems performed better than unmodified membrane with unmodified feed spacer but that the antimicrobial activity on the modified feed spacer lasted longer (H.-L. Yang et al., 2009). With the coated spacer, nearly no multiplication of cells was observed during testing. However, we note that fouling was observed as permeate flux decline and salt rejection, which we have noted are not accurate or early indicators of fouling extent and type. A decade later, Thamaraiselvan et al. modified the surface of PP feed spacers by growing ZnO nanorods (Thamaraiselvan et al., 2019). The antibacterial activity increased when the modification layer consisted of nanorods as opposed to nanoparticles, indicating the role of morphology on antibacterial behavior: The crystalline nanorods had an abrasive surface, combined with hydrophobic repulsion and oxidation toward biofoulants (Thamaraiselvan et al., 2019). The modified spacers showed a 40% reduction in biofouling compared to unmodified spacers. Figure 6.7 shows SEM images of the chemical-bath-deposited ZnO nanorods grown on PP spacers. Their group also applied superhydrophobic candle soot as a low fouling stable coating on membrane feed spacers (Thamaraiselvan et al., 2021). They coated a PP feed spacer in two steps: first with polydimethylsiloxane (PDMS) and then with candle soot nanoparticles embedded into a PDMS layer. The coated membranes showed almost complete removal of biofilm growth, which was confirmed through confocal laser scanning microscopy (CLSM) and SEM. The superhydrophobic nanostructured surface acts to trap gases and form a stable plastron.

Ronen et al. used two approaches to modify spiral wound PP feed spacers: The first involved covalent binding of polymeric quaternary ammonium groups (pQAs) to the spacer, and the second was embedding silver nanoparticles in the polymeric spacer via sonochemical deposition (Ronen et al., 2015; Ronen, Resnick et al., 2016). While both modified spacers prevented biofouling in membranes, pQA grafted spacer showed localized antibacterial activity through direct contact, whereas Ag-coated spacers demonstrated long-distance antibacterial activity, resulting in a steadier permeate flux profile. Previously, their group used ZnO to modify feed spacers to suppress flux decline from biofouling (Ronen et al., 2013). Membrane flux was five times

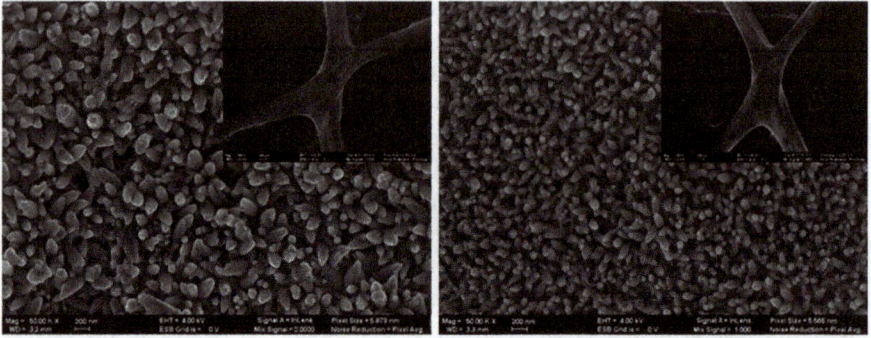

FIGURE 6.7 SEM images of the chemical bath deposited ZnO nanorods grown on PP spacers. Left: ZnO nanorods; right: ZnO nanorods with hydrophobic coating. Insets show intersection of spacer strands.

Source: Thamaraiselvan, C., Carmiel, Y., Eliad, G., Sukenik, C. N., Semiat, R., & Dosoretz, C. G. (2019). Modification of a polypropylene feed spacer with metal oxide-thin film by chemical bath deposition for biofouling control in membrane filtration. *Journal of Membrane Science, 573,* 511–519. Reprinted with permission.

higher in the case of ZnO-modified spacers due to the disruption of foulant accumulation with coated spacers, as shown in Figure 6.8. However, long-term antifouling is still difficult to achieve (Siddiqui et al., 2017a).

6.3.1.2 Plasma Treatment

Plasma treatment is a common approach used to modify the surface of polymeric materials. It consists of exposing surfaces to an electrical discharge or plasma to alter their surface chemistry. Plasma treatment has been extensively employed on various polymers with the objective of improving adhesion of coatings, enhancing wetting, antifouling, and developing biomaterials (Hegemann et al., 2003). Plasma treatment is versatile in that the end surface properties can be tailored by controlling gas flow, pressure, and treatment time. Depending on plasma conditions, the surface is modified by ions, electrons, fast neutrals, radicals, and vacuum ultraviolet (VUV) radiation (Fozza et al., 1999; Liston et al., 1993).

Plasma treatment has been used on feed spacers in membrane processes to improve their resistance to fouling. Nikkola and coworkers reported the surface modification of polypropylene (PP) and low-density polyethylene (LDPE) spacer surfaces using alkoxysilane coatings prepared by the sol–gel method, followed by atmospheric plasma treatment to improve surface hydrophilicity and alkoxysilane adhesion (Nikkola et al., 2012). Plasma pretreatment increases the surface energy of both spacer materials while the alkoxysilane coating lowers surface roughness. The functionalized coating imparted antimicrobial activity against *E. coli* and *S. aureus.*

In another approach to mitigate biofouling, Wibisono et al. grafted neutral, cationic, and anionic hydrogels onto PP feed spacers using plasma-mediated UV polymerization (Wibisono et al., 2015). The antibiofouling properties were investigated using *E. coli* during nanofiltration experiments; the authors found that the highly

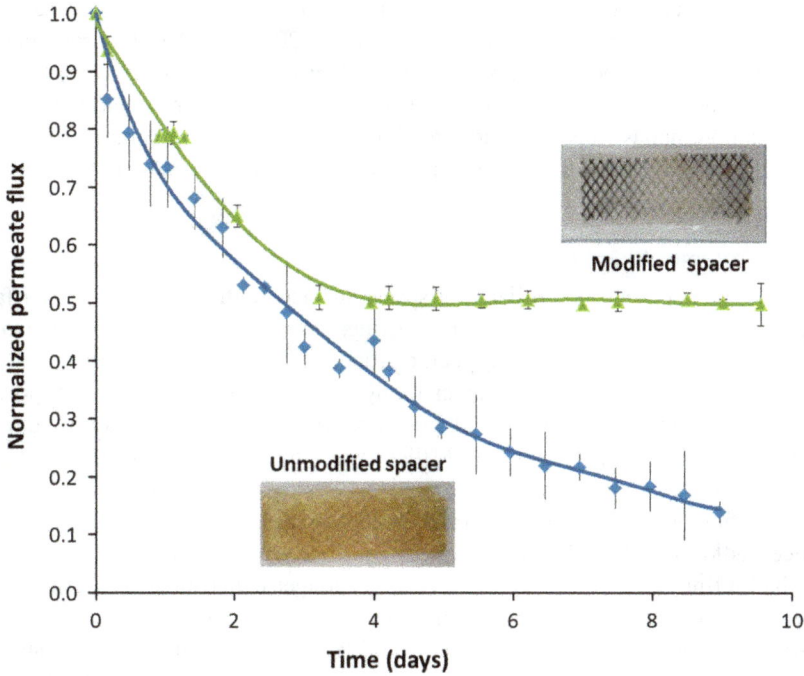

FIGURE 6.8   Flux profile of a polysulfone membrane during 230 hour experiment in a planar flow cell with exposure to mixed bacterial cultures.

Source: Ronen, A., Lerman, S., Ramon, G. Z., & Dosoretz, C. G. (2015). Experimental characterization and numerical simulation of the anti-biofouling activity of nanosilver-modified feed spacers in membrane filtration. *Journal of Membrane Science, 475,* 320–329. Reprinted with permission.

hydrophilic and negatively charged anionic-coated spacer exhibited reduced biofouling and delayed biofilm growth. Katherine Reid et al. used plasma polymerization to modify PP spacers with diglyme monomer and correlated the energy density of the treatment with biofouling (Reid et al., 2014). They found that low energy density caused insufficient cross-linking of the monomer to the spacer surface, which resulted in increased biofouling as compared to the unmodified spacer. Optimal plasma energy density between 120–170 J cm^{-3} yielded stable coatings with reduced biomass attachment (Reid et al., 2014). Plasma polymerization is used to improve hydrophilicity. Although hydrophilic surfaces are resistant to protein adsorption and cell adhesion (Krishnan et al., 2008), hydrophobic surfaces are known for their innate self-cleaning ability (Bixler & Bhushan, 2012; Marmur, 2006). This only points to the importance of understanding the type of foulants involved and of using accurate tools to measure atomic-scale interaction forces and interfacial energy values to devise antifouling characteristics. Many studies have attempted to explain fouling interactions through atomic force microscopy (Bowen & Hilal, 2009; Hilal et al., 2002), which is out of the scope of the current text. Jabłońska et al. also pretreated

HDPE/PP spacers with plasma before coating them with zwitterionic poly(sulfobe-taine methacrylate) (pSBMA) (Jabłońska et al., 2020). Antiadhesive properties of the modified membrane delayed short-term EPS production by adhered bacteria by lowering adhesion of *P. fluorescens* by 70%; however, biofouling could not be pre-vented in continuous runs. Intensity and duration are important parameters during plasma treatment, as fast interaction with radicals or ions may cause degradation of polymer surfaces.

6.3.1.3 Polymer Blend Coatings

Polymer blend coatings are utilized for spacer surface modification to tailor surface properties and mitigate fouling. Blend coatings combine the properties of two or more different polymers to obtain specific properties in the end material.

Miller et al. evaluated the impact of hydrophilic modification using polydo pamine-*g*-poly(ethylene glycol) and polydopamine feed spacer coatings during UF and NF on fouling propensity (Miller et al., 2012). Short-term UF tests and long-term NF tests showed that the coated spacers reduced adhesion of *P. aerugi-nosa* and bovine serum albumin (BSA). Both coatings demonstrated considerably reduced adhesion of BSA and *P. aeruginosa* in short-term adhesion tests, how-ever biofouling reduction was not observed in longer experiments using modified feed spacers (Figure 6.6). Although their revelation casts a doubt on the long-term performance of antifouling feed spacer modifications and can be somewhat dis-couraging, it should emphasize the need for long-term tests for novel feed spacer materials and cleaning strategies. The same group later reported that feed spacer or membrane modification through surface coating may not be an effective strategy for biofouling control (Araújo et al., 2012).

Rice and coworkers compared the effect of four different coatings on the biofoul-ing resistance of RO feed spacers: biocidal silver coating, hydrophilic SiO_2 coating, superhydrophobic antiadhesive coating with trimethoxy(propyl)silane (TMPSi)-TiO_2, and hybrid hydrophilic-biocidal coating with graphene oxide (Rice et al., 2018). Adhesion assays, viability tests, and flux decline were all used as measurements for biofouling control. Interestingly, the unmodified spacer performed well in terms of bacterial deposition during dynamic RO fouling experiments due to its near ideal surface free energy (SFE), second only to the biocidal silver coating (Rice et al., 2018). Their study implies that the improvement of spacer materials using coatings should focus on engineered biocidal coatings rather than on antiadhesive coatings.

6.3.2 3D-Printed Spacers

Engineering the geometry of feed spacers to optimize feed flow properties and foul-ing propensity requires the use of advanced fabrication technologies that are able to generate complex geometries with relatively low material cost and time. The devel-opment of 3D printing in recent years has been particularly promising for realizing innovative spacer designs. The flexibility of 3D printing for fabricating complex geom-etries makes them an attractive candidate for feed spacer fabrication. Benefits associ-ated with 3D printing include flexibility of design and material, reduced cycle time, and reduced material wastage (Ngo et al., 2018). In the area of membrane technology,

3D printing is being used to fabricate membrane and module components, including modified patterned polymeric membranes (Issac & Kandasubramanian, 2020), ceramic membranes with controlled morphology (Dommati et al., 2019), and other module components with innovative geometries (Lee et al., 2016). Due to resolution limits, 3D printing is currently more suitable for the production of spacer materials rather than microporous membranes. Eventually, the development of apparatus with increased resolution and speed output will extend this technology to the fabrication of membranes with submicron pores (Lee et al., 2016). Figure 6.9 shows key milestones of 3D printing in membrane module design.

Contrary to conventional manufacturing methods, 3D printing or additive manufacturing allows the precise construction of free-form objects layer by layer using a wide variety of materials including metals, polymers, composites, and other materials (Lee et al., 2016; W. S. Tan et al., 2017). It facilitates engineering patterns in designing complex objects. 3D printing is used in various applications including building and construction (Ding et al., 2014), marine industries (Kim et al., 2014), food industry (Sun et al., 2015), biomedical devices (M. Wang et al., 2015), and more recently water treatment systems (W. S. Tan et al., 2017)

Li and coworkers were the first to design 3D-printed feed spacers using a powder-based selective laser sintering (SLS) process (F Li et al., 2005). Their work highlighted the difference in fluid flow patterns by changing the configuration between modified strand, helical, and multilayered feed spacers. Balster et al. also investigated geometrical factors in feed spacers for an electrodialysis (ED) plant to reduce concentration polarization (Balster et al., 2006). Another 3D printing technique, known as stereolithography, was applied by Liu et al. to develop feed spacers for static mixing in ultrafiltration (Liu et al., 2013). Fritzmann et al. used PolyJet printing, one of the most common types of 3D printing, to fabricate feed spacers for submerged membrane modules (Fritzmann et al., 2013). Fused deposition modeling (FDM), SLS, stereolithography, and PolyJet printing are among the commonly used 3D printing techniques for feed spacer engineering (Siddiqui et al., 2016).

Tan et al. investigated the impact of 3D printing technique, geometry, and surface morphology on membrane performance and fouling detachment (W. S. Tan et al., 2017). As predicted, surface modification strongly influences bacterial attachment.

Castillo et al. develop 3D-printed triply periodic minimal surfaces (TPMS) spacers for organic fouling mitigation in direct contact membrane distillation (DCMD) in two geometries: Schwarz transverse crossed layer of parallel (tCLP) and Schoen gyroid (gyroid) (Castillo et al., 2019). Figure 6.10a and c show the design features and mechanism of fouling during DCMD of both 3D-printed geometries. Compared to commercial spacers, the 3D-printed spacer with smaller hydraulic diameter can enhance flux by 50–65% due to specific geometrical design features such as tortuous design. Their study confirmed that the 3D-printed spacers performed better than commercial and no-spacer scenarios with less foulant accumulated on the membrane, as shown by liquid chromatography with organic carbon detection (LC-OCD) analysis (Figure 6.10b). Stability of novel membrane-spacer systems relies on suitable pretreatment, as well as on cleaning-in-place (CIP) methods in conjunction with appropriate spacer material and geometry, to ensure effective long-term performance (Castillo et al., 2019). Sreedhar et al. functionalized 3D-printed UF feed spacers with

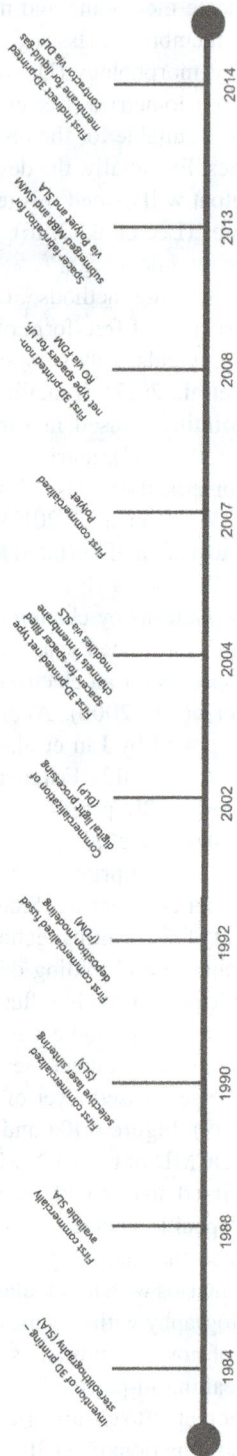

FIGURE 6.9   Milestones of 3D printing in membrane modules.

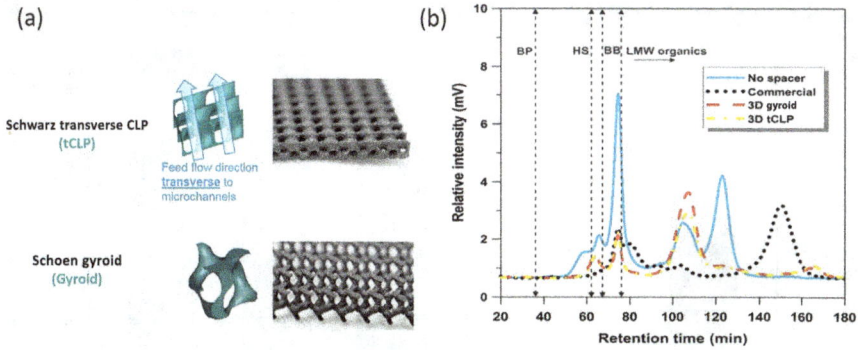

FIGURE 6.10 (a) Design features of two 3D-printed spacer geometries: Schwarz transverse CLP (tCLP) (top) and Schoen gyroid (bottom); (b) LC-OCD chromatogram of membrane foulant composition with and without commercial and 3D-printed spacers.

Source: Castillo, E. H. C., Thomas, N., Al-Ketan, O., Rowshan, R., Abu Al-Rub, R. K., Nghiem, L. D., & Naidu, G. (2019). 3D printed spacers for organic fouling mitigation in membrane distillation. *Journal of Membrane Science*, *581*, 331–343. Reprinted with permission.

photocatalytic ß-FeOOH nanorods for pollutant degradation and membrane cleaning (Sreedhar et al., 2022). For three organic foulants, namely HA, SA, and BSA, membrane cleaning resulted in a flux recovery of 92, 60, and 54%, respectively. Figure 6.3 shows the effect of various 3D-printed spacer geometries on organic and inorganic fouling (Khalil et al., 2021).

Siddiqui et al. attempted to elucidate the relationship between commercial spacers with and without geometric modification for biofouling resistance in membrane fouling simulators, but the manufacturers who provided the spacers did not specify which aspects of the feed spacer geometry were changed (Siddiqui et al., 2017b).

While 3D printing has shown promise in feed spacer design, especially in adopting fouling resistant geometries, limitations in the form of expensive raw materials and in the range of suitable materials still need to be overcome.

6.3.3 ELECTRICALLY CONDUCTIVE SPACERS FOR FOULING MITIGATION

The role of electrically conductive materials for fouling mitigation has been described in Section 5.2. Here, we assess recent literature in electrically conductive spacers for fouling mitigation.

Baek et al. applied electrically conductive Ti feed spacers to control biofouling in a lab-scale cross-flow filtration system using a SWM (Baek et al., 2014). Electrical polarization for 30 minutes after a period of 24 hours of fouling recovered permeate flux by 33–44% without any noticeable membrane damage (Baek et al., 2014). Figure 6.11 shows permeate flux recovery with electrical polarization. Permeate flux data was supported by measurement of bacterial concentration and microscopic images of bacteria or biofilms on both the membrane and the spacer surfaces. The low electrical potentials used ensure that bacterial detachment from the spacer surface,

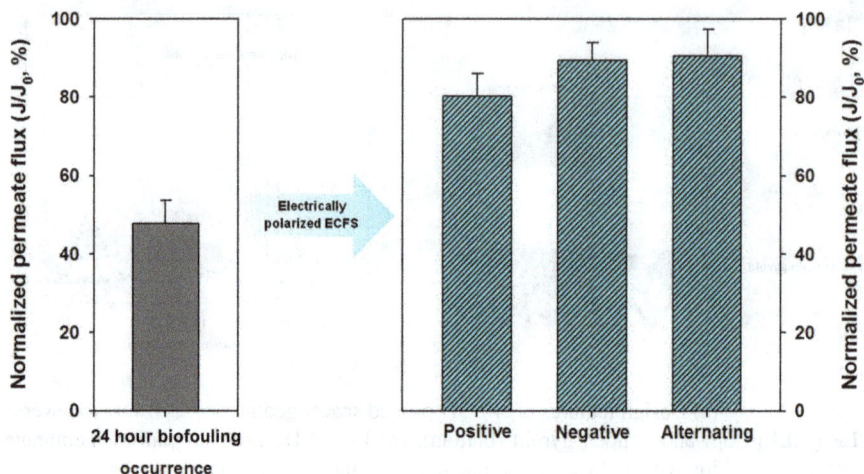

FIGURE 6.11 Flux regeneration through the application of +1 V, −1 V, and alternating potential on Ti mesh spacer for 30 minutes after 24 hours of biofouling in a cross-flow membrane system; *P. aeruginosa* PA01 is used as biofoulant in a feed solution containing 10 mM NaCl, 10 mM sodium citrate, and 0.1% tryptic soy broth; initial flux: 40 L m^{-2} h^{-1}; cross-flow velocity: 7.6 cm s^{-1}; temperature: 25 °C).

Source: Reprinted with permission from *Environ. Sci. Technol. Lett.* 2014, *1*(2), 179–184. Copyright © 2014 American Chemical Society. https://doi.org/10.1021/ez400206d.

rather than inactivation through oxidant generation, is the dominant mechanism for bacterial removal. Bacterial removal is attributed to electrostatic forces between the feed spacer and attached *P. aeruginosa*.

In another study, building on periodic electrolytic cleaning methods for membranes, Abid and coworkers developed electrically conducting feed spacers by coating a polymeric spacer with an ink consisting of graphene nanoplates (GNP) (Abid et al., 2017). GNPs were chosen over nanosilver ink due to the observed leaching of the latter. Bubble generation and sufficient current was observed only when a voltage of −6 V was applied to the GNP-coated spacer, so periodic electrolysis was carried out at this potential.

A potential of −6 V was applied for 2 minutes after each cycle of fouling of PVDF MF membranes with sodium alginate (SA) such that the generation of hydrogen microbubbles resulted in lifting the foulant layer. Flux regeneration via electrolytic spacer cleaning was more effective when carried out after shorter filtration cycles (Abid et al., 2017). Figure 6.12 shows permeate flux regeneration during filtration of SA using both GNP-coated spacers and Ti-coated spacers. Electrochemical cleaning using conducting spacers is a chemical-free method to address organic fouling in membrane processes.

The material of the metal mesh used as spacer strongly affects its performance for electrolytic self-cleaning. Abid et al. compared titanium and steel spacers to evaluate the effect of material, foulant concentration in feed, and filtration interval

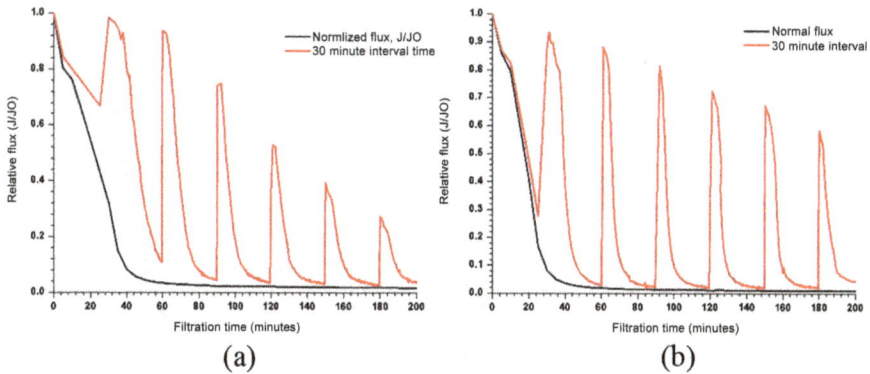

FIGURE 6.12 Permeate flux regeneration with respect to initial flux for filtration of sodium alginate suspension using (a) GNP-coated polymer spacer and (b) Ti spacer.

Source: Abid, H. S., Lalia, B. S., Bertoncello, P., Hashaikeh, R., Clifford, B., Gethin, D. T., & Hilal, N. (2017). Electrically conductive spacers for self-cleaning membrane surfaces via periodic electrolysis. *Desalination, 416*, 16–23. Reprinted with permission.

before cleaning on permeate flux when using the spacer as cathodes (Abid, 2019). They used four initial SA feed concentrations (20, 60, 100, and 150 mg L^{-1}), and four HA concentrations (8, 12, 16, and 20 mg L^{-1}). Commercial MF membranes were used, and feed flow rate was kept at 0.58 L min^{-1}. Graphite was employed as the counter-electrode in these experiments. Stainless steel as anode resulted in a noticeable rust layer.

Figure 6.13 shows linear sweep voltammograms of both metal meshes. Comparing the electrochemical performance of both spacers with the Ag/AgCl reference electrode, they found that microbubble evolution starts at −1 V for the stainless steel spacer and at −0.5 V for the Ti mesh.

6.3.3.1 Effect of Applied Voltage on in Situ Cleaning Using Electrically Conductive Spacers

The effect of applied voltage was studied by filtering SA/NaCl solutions in an electrochemical cross-flow cell at an operating pressure of 0.5 bar and flow rate of 0.58 L min^{-1}. No bubbles were observed when potentials of −1 V and −2 V were applied for 2 minutes each. At −2 V, the initial measured current was 10 mA, which quickly dropped over 2 minutes. At −3 V, a constant current of 10 mA was measured, and noticeable bubbles evolved at the feed spacer strands (Figure 6.14). The formation of gas bubbles at high voltages can block the reactive surfaces and increase overpotential and ohmic voltage drop. Cho et al. tackle this problem by applying ultrasound during electrolysis to reduce overpotential and save energy for HER on a stainless steel plate (Cho et al., 2021). Electrically conductive spacers is still a young field that yields the possibility of optimizing membrane separation characteristics and spacer fouling control separately. Although this work illustrates the potential for electrically conductive spacers, the field is deterred by a lack of understanding of

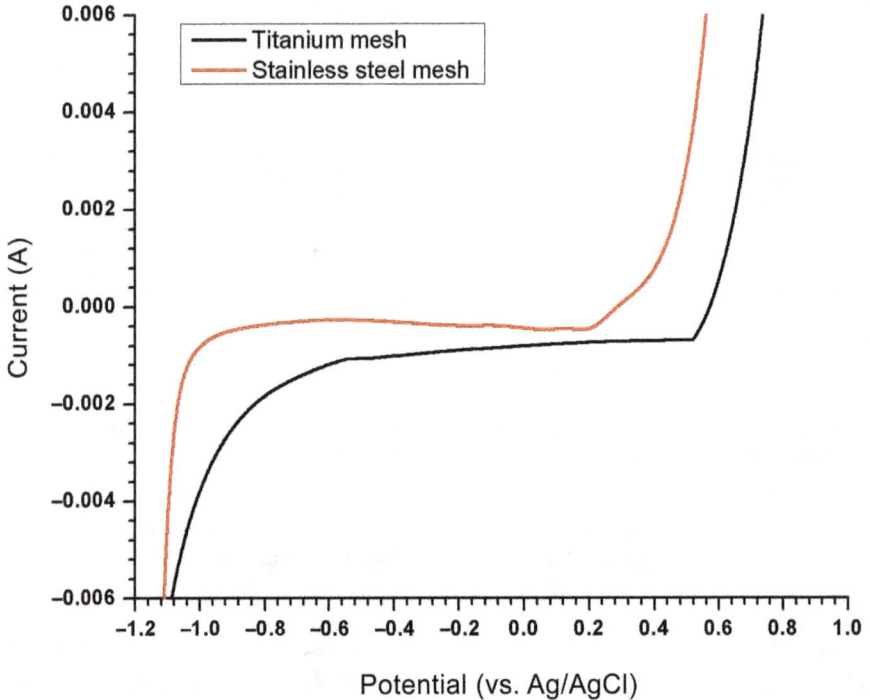

FIGURE 6.13 Linear sweep voltammograms of electrically conductive titanium and stainless steel feed spacers.

Source: Abid, H. S. (2019). *Periodic electrolysis for advanced water treatment* (Doctor of Philosophy). Swansea University, United Kingdom.

the fundamental relationship between gas-evolving electrode surface characteristics, bubble dynamics, and overpotential (Iwata et al., 2021). Investigating voltage, time, as well as the size and uniformity of bubbles formed is critical to optimizing conditions for electrolytic cleaning.

6.3.3.2 Effect of Filtration Interval and Foulant Concentration on Electrolytic Self-Cleaning

The efficacy of in situ cleaning of Ti spacers was evaluated with and without periodic electrolysis (Abid, 2019). Figure 6.15 shows the change in flux over time for varying SA concentrations and filtration intervals with and without an applied potential of −3 V. Cleaning was carried out for 6, 4, and 3 cycles, at filtration times of 30, 45, and 60 minutes, respectively. Initial flux decline was attributed to foulant accumulation. In all cases, each cycle of applied potential helped regenerate flux, although flux recovery dropped with the number of cycles. After the first cycle of electrolytic cleaning carried out after 30 minutes of filtration with 20 ppm SA, flux improved from 1006 L m⁻² h⁻¹ to 3400 L m⁻² h⁻¹. Similarly, following the first cycle of electrolytic cleaning carried out after 45 minutes of filtration, flux improved from 724 L

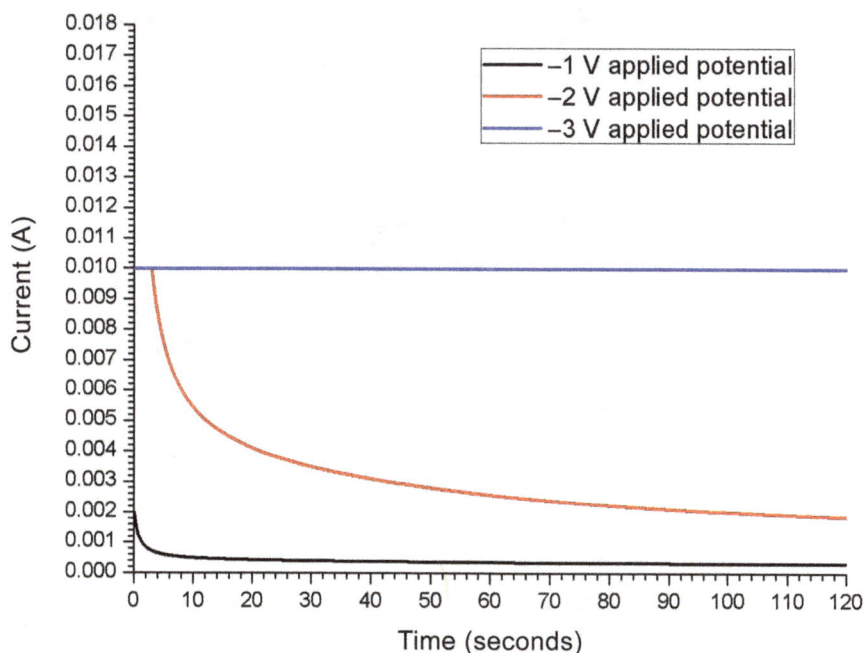

FIGURE 6.14 Effect of applied voltage on current through electrochemical cell using titanium feed spacer as cathode and graphite as anode.

Source: Abid, H. S. (2019). *Periodic electrolysis for advanced water treatment* (Doctor of Philosophy). Swansea University, United Kingdom.

$m^{-2}\,h^{-1}$ to 3300 L $m^{-2}\,h^{-1}$. When periodic electrolysis was carried out after a filtration time of 60 minutes, the first cycle improved flux from 138 to 3100 L $m^{-2}\,h^{-1}$. The longer the filtration time, the more the flux decline was due to fouling because there was more time for foulants to accumulate. Hydrogen bubbles formed during electrolysis detached SA molecules from the membrane surface back into the feed stream; this reduction of fouling is reflected as flux recovery. They noted that flux recovery dropped as initial SA concentration increased, as higher foulant concentration makes cleaning less efficient. The adhesion of foulants is dictated by various phenomena including hydrophobic interactions, electrostatic interactions, electrostatic interactions, hydrogen bonding, and van der Waals attractions. Higher concentration leads to greater buildup close to the feed spacer filaments, making physical removal more difficult.

Due to the higher overpotential for HER using stainless steel, the appropriate applied potential during cross-flow filtration was set at −4 V as dense bubbles were not found to evolve at lower potentials. Figure 6.16 shows flux decline with and without applied potential using stainless steel as spacer instead. The trend is similar to that observed for titanium, and an improvement with applied potential is observed when periodic electrolysis is carried out after 30, 45, and 60 minute filtration intervals. Energy consumption was slightly higher when titanium was used due to the

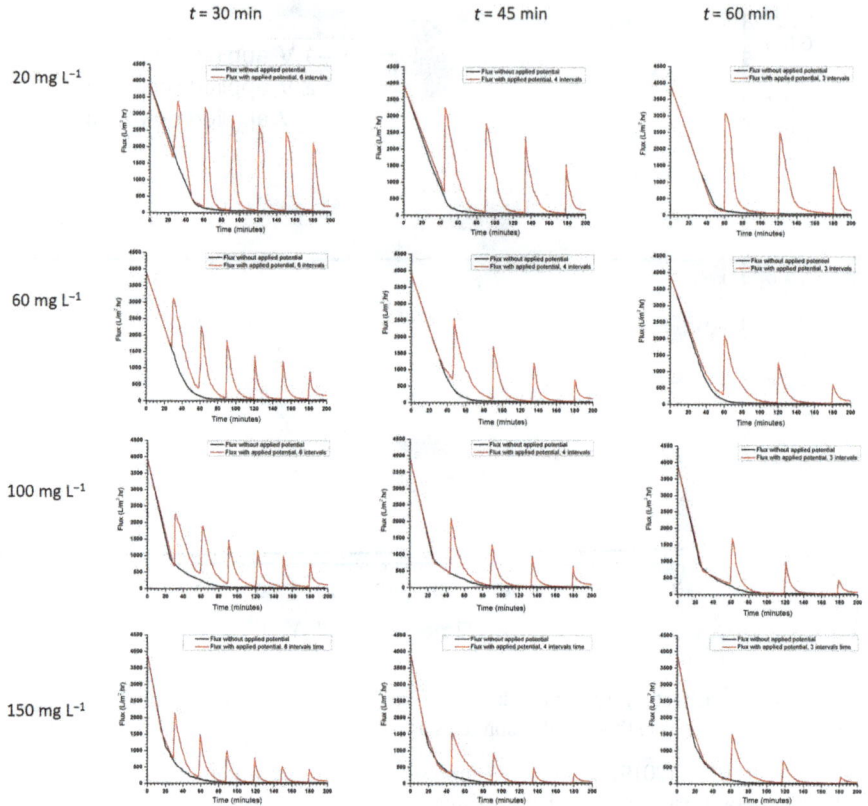

FIGURE 6.15　Flux with time for filtration using Ti mesh spacer for varying SA concentrations and filtration intervals with and without applied potential of –3 V.

Source: Abid, H. S. (2019). *Periodic electrolysis for advanced water treatment* (Doctor of Philosophy). Swansea University, United Kingdom.

greater potential that had to be applied for HER. Titanium is more hydrophilic than stainless steel, which makes its higher overpotential surprising, as Iwata et al. found that higher wettability leads to lower overpotentials for bubble formation on electrodes (Iwata et al., 2021).

Later, the same group investigated the effect of spacer aperture size on in situ electrolytic fouling mitigation (Abid et al., 2018). The same dip-coating method was used to coat a carbon-based ink on two polypropylene diamond-shaped meshes with aperture sizes of 2 mm and 3 mm in conjunction with a PVDF MF membrane, where HA at different concentrations was used as foulant. They tested both spacers for fouling mitigation and recirculated the concentrate streams into the feed tank. An SA solution with 10 mg L^{-1} NaCl was used as electrolyte, and in situ periodic electrolysis was carried out at intervals of either 30, 45, or 60 minutes of filtration time, using the spacer as cathode. The applied potential helped mitigate SA fouling by influencing the formation of cake layer.

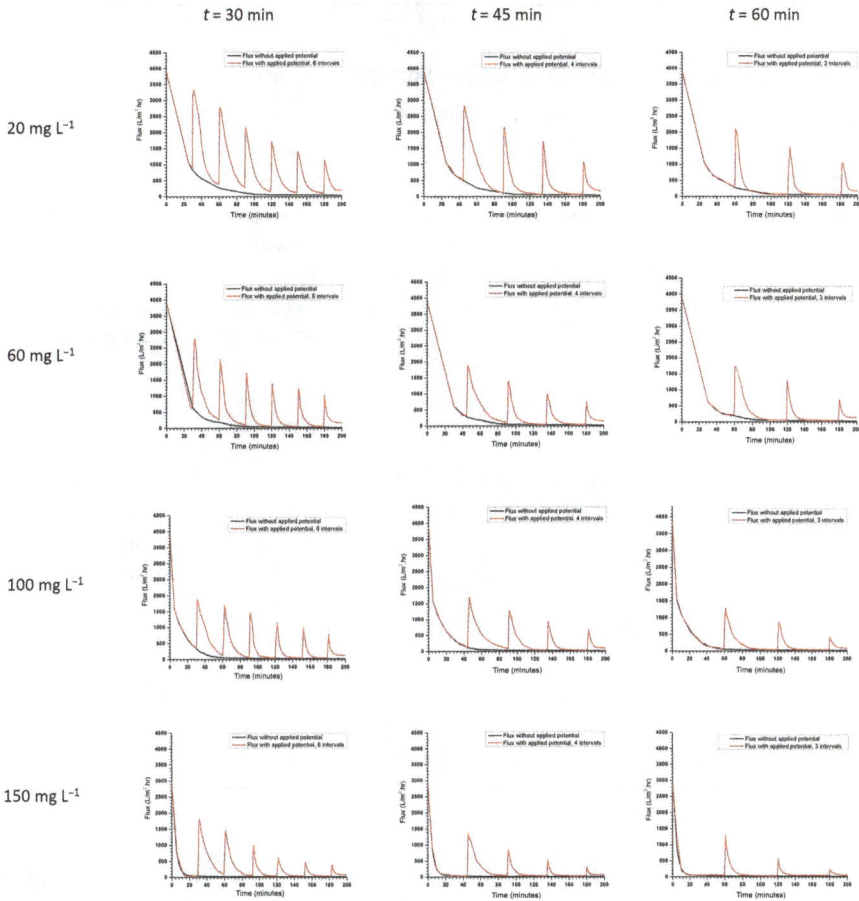

FIGURE 6.16   Flux with time for filtration using stainless steel mesh spacer for varying SA concentrations and filtration intervals with and without applied potential of −4 V.

Source: Abid, H. S. (2019). *Periodic electrolysis for advanced water treatment* (Doctor of Philosophy). Swansea University, United Kingdom.

They found that the smaller aperture size resulted in improved flux and lower energy consumption; this improvement was attributed to the greater conductive area for electrolytic cleaning in the spacer with a smaller aperture size (Abid et al., 2018). However, more work should be carried out to find the optimum aperture size at which electrolytic cleaning is effective and spacer-enhanced flux is also noticeable.

Yanar and coworkers also developed electrically conductive spacers by blending graphene with poly(lactic acid) PLA to develop 3D-printed spacers; they applied electrical polarization to control organic fouling ("Electrically Polarized Graphene-Blended Spacers for Organic Fouling Reduction in Forward Osmosis," 2021). Electrostatic forces on the modified membrane helped reduce membrane fouling by 70%, with a flux decline of only 14% after 5 hours of fouling.

TABLE 6.1

Application of Electrically Conductive Spacers to Control Fouling.

Electrically Conducting Component	Material	Type of Fouling	Mechanism of Fouling Control	Improvement	Reference
Spacer	Titanium	Biofouling	Electrostatic forces	33–44% flux recovery	(Baek et al., 2014)
Spacer	GNP-PVDF; titanium	Organic (SA)	Hydrogen bubble evolution	30% (GNP-PVDF) and 60% (Ti) flux recovery after six cycles of periodic cleaning	(Abid, Johnson et al., 2018; Abid, Lalia et al., 2017)
Spacer	Graphene/PLA	Organic (SA)	Electrostatic forces	70%	("Electrically Polarized Graphene-Blended Spacers for Organic Fouling Reduction in Forward Osmosis," 2021)
Spacer	Graphite/Ag nanoparticles/PES	Organic (HA)	Hydrogen and chlorine bubble evolution	90%	(Khalil et al., 2022)

Khalil and coworkers used 3D printing to fabricate electrically conductive inter-digitated spacers directly on commercial UF membranes from a paste of graphite powder, silver nanoparticles, PES, and surfactant (Khalil et al., 2022). The multifunctional membrane fabricated as such also acts as spacer and electrode for self-cleaning, as shown in Figure 6.17. Three-dimensional printing on the membrane works to increase turbulence and eliminates the need for counter-electrodes, as is the case with other electrochemical filtration systems reported in the literature. With an applied potential in the range of 5–8 V, hydrogen evolution at the cathode helps detach HA, while at the anode, the generation of chlorine may provide disinfection as the overall flux is recovered to 90% of its original value in the first few cycles (Khalil et al., 2022).

6.4 IN SITU CHARACTERIZATION OF FEED SPACER FOULING

In Chapter 3, we discussed techniques used for in situ detection of membrane fouling. Some of these techniques have been extended to real-time monitoring of feed spacer channel fouling, as we will see in this section.

Optical coherence tomography (OCT) is an advanced optical imaging technique in which light is introduced into the feed channel through an optical window and an

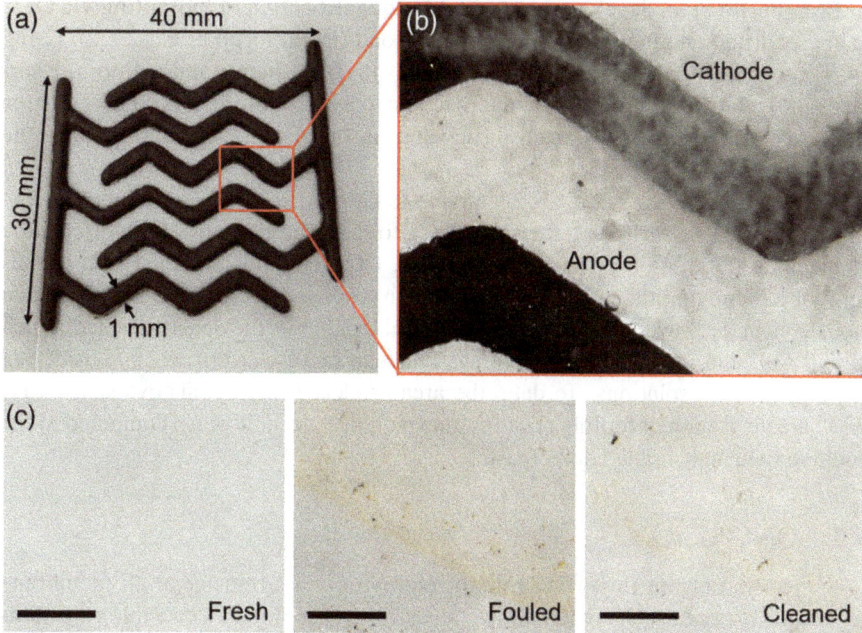

FIGURE 6.17   Cleaning effect of electrically conductive interdigitated spacer: (a) photograph of 3D-printed spacer on commercial UF membrane; (b) optical microscope image showing hydrogen evolution at cathodic branch under 5V; (c) optical microscope images showing pristine, fouled, and cleaned membranes with electrically conductive 3D-printed spacer.

OCT signal is generated when light backscattered from the sample, e.g. foulant layer, membrane, etc., is combined with light reflected from a reference. To characterize fouling, the 3D OCT scan can be carried out continuously at time intervals of 30 or 60 seconds.

West and coworkers were the first to apply in situ non-invasive OCT for the 3D visualization and quantification of feed spacer channel fouling in RO modules (West et al., 2016). They used OCT to study the effect of feed spacer geometry and organic foulant load on biofilm development. Their work demonstrated that OCT is a suitable tool to monitor the evolution of biofilm formation on the membrane feed spacer without interrupting the process. OCT data correlated with macroscopic measurements in feed channel pressure drop while offering the ability to deconstruct and visualize the structure of biofilm formation, which is not possible by simple determination of feed channel pressure drop. The following year, Fortunato et al. demonstrated spatially resolved in situ quantification of biofouling using OCT in a spacer filled channel of a UF membrane module (Fortunato et al., 2017). They used OCT to evaluate spatial biomass distribution, distinguishing total biomass volume among the different components of the module: membrane, feed spacer, and glass cell window. Such 3D biofilm thickness mapping can be useful in visualizing biofilm deposition in situ (Fortunato & Leiknes, 2017).

Pratofiorito et al. developed a method for quantification of biofouling via OCT in spacer filled channels with high organic load (Pratofiorito et al., 2022). Using OCT, they distinguished biofilm growth on spacer filaments and membrane surface and related biofouling to macroscopic process parameters. Their results showed that biofilm first started to develop on the feed spacer and then on the membrane surface.

Liu et al. used the same OCT technique to determine local deposition rates (X. Liu et al., 2017). In particular, they used OCT to show the effect of spacer orientation of membrane fouling by particulate foulants. OCT has thus emerged as a powerful tool for studying the evolution of fouling in response to the microhydrodynamic environment. However, since OCT is an optical technique, the opacity of media such as cell wall can interfere with imaging. This can be of concern when considering challenging feed solutions. To date, the area of electrically conductive spacers for the characterization of fouling remains unexplored, especially when compared to the boom in their membrane counterpart.

6.5 CONCLUSION

Feed spacers play an important role in improving mass transfer in all membrane separation processes. Currently, ample research activity focuses on exploiting modified feed spacer performance. Little has been done to mitigate membrane fouling from the perspective of spacer design. However, since spacers are typically the first to foul, there is growing interest in developing antifouling spacers as well as electrically conducting spacers for fouling mitigation. This is also accompanied by the effect of spacer parameters on fouling dynamics, into which in situ microscopic monitoring techniques such as OCT can give a deeper insight. Understanding fouling on membrane spacers will eventually lead to design improvements and more efficient membrane modules. Today, as membrane development has reached new heights, considering module improvements in other components such as feed spacers is more important than ever in enhancing performance. Electrically conducting spacers may have certain advantages over electrically conducting membranes in that often the electrically conducting modification of membrane materials compromises key membrane properties such as selectivity; it is still a challenge to tailor electrically conducting membranes with a certain permeate flux and solute rejection. Similar to conducting membranes, electrically conducting spacers can be used for electrolytic self-cleaning to control fouling and flux recovery. Nevertheless, the simplified system of electrically conductive membranes or membranes with integrated spacers provides the most direct route to electrical fouling mitigation at the membrane surface. The ample progress and widespread knowledge in membrane fabrication techniques, as opposed to the limitations in spacer material fabrication, has allowed breakthroughs in conductive membrane fabrication at a much faster rate than conductive spacer fabrication, as shown by the literature. Instead, an increased interest in developing electrically conducting spacers to mitigate fouling is thus not surprising and will help enhance the overall process for membrane technologies in desalination and water treatment. However, most studies are limited to the laboratory scale. As shown in this section, the fouling control of coatings during long-term testing may differ significantly from short-term tests; hence more effort should be directed toward the long term and studies on a larger scale. Additionally, the energy consumption of electrically conducting spacers is unclear, although approaches have involved conductive coatings and the use of metal meshes directly as feed spacers. Commercialization of these technologies is still pending, but research so far has provided valuable insight into novel spacer-based strategies for overcoming fouling and related challenges in membrane modules. Other strategies that have been adopted to optimize engineered feed spacers for fouling mitigation involve 3D-printed spacers, surface coatings, and modified geometry. Modification of novel feed spacers has emerged as a promising area, especially as new research in nanomaterials continues to take form on a daily basis.

BIBLIOGRAPHY

Abid, H. S. (2019). *Periodic electrolysis for advanced water treatment* (Doctor of Philosophy). Swansea University, United Kingdom.

Abid, H. S., Johnson, D. J., Clifford, B., Gethin, D. T., Bertoncello, P., Hashaikeh, R., & Hilal, N. (2018). Periodic electrolysis technique for in situ fouling control and removal with low-pressure membrane filtration. *Desalination, 433*, 10–24. https://doi.org/10.1016/j.desal.2018.01.019

Abid, H. S., Lalia, B. S., Bertoncello, P., Hashaikeh, R., Clifford, B., Gethin, D. T., & Hilal, N. (2017). Electrically conductive spacers for self-cleaning membrane surfaces via periodic electrolysis. *Desalination, 416*, 16–23. https://doi.org/10.1016/j.desal.2017.04.018

Ahmad, A. L., Lau, K. K., & Abu Bakar, M. Z. (2005). Impact of different spacer filament geometries on concentration polarization control in narrow membrane channel. *Journal of Membrane Science, 262*(1), 138–152. https://doi.org/10.1016/j.memsci.2005.06.056

Al Ashhab, A., Gillor, O., & Herzberg, M. (2014). Biofouling of reverse-osmosis membranes under different shear rates during tertiary wastewater desalination: Microbial community composition. *Water Research, 67*, 86–95.

Ang, W. L., & Mohammad, A. W. (2015). 12—Mathematical modeling of membrane operations for water treatment. In A. Basile, A. Cassano, & N. K. Rastogi (Eds.), *Advances in membrane technologies for water treatment* (pp. 379–407). Woodhead Publishing.

Araújo, P. A., Kruithof, J. C., Van Loosdrecht, M. C. M., & Vrouwenvelder, J. S. (2012). The potential of standard and modified feed spacers for biofouling control. *Journal of Membrane Science, 403–404*, 58–70. https://doi.org/10.1016/j.memsci.2012.02.015

Araújo, P. A., Miller, D. J., Correia, P. B., van Loosdrecht, M. C. M., Kruithof, J. C., Freeman, B. D., . . . Vrouwenvelder, J. S. (2012). Impact of feed spacer and membrane modification by hydrophilic, bactericidal and biocidal coating on biofouling control. *Desalination, 295*, 1–10. https://doi.org/10.1016/j.desal.2012.02.026

Baek, Y., Yoon, H., Shim, S., Choi, J., & Yoon, J. (2014). Electroconductive feed spacer as a tool for biofouling control in a membrane system for water treatment. *Environmental Science & Technology Letters, 1*(2), 179–184. doi:10.1021/ez400206d

Baker, J. S., & Dudley, L. Y. (1998). Biofouling in membrane systems—a review. *Desalination, 118*(1–3), 81–89.

Balster, J., Pünt, I., Stamatialis, D. F., & Wessling, M. (2006). Multi-layer spacer geometries with improved mass transport. *Journal of Membrane Science, 282*(1), 351–361. https://doi.org/10.1016/j.memsci.2006.05.039

Bartels, C., Hirose, M., & Fujioka, H. (2008). Performance advancement in the spiral wound RO/NF element design. *Desalination, 221*(1), 207–214. https://doi.org/10.1016/j.desal.2007.01.077

Ben-Sasson, M., Zodrow, K. R., Genggeng, Q., Kang, Y., Giannelis, E. P., & Elimelech, M. (2014). Surface functionalization of thin-film composite membranes with copper nanoparticles for antimicrobial surface properties. *Environmental Science & Technology*, *48*(1), 384–393.

Bixler, G. D., & Bhushan, B. (2012). Biofouling: Lessons from nature. *Philosophical Transactions of the Royal Society A: Mathematical, Physical and Engineering Sciences*, *370*(1967), 2381–2417. doi:10.1098/rsta.2011.0502

Bowen, R., & Hilal, N. (2009). *Atomic force microscopy in process engineering: An introduction to AFM for improved processes and products*. Elsevier Science.

Bucs, S. S., Radu, A. I., Lavric, V., Vrouwenvelder, J. S., & Picioreanu, C. (2014). Effect of different commercial feed spacers on biofouling of reverse osmosis membrane systems: A numerical study. *Desalination*, *343*, 26–37.

Casini, M. (2022). Chapter 7—Advanced construction materials. In M. Casini (Ed.), *Construction 4.0* (pp. 337–404). Woodhead Publishing.

Castillo, E. H. C., Thomas, N., Al-Ketan, O., Rowshan, R., Abu Al-Rub, R. K., Nghiem, L. D., . . . Naidu, G. (2019). 3D printed spacers for organic fouling mitigation in membrane distillation. *Journal of Membrane Science*, *581*, 331–343. https://doi.org/10.1016/j.memsci.2019.03.040

Cho, K. M., Deshmukh, P. R., & Shin, W. G. (2021). Hydrodynamic behavior of bubbles at gas-evolving electrode in ultrasonic field during water electrolysis. *Ultrason Sonochem*, *80*, 105796. doi:10.1016/j.ultsonch.2021.105796

Chong, T., & Fane, A. (2009). Implications of critical flux and cake enhanced osmotic pressure (CEOP) on colloidal fouling in reverse osmosis: Modeling approach. *Desalination and Water Treatment*, *8*(1–3), 68–90.

Da Costa, A. R., & Fane, A. G. (1994). Net-type spacers: Effect of configuration on fluid flow path and ultrafiltration flux. *Industrial & Engineering Chemistry Research*, *33*(7), 1845–1851.

Da Costa, A. R., Fane, A. G., Fell, C., & Franken, A. (1991). Optimal channel spacer design for ultrafiltration. *Journal of Membrane Science*, *62*(3), 275–291.

Ding, L., Wei, R., & Che, H. (2014). Development of a BIM-based automated construction system. *Procedia Engineering*, *85*, 123–131. https://doi.org/10.1016/j.proeng.2014.10.536

Dommati, H., Ray, S. S., Wang, J. C., & Chen, S. S. (2019). A comprehensive review of recent developments in 3D printing technique for ceramic membrane fabrication for water purification. *RSC Advances*, *9*(29), 16869–16883.

Electrically polarized graphene-blended spacers for organic fouling reduction in forward osmosis. (2021). *Membranes*, *11*(1), 36. http://dx.doi.org/10.3390/membranes11010036

Field, R. W., & Wu, J. J. (2018). On boundary layers and the attenuation of driving forces in forward osmosis and other membrane processes. *Desalination*, *429*, 167–174. https://doi.org/10.1016/j.desal.2017.12.001

Fimbres-Weihs, G. A., & Wiley, D. E. (2007). Numerical study of mass transfer in three-dimensional spacer-filled narrow channels with steady flow. *Journal of Membrane Science*, *306*(1), 228–243. https://doi.org/10.1016/j.memsci.2007.08.043

Fimbres-Weihs, G. A., & Wiley, D. E. (2010). Review of 3D CFD modeling of flow and mass transfer in narrow spacer-filled channels in membrane modules. *Chemical Engineering and Processing: Process Intensification*, *49*(7), 759–781. https://doi.org/10.1016/j.cep.2010.01.007

Fortunato, L., Bucs, S., Linares, R. V., Cali, C., Vrouwenvelder, J. S., & Leiknes, T. (2017). Spatially-resolved in-situ quantification of biofouling using optical coherence tomography (OCT) and 3D image analysis in a spacer filled channel. *Journal of Membrane Science*, *524*, 673–681. https://doi.org/10.1016/j.memsci.2016.11.052

Fortunato, L., & Leiknes, T. (2017). In-situ biofouling assessment in spacer filled channels using optical coherence tomography (OCT): 3D biofilm thickness mapping. *Bioresource Technology*, *229*, 231–235. https://doi.org/10.1016/j.biortech.2017.01.021

Fozza, A. C., Klemberg-Sapieha, J. E., & Wertheimer, M. R. (1999). Vacuum ultraviolet irradiation of polymers. *Plasmas and Polymers*, *4*(2), 183–206. doi:10.1023/a:1021853 026619

Fritzmann, C., Hausmann, M., Wiese, M., Wessling, M., & Melin, T. (2013). Microstructured spacers for submerged membrane filtration systems. *Journal of Membrane Science*, *446*, 189–200. https://doi.org/10.1016/j.memsci.2013.06.033

Gu, B., Adjiman, C. S., & Xu, X. Y. (2017). The effect of feed spacer geometry on membrane performance and concentration polarisation based on 3D CFD simulations. *Journal of Membrane Science*, *527*, 78–91.

Guo, W., Ngo, H.-H., & Li, J. (2012). A mini-review on membrane fouling. *Bioresource Technology*, *122*, 27–34. https://doi.org/10.1016/j.biortech.2012.04.089

Haaksman, V. A., Siddiqui, A., Schellenberg, C., Kidwell, J., Vrouwenvelder, J. S., & Picioreanu, C. (2017). Characterization of feed channel spacer performance using geometries obtained by X-ray computed tomography. *Journal of Membrane Science*, *522*, 124–139. https://doi.org/10.1016/j.memsci.2016.09.005

Hai, F. I., Yamamoto, K., Fukushi, K., & Nakajima, F. (2008). Fouling resistant compact hollow-fiber module with spacer for submerged membrane bioreactor treating high strength industrial wastewater. *Journal of Membrane Science*, *317*(1), 34–42. https://doi.org/10.1016/j.memsci.2007.06.026

Haidari, A. H., Heijman, S. G. J., & van der Meer, W. G. J. (2018). Optimal design of spacers in reverse osmosis. *Separation and Purification Technology*, *192*, 441–456. https://doi.org/10.1016/j.seppur.2017.10.042

Hartinger, M., Napiwotzki, J., Schmid, E.-M., Hoffmann, D., Kurz, F., & Kulozik, U. (2020). Influence of spacer design and module geometry on the filtration performance during skim milk microfiltration with flat sheet and spiral-wound membranes. *Membranes*, *10*(4), 57.

Hausman, R., & Escobar, I. C. (2011). A Fourier transform infrared spectroscopic based biofilm characterization technique and its use to show the effect of copper-charged polypropylene feed spacers in biofouling control. In *Modern applications in membrane science and technology* (Vol. 1078, pp. 225–237). American Chemical Society.

Hausman, R., & Escobar, I. C. (2013). A comparison of silver- and copper-charged polypropylene feed spacers for biofouling control. *Journal of Applied Polymer Science*, *128*(3), 1706–1714. https://doi.org/10.1002/app.38164

Hausman, R., Gullinkala, T., & Escobar, I. C. (2009). Development of low-biofouling polypropylene feedspacers for reverse osmosis. *Journal of Applied Polymer Science*, *114*(5), 3068–3073. https://doi.org/10.1002/app.30755

Hegemann, D., Brunner, H., & Oehr, C. (2003). Plasma treatment of polymers for surface and adhesion improvement. *Nuclear Instruments and Methods in Physics Research Section B: Beam Interactions with Materials and Atoms*, *208*, 281–286. https://doi.org/10.1016/S0168-583X(03)00644-X

Hilal, N., Bowen, W. R., Lovitt, R. W., & Wright, C. J. (2002). (Bio)fouling of polymeric membranes: Atomic force microscope study. *Engineering in Life Sciences*, *2*(5), 131–135. https://doi.org/10.1002/1618-2863(200205)2:5<131::AID-ELSC131>3.0.CO;2-G

Hoek, E. M., & Elimelech, M. (2003). Cake-enhanced concentration polarization: A new fouling mechanism for salt-rejecting membranes. *Environmental Science & Technology*, *37*(24), 5581–5588.

Issac, M. N., & Kandasubramanian, B. (2020). Review of manufacturing three-dimensional-printed membranes for water treatment. *Environmental Science and Pollution Research*, *27*(29), 36091–36108.

Iwata, R., Zhang, L., Wilke, K. L., Gong, S., He, M., Gallant, B. M., & Wang, E. N. (2021). Bubble growth and departure modes on wettable/non-wettable porous foams in alkaline water splitting. *Joule*, *5*(4), 887–900. https://doi.org/10.1016/j.joule.2021.02.015

Jabłońska, M., Menzel, M., Hirsch, U., & Heilmann, A. (2020). Assessment of anti-bacterial adhesion and anti-biofouling potential of plasma-mediated poly(sulfobetaine methacrylate) coatings of feed spacer. *Desalination*, *493*, 114664. https://doi.org/10.1016/j. desal.2020.114664

Kavianipour, O., Ingram, G. D., & Vuthaluru, H. B. (2017). Investigation into the effectiveness of feed spacer configurations for reverse osmosis membrane modules using Computational Fluid Dynamics. *Journal of Membrane Science*, *526*, 156–171.

Kerdi, S., Qamar, A., Alpatova, A., & Ghaffour, N. (2019). An in-situ technique for the direct structural characterization of biofouling in membrane filtration. *Journal of Membrane Science*, *583*, 81–92.

Kerdi, S., Qamar, A., Alpatova, A., Vrouwenvelder, J. S., & Ghaffour, N. (2020). Membrane filtration performance enhancement and biofouling mitigation using symmetric spacers with helical filaments. *Desalination*, *484*, 114454. https://doi.org/10.1016/j. desal.2020.114454

Kerdi, S., Qamar, A., Vrouwenvelder, J. S., & Ghaffour, N. (2018). Fouling resilient perforated feed spacers for membrane filtration. *Water Research*, *140*, 211–219. https://doi. org/10.1016/j.watres.2018.04.049

Khalil, A., Ahmed, F. E., Hashaikeh, R., & Hilal, N. (2022). 3D printed electrically conductive interdigitated spacer on ultrafiltration membrane for electrolytic cleaning and chlorination. *Journal of Applied Polymer Science*, 52292.

Khalil, A., Ahmed, F. E., & Hilal, N. (2021). The emerging role of 3D printing in water desalination. *Science of the Total Environment*, *790*, 148238. https://doi.org/10.1016/j. scitotenv.2021.148238

Kim, T. B., Yue, S., Zhang, Z., Jones, E., Jones, J. R., & Lee, P. D. (2014). Additive manufactured porous titanium structures: Through-process quantification of pore and strut networks. *Journal of Materials Processing Technology*, *214*(11), 2706–2715. https://doi. org/10.1016/j.jmatprotec.2014.05.006

Koo, J. W., Ho, J. S., An, J., Zhang, Y., Chua, C. K., & Chong, T. H. (2021). A review on spacers and membranes: Conventional or hybrid additive manufacturing? *Water Research*, *188*, 116497. https://doi.org/10.1016/j.watres.2020.116497

Koutsou, C., Yiantsios, S., & Karabelas, A. (2007). Direct numerical simulation of flow in spacer-filled channels: Effect of spacer geometrical characteristics. *Journal of Membrane Science*, *291*(1–2), 53–69.

Krishnan, S., Weinman, C. J., & Ober, C. K. (2008). Advances in polymers for anti-biofouling surfaces. *Journal of Materials Chemistry*, *18*(29), 3405–3413. doi:10.1039/ B801491D

Lansdown, A. (2006). Silver in health care: Antimicrobial effects and safety in use. *Current Problems in Dermatology*, *33*, 17–34. doi:10.1159/000093928

Lee, J.-Y., Tan, W. S., An, J., Chua, C. K., Tang, C. Y., Fane, A. G., & Chong, T. H. (2016). The potential to enhance membrane module design with 3D printing technology. *Journal of Membrane Science*, *499*, 480–490. https://doi.org/10.1016/j.memsci.2015.11.008

Li, F., Meindersma, W., de Haan, A. B., & Reith, T. (2002). Optimization of commercial net spacers in spiral wound membrane modules. *Journal of Membrane Science*, *208*(1), 289–302. https://doi.org/10.1016/S0376-7388(02)00307-1

Li, F., Meindersma, W., De Haan, A. B., & Reith, T. (2005). Novel spacers for mass transfer enhancement in membrane separations. *Journal of Membrane Science*, *253*(1–2), 1–12.

Lin, W., Wang, Q., Sun, L., Wang, D., Cabrera, J., Li, D., . . . Huang, X. (2022). The critical role of feed spacer channel porosity in membrane biofouling: Insights and implications. *Journal of Membrane Science*, *649*, 120395. https://doi.org/10.1016/j.memsci.2022.120395

Linares, R. V., Bucs, S. S., Li, Z., AbuGhdeeb, M., Amy, G., & Vrouwenvelder, J. S. (2014). Impact of spacer thickness on biofouling in forward osmosis. *Water Research*, *57*, 223–233.

Liston, E. M., Martinu, L., & Wertheimer, M. R. (1993). Plasma surface modification of polymers for improved adhesion: A critical review. *Journal of Adhesion Science and Technology, 7*(10), 1091–1127. doi:10.1163/156856193X00600

Liu, J., Iranshahi, A., Lou, Y., & Lipscomb, G. (2013). Static mixing spacers for spiral wound modules. *Journal of Membrane Science, 442*, 140–148.

Liu, J., Liu, Z., Xu, X., & Liu, F. (2015). Saw-tooth spacer for membrane filtration: Hydrodynamic investigation by PIV and filtration experiment validation. *Chemical Engineering and Processing: Process Intensification, 91*, 23–34. https://doi.org/10.1016/j.cep.2015.03.013

Liu, X., Li, W., Chong, T. H., & Fane, A. G. (2017). Effects of spacer orientations on the cake formation during membrane fouling: Quantitative analysis based on 3D OCT imaging. *Water Research, 110*, 1–14. https://doi.org/10.1016/j.watres.2016.12.002

Marmur, A. (2006). Super-hydrophobicity fundamentals: Implications to biofouling prevention. *Biofouling, 22*(2), 107–115. doi:10.1080/08927010600562328

Matsuura, T. (2020). *Synthetic membranes and membrane separation processes.* CRC press.

Miller, D. J., Araújo, P. A., Correia, P. B., Ramsey, M. M., Kruithof, J. C., van Loosdrecht, M. C. M., ... Vrouwenvelder, J. S. (2012). Short-term adhesion and long-term biofouling testing of polydopamine and poly(ethylene glycol) surface modifications of membranes and feed spacers for biofouling control. *Water Research, 46*(12), 3737–3753. https://doi.org/10.1016/j.watres.2012.03.058

Ngene, I. S., Lammertink, R. G. H., Kemperman, A. J. B., van de Ven, W. J. C., Wessels, L. P., Wessling, M., & Van der Meer, W. G. J. (2010). CO2 nucleation in membrane spacer channels remove biofilms and fouling deposits. *Industrial & Engineering Chemistry Research, 49*(20), 10034–10039. doi:10.1021/ie1011245

Ngo, T. D., Kashani, A., Imbalzano, G., Nguyen, K. T., & Hui, D. (2018). Additive manufacturing (3D printing): A review of materials, methods, applications and challenges. *Composites Part B: Engineering, 143*, 172–196.

Nikkola, J., Nättinen, K., Mannila, J., Vartiainen, J., Alakomi, H. L., & Tang, C. Y. (2012). Plasma-assisted hybrid coatings as low-fouling surface treatment of membrane spacer materials. *Procedia Engineering, 44*, 1479–1480. https://doi.org/10.1016/j.proeng.2012.08.835

Park, H.-G., Cho, S.-G., Kim, K.-J., & Kwon, Y.-N. (2016). Effect of feed spacer thickness on the fouling behavior in reverse osmosis process—A pilot scale study. *Desalination, 379*, 155–163. https://doi.org/10.1016/j.desal.2015.11.011

Picioreanu, C., Vrouwenvelder, J. S., & van Loosdrecht, M. C. M. (2009). Three-dimensional modeling of biofouling and fluid dynamics in feed spacer channels of membrane devices. *Journal of Membrane Science, 345*(1), 340–354. https://doi.org/10.1016/j.memsci.2009.09.024

Pratofiorito, G., Horn, H., & Saravia, F. (2022). Differentiating fouling on the membrane and on the spacer in low-pressure reverse-osmosis under high organic load using optical coherence tomography. *Separation and Purification Technology, 291*, 120885. https://doi.org/10.1016/j.seppur.2022.120885

Radu, A. I., Bergwerff, L., van Loosdrecht, M. C. M., & Picioreanu, C. (2014). A two-dimensional mechanistic model for scaling in spiral wound membrane systems. *Chemical Engineering Journal, 241*, 77–91. https://doi.org/10.1016/j.cej.2013.12.021

Rahmawati, R., Bilad, M. R., Nawi, N. I. M., Wibisono, Y., Suhaimi, H., Shamsuddin, N., & Arahman, N. (2021). Engineered spacers for fouling mitigation in pressure driven membrane processes: Progress and projection. *Journal of Environmental Chemical Engineering, 9*(5), 106285. https://doi.org/10.1016/j.jece.2021.106285

Ranade, V. V., & Kumar, A. (2006). Fluid dynamics of spacer filled rectangular and curvilinear channels. *Journal of Membrane Science, 271*(1), 1–15. https://doi.org/10.1016/j.memsci.2005.07.013

Reid, K., Dixon, M., Pelekani, C., Jarvis, K., Willis, M., & Yu, Y. (2014). Biofouling control by hydrophilic surface modification of polypropylene feed spacers by plasma polymerisation. *Desalination, 335*(1), 108–118. https://doi.org/10.1016/j.desal.2013.12.017

Rice, D., Barrios, A. C., Xiao, Z., Bogler, A., Bar-Zeev, E., & Perreault, F. (2018). Development of anti-biofouling feed spacers to improve performance of reverse osmosis modules. *Water Research, 145*, 599–607. https://doi.org/10.1016/j.watres.2018.08.068

Ronen, A., Lerman, S., Ramon, G. Z., & Dosoretz, C. G. (2015). Experimental characterization and numerical simulation of the anti-biofueling activity of nanosilver-modified feed spacers in membrane filtration. *Journal of Membrane Science, 475*, 320–329. https://doi.org/10.1016/j.memsci.2014.10.042

Ronen, A., Resnick, A., Lerman, S., Eisen, M. S., & Dosoretz, C. G. (2016). Biofouling suppression of modified feed spacers: Localized and long-distance antibacterial activity. *Desalination, 393*, 159–165. https://doi.org/10.1016/j.desal.2015.07.004

Ronen, A., Semiat, R., & Dosoretz, C. G. (2013). Impact of ZnO embedded feed spacer on biofilm development in membrane systems. *Water Research, 47*(17), 6628–6638.

Sablani, S. S., Goosen, M. F. A., Al-Belushi, R., & Wilf, M. (2001). Concentration polarization in ultrafiltration and reverse osmosis: A critical review. *Desalination, 141*(3), 269–289. https://doi.org/10.1016/S0011-9164(01)85005-0

Saeed, A., Vuthaluru, R., & Vuthaluru, H. B. (2015a). Impact of feed spacer filament spacing on mass transport and fouling propensities of RO membrane surfaces. *Chemical Engineering Communications, 202*(5), 634–646. doi:10.1080/00986445.2013.860525

Saeed, A., Vuthaluru, R., & Vuthaluru, H. B. (2015b). Investigations into the effects of mass transport and flow dynamics of spacer filled membrane modules using CFD. *Chemical Engineering Research and Design, 93*, 79–99. https://doi.org/10.1016/j.cherd.2014.07.002

Schwinge, J., Neal, P. R., Wiley, D. E., Fletcher, D. F., & Fane, A. G. (2004). Spiral wound modules and spacers: Review and analysis. *Journal of Membrane Science, 242*(1), 129–153. https://doi.org/10.1016/j.memsci.2003.09.031

Shakaib, M. (2009). *Pressure and concentration gradients in membrane feed channels numerical and experimental investigations.* NED University of Engineering & Technology.

Shirazi, S., Lin, C.-J., & Chen, D. (2010). Inorganic fouling of pressure-driven membrane processes—a critical review. *Desalination, 250*(1), 236–248. https://doi.org/10.1016/j.desal.2009.02.056

Siddiqui, A., Farhat, N., Bucs, S. S., Linares, R. V., Picioreanu, C., Kruithof, J. C., . . . Vrouwenvelder, J. S. (2016). Development and characterization of 3D-printed feed spacers for spiral wound membrane systems. *Water Res, 91*, 55–67. doi:10.1016/j.watres.2015.12.052

Siddiqui, A., Lehmann, S., Bucs, S. S., Fresquet, M., Fel, L., Prest, E. I. E. C., . . . Vrouwenvelder, J. S. (2017a). Predicting the impact of feed spacer modification on biofouling by hydraulic characterization and biofouling studies in membrane fouling simulators. *Water Research, 110*, 281–287. https://doi.org/10.1016/j.watres.2016.12.034

Siddiqui, A., Lehmann, S., Haaksman, V., Ogier, J., Schellenberg, C., van Loosdrecht, M. C. M., . . . Vrouwenvelder, J. S. (2017b). Porosity of spacer-filled channels in spiral-wound membrane systems: Quantification methods and impact on hydraulic characterization. *Water Research, 119*, 304–311. https://doi.org/10.1016/j.watres.2017.04.034

Sousa, P., Soares, A., Monteiro, E., & Rouboa, A. (2014). A CFD study of the hydrodynamics in a desalination membrane filled with spacers. *Desalination, 349*, 22–30. https://doi.org/10.1016/j.desal.2014.06.019

Sreedhar, N., Kumar, M., Al Jitan, S., Thomas, N., Palmisano, G., & Arafat, H. A. (2022). 3D printed photocatalytic feed spacers functionalized with β-FeOOH nanorods inducing pollutant degradation and membrane cleaning capabilities in water treatment. *Applied Catalysis B: Environmental, 300*, 120318. https://doi.org/10.1016/j.apcatb.2021.120318

Sun, J., Peng, Z., Yan, L., Fuh, J. Y. H., & Hong, G. S. (2015). 3D food printing an innovative way of mass customization in food fabrication. *International Journal of Bioprinting*, *1*(1).

Tan, W. S., Suwarno, S. R., An, J., Chua, C. K., Fane, A. G., & Chong, T. H. (2017). Comparison of solid, liquid and powder forms of 3D printing techniques in membrane spacer fabrication. *Journal of Membrane Science*, *537*, 283–296. https://doi.org/10.1016/j.memsci.2017.05.037

Tan, Y. Z., Mao, Z., Zhang, Y., Tan, W. S., Chong, T. H., Wu, B., & Chew, J. W. (2019). Enhancing fouling mitigation of submerged flat-sheet membranes by vibrating 3D-spacers. *Separation and Purification Technology*, *215*, 70–80. https://doi.org/10.1016/j.seppur.2018.12.085

Thamaraiselvan, C., Carmiel, Y., Eliad, G., Sukenik, C. N., Semiat, R., & Dosoretz, C. G. (2019). Modification of a polypropylene feed spacer with metal oxide-thin film by chemical bath deposition for biofouling control in membrane filtration. *Journal of Membrane Science*, *573*, 511–519. https://doi.org/10.1016/j.memsci.2018.12.033

Thamaraiselvan, C., Manderfeld, E., Kleinberg, M. N., Rosenhahn, A., & Arnusch, C. J. (2021). Superhydrophobic candle soot as a low fouling stable coating on water treatment membrane feed spacers. *ACS Applied Bio Materials*, *4*(5), 4191–4200. doi:10.1021/acsabm.0c01677

Toh, K. Y., Liang, Y. Y., Lau, W. J., & Fimbres Weihs, G. A. (2020). A review of CFD modelling and performance metrics for osmotic membrane processes. *Membranes*, *10*(10), 285. doi:10.3390/membranes10100285

Vrouwenvelder, J. S., Graf von der Schulenburg, D. A., Kruithof, J. C., Johns, M. L., & van Loosdrecht, M. C. M. (2009). Biofouling of spiral-wound nanofiltration and reverse osmosis membranes: A feed spacer problem. *Water Research*, *43*(3), 583–594. https://doi.org/10.1016/j.watres.2008.11.019

Wang, F., & Tarabara, V. V. (2007). Coupled effects of colloidal deposition and salt concentration polarization on reverse osmosis membrane performance. *Journal of Membrane Science*, *293*(1), 111–123. https://doi.org/10.1016/j.memsci.2007.02.003

Wang, M., He, J., Liu, Y., Li, M., Li, D., & Jin, Z. (2015). The trend towards in vivo bioprinting. *International Journal of Bioprinting*, *1*(1).

West, S., Wagner, M., Engelke, C., & Horn, H. (2016). Optical coherence tomography for the in situ three-dimensional visualization and quantification of feed spacer channel fouling in reverse osmosis membrane modules. *Journal of Membrane Science*, *498*, 345–352. https://doi.org/10.1016/j.memsci.2015.09.047

Wibisono, Y., Yandi, W., Golabi, M., Nugraha, R., Cornelissen, Emile, R., Kemperman, A. J. B., . . . Nijmeijer, K. (2015). Hydrogel-coated feed spacers in two-phase flow cleaning in spiral wound membrane elements: A novel platform for eco-friendly biofouling mitigation. *Water Research*, *71*, 171–186. https://doi.org/10.1016/j.watres.2014.12.030

Winograd, Y., Solan, A., & Toren, M. (1973). Mass transfer in narrow channels in the presence of turbulence promoters. *Desalination*, *13*(2), 171–186. https://doi.org/10.1016/S0011-9164(00)82043-3

Wu, B., Zhang, Y., Mao, Z., Tan, W. S., Tan, Y. Z., Chew, J. W., . . . Fane, A. G. (2019). Spacer vibration for fouling control of submerged flat sheet membranes. *Separation and Purification Technology*, *210*, 719–728. https://doi.org/10.1016/j.seppur.2018.08.062

Yang, H.-L., Lin, J. C.-T., & Huang, C. (2009). Application of nanosilver surface modification to RO membrane and spacer for mitigating biofouling in seawater desalination. *Water Research*, *43*(15), 3777–3786. https://doi.org/10.1016/j.watres.2009.06.002

Yang, W., Son, M., Xiong, B., Kumar, M., Bucs, S., Vrouwenvelder, J. S., & Logan, B. E. (2019). Effective biofouling control using periodic H2O2 cleaning with CuO modified and polypropylene spacers. *ACS Sustainable Chemistry & Engineering*, *7*(10), 9582–9587. doi:10.1021/acssuschemeng.9b01086

Zhang, H., Cheng, S., & Yang, F. (2014). Use of a spacer to mitigate concentration polarization during forward osmosis process. *Desalination*, *347*, 112–119.

7 Electrically Conductive Systems in Membrane Distillation

7.1 INTRODUCTION

Membrane distillation (MD) is a separation process that was first introduced in the 1960s but could not achieve its commercial potential as a desalination process, largely due to the lack of suitable materials and relatively high energy consumption (Alklaibi & Lior, 2005). Today, however, increasing interest in treating highly saline wastewater and growing acknowledgment of the limitations of reverse osmosis (RO) have led to a resurgence in MD research activity. While much of this activity focuses on finding materials with special wettability (Z. Wang & Lin, 2017; Yao et al., 2020), improving the energy efficiency of MD systems has taken the form of novel configurations, improved fluid flow channel design, and multifunctional materials. In this section, we will look at some of the challenges associated with MD and the role of electrically conducting surfaces in addressing those issues.

Membrane distillation (MD) is a thermally driven separation process. A temperature difference across a porous hydrophobic (typically polymeric) membrane is induced by heated saline feed flowing along one side. The resulting vapor pressure gradient causes volatile compounds to vaporize and transport through the membrane pores. On the other side of the membrane, vapor is condensed and collected, the methods for which differ according to the selected MD configuration (Drioli et al., 2015). MD is a versatile technique with applications in the treatment of pharmaceutical compounds, separation of dairy compounds, juices, desalination, as well as oily wastewater treatment (Drioli et al., 2015). Within desalination, MD has the potential to compete with RO in niche applications such as brine concentration (Adham et al., 2013; Amy et al., 2017; Kesieme et al., 2013; Minier-Matar et al., 2014) or off-grid remote desalination systems using waste heat or renewable energy (F. Banat et al., 2002; Curcio & Drioli, 2005; Schneider et al., 1988). RO membranes cannot sustain the high pressures needed for brine treatment, nor would RO be energetically feasible as its energy consumption increases with increasing feed osmotic pressure, which can be extremely high for brine streams. In commercial desalination, MD lags behind RO due to the high cost of product water resulting from the high energy consumption of MD (Zaragoza et al., 2018).

There are four common MD configurations: direct contact membrane distillation (DCMD), air gap membrane distillation (AGMD), sweeping gas membrane

DOI: 10.1201/9781003144991-7

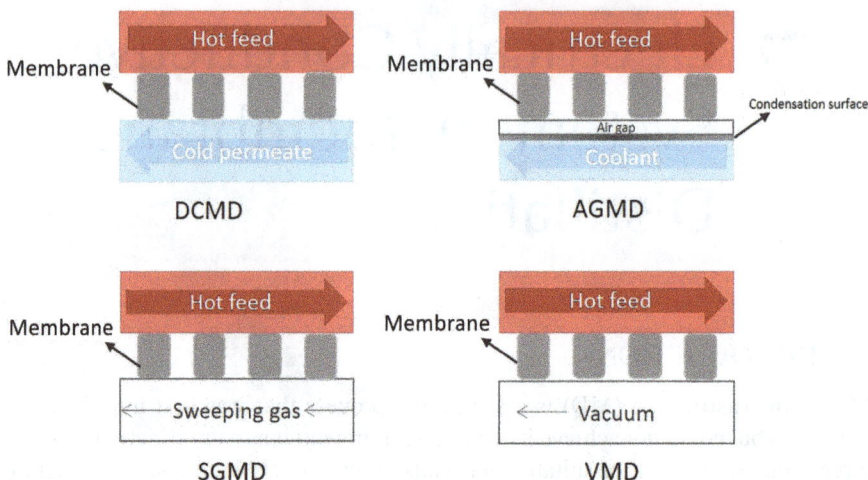

FIGURE 7.1 Schematic of four main MD configurations.

distillation (SGMD), and vacuum membrane distillation (VMD), as shown in Figure 7.1 In all MD configurations, the membrane is in contact with a heated feed solution. Ease and simplicity of operation make DCMD the preferred choice for most lab-scale MD studies (Drioli et al., 2011). However, recent research consciously focuses on more energy-efficient configurations to other configurations, namely AGMD and VMD, simply due to their greater thermal energy efficiency (Benyahia, 2019; Summers et al., 2012). For this reason, AGMD and VMD have been the most suitable configuration for pilot installations (Drioli et al., 2015), as we will see later in this chapter.

Thermal energy used to heat the feed is the largest contributor of energy in MD. However, one must note that MD is still considered a low temperature distillation process as the temperature required, typically 60–80 °C, is significantly less than that required for other thermal desalination processes such as multistage flash (MSF). Another important parameter in MD is thermal energy efficiency, which takes into account heat energy in the form of latent heat and conductive heat. Energy costs associated with MD can be kept low if waste heat is used to supply thermal energy. However, in cases where waste heat is not available, MD systems should operate at maximum thermal energy efficiency. Energy consumption in MD depends strongly on many factors such as configuration, temperature, flow rate, etc. Table 7.1 shows the pros and cons of the four main MD configurations, as illustrated by Drioli et al. (Drioli et al., 2015). Non-standardized conditions and different module sizes have made it difficult to draw direct comparisons among various configurations.

The reemergence of MD is characterized by increased research activity and progress in MD membrane materials (Hung Cong Duong et al., 2018; Guo et al., 2015; Huang et al., 2017; Lalia et al., 2013; Leitch et al., 2016; Li et al., 2016; Liao et al., 2014; Lu et al., 2017; C. Su et al., 2017; Tijing et al., 2014; Zuo & Wang, 2013). Yet

TABLE 7.1

Pros and Cons of the Four Major MD Configurations.

Configuration	Pros	Cons
DCMD	The easiest and simplest configuration to realize practically, flux is more stable than VMD for the feeds with fouling tendency, high gained output ratio, it might be the most appropriate configuration for removal of volatile components.	Flux obtained is relatively lower than vacuum configurations under the identical operating conditions, thermal polarization is highest among all the configurations, flux is relatively more sensitive to feed concentration, the permeate quality is sensitive to membrane wetting, suitable mainly for aqueous solutions.
VMD	High flux, can be used for recovery of aroma compounds and related substances, the permeate quality is stable despite some wetting, no possibility of wetting from distillate side, thermal polarization is very low.	Higher probability of pore wetting, higher fouling, minimum selectivity of volatile components, require vacuum pump and external condenser.
AGMD	Relatively high flux, low thermal losses, no wetting on permeate side, less fouling tendency.	Air gap provides an additional resistance to vapors, difficult module designing, difficult to model due to the involvement of too many variables, lowest gained output ratio.
SGMD	Thermal polarization is lower, no wetting from permeate side, permeate quality independent of membrane wetting	

Source: Drioli, E., Ali, A., & Macedonio, F. (2015). Membrane distillation: Recent developments and perspectives. *Desalination, 356,* 56–84. Reprinted with permission.

the specific energy consumption (SEC) of practical MD systems remains unclear with a wide range of values reported in the literature, as shown in Table 7.2. Khayet et al. suggest the implementation of standard testing methods to calculate energy and subsequent water production costs using MD. This step is crucial in MD studies as the technology inches toward commercialization (Ahmed et al., 2021). They also state that operating conditions such as permeate recovery rate are often not optimized. Studies suggest that countercurrent flow between feed and permeate leads to high energy efficiency when considering relative flow directions (Swaminathan et al., 2016), followed by cross-flow and parallel.

7.2 ELECTRICALLY CONDUCTING MATERIALS FOR REAL-TIME MONITORING

7.2.1 WETTING

Pore wetting reduces the efficiency of the MD process. When the pore wets completely, saline water from the feed side is allowed to penetrate into the permeate, thus causing the permeate water quality to deteriorate. Membrane wetting causes a

TABLE 7.2

Specific Energy Consumption (SEC) of Selected MD Systems.

Configuration	Membrane Characteristics	Operating Conditions		Feed Type	SEC (kWh m⁻³)	Plant Capacity (m³ h⁻¹)	Reference
		T_i (°C)	T_p (°C)				
DCMD	Spiral wound PTFE (*SEP GmbH*), *pore size 0.2 μ, porosity 80%*	35–80	5–30	Radioactive solution	600–1600	0.05	(Zakrzewska-Trznadel et al., 1999)
AGMD	PTFE, *pore size 0.2 μ*	60–85	—	Seawater	140–200	0.2–20	(Koschikowski et al., 2003)
AGMD		313–343	—	Brackish water	30.8		(Bouguecha et al., 2005)
AGMD	PTFE, *pore size 0.2 μ, porosity 80%*	—	—	Seawater	200–300	3.46–19	(Fawzi Banat 2007)
DCMD in hybrid systems	PP modules from Microdyn Nadir, *pore size 0.2 μ, porosity 73%*	—	—	Seawater	1.6–27.5	931 (overall)	(Macedonio et al., 2007)
DCMD	Commercial membranes from Membrana with *pore size 0.2 μ, thickness 91 μ*	39.8–59	13.4–14.4	Distilled water	3550–4580	—	(Alessandra Criscuoli et al., 2008)
VMD	PP, thickness 35 μ, *pore size 0.1 μ*	15–22		Underground water	8100.8–9089.5	2.67–6.94	(X. Wang et al., 2009)
AGMD	LDPE, thickness 76 μ, *Pore size 0.3 μm, porosity 85%, Am 7.4 m²*	50–70		Tap water, synthetic seawater	~65 to ~127	0.0034–0.0094	(Hung C. Duong et al., 2016)
VMD	Flat-sheet PP, thickness 400 μm, *pore size 0.1 μ, porosity 70%, Am 5 m²*	80		Distilled water	130		(A. Criscuoli et al., 2013)
DCMD	PVDF hollow fiber, thickness 240 μm	80	30	Simulated reverse osmosis brine	~130–1700		(Guan et al., 2014)
DCMD	PTFE with PP support, mean pore size 0.5 ± 0.08, porosity 91 ± 0.5, active layer thickness 46 ± 1 μm, Am 0.67 m²	60	18–21	Wastewater	1500	3.85	(Dow et al., 2016)
DCMD	Several commercial membranes with different characteristics	85	20	Seawater	~697–10,457		(Ali et al., 2012)

Source: Jantaporn, W., Ali, A., & Aimar, P. (2017). Specific energy requirement of direct contact membrane distillation. *Chemical Engineering Research and Design, 128*, 15–26. Reprinted with permission.

decline in salt rejection over time (Gryta & Barancewicz, 2010). The phenomenon of wetting has been discussed in Chapter 2. A suitable MD membrane will show high resistance to pore wetting. Progressive wetting prevents separation. Long-term lack of stability of MD membrane materials often leads to wetting and is a major factor in preventing MD from competing with other industrial technologies. Apart from reduced salt rejection, membrane wetting is also closely related to fouling and scaling (Gryta, 2005).

Wetting occurs when the transmembrane pressure is greater than the liquid entry pressure of the membrane for the given feed solution. In its simplest form, LEP depends on pore size, hydrophobicity, and pore geometry. It is now understood that wetting is a complex dynamic phenomenon. Wetting and factors affecting pore wetting are discussed in detail in Section 2.4.

At present, two types of wetting detection techniques are used in MD systems. The first and most common is monitoring permeate water quality. However, by the time a noticeable increase in permeate conductivity is observed, failure in the form of complete pore wetting has already occurred, and reversing this phenomenon would be difficult. The other membrane-based detection method involves studying variation of the membrane itself. A common membrane-based technique involves examination of possible salt deposits on the permeate side of the membrane surface. However, this type of membrane examination, known as membrane autopsy, is not real-time and can only be carried out once the membrane has been removed from the cell.

7.2.1.1 Electrically Conducting Membrane Surface for Wetting Detection

Our group modified an electrospun polyvinylidene fluoride-*co*-hexafluoropropylene (PVDF-HFP) membrane with a conductive layer of carbon cloth to facilitate rapid electrolytic wetting detection (Ahmed et al., 2017). The conductive membrane acted as an electrode, allowing for feedback and control. Electrospinning is a fabrication process in which a polymer solution in or melt of suitable viscosity in a syringe needle is subjected to a high electric field; the charged liquid is ejected toward the collector once electrostatic and surface forces form a conical structure at the needle tip, depositing nanofibers on the collector surface (Zheng, 2019). For a detailed review on electrospinning as a membrane fabrication process, including solution parameters, operating conditions, and post-treatment, the reader is referred to (Ahmed et al., 2015; Francis et al., 2022). Electrospun membranes have been successfully applied as MD membranes due to their high porosity, interconnected pore structure, and ability to tailor morphology and material (Tijing et al., 2014). Electrospun PVDF-HFP membranes have been reported in several MD studies due to the hydrophobicity and ease of electrospinning of the polymer using organic solvents at room temperature with the ability to tailor properties to MD or MF (C.-I. Su et al., 2012). The membrane can subsequently be post-treated to reduce pore size or introduce electrical conductivity, etc. depending on end use. We used heat pressing to fuse a layer of carbon cloth on the permeate side of the electrospun PVDF-co-HFP membrane; the conductive membrane acted as an electrode in the combined MD-electrochemical cell (Ahmed et al., 2017). Heat pressing also causes the nanofibers to fuse together at the interconnections, increasing overall mechanical strength (Ejaz Ahmed et al.,

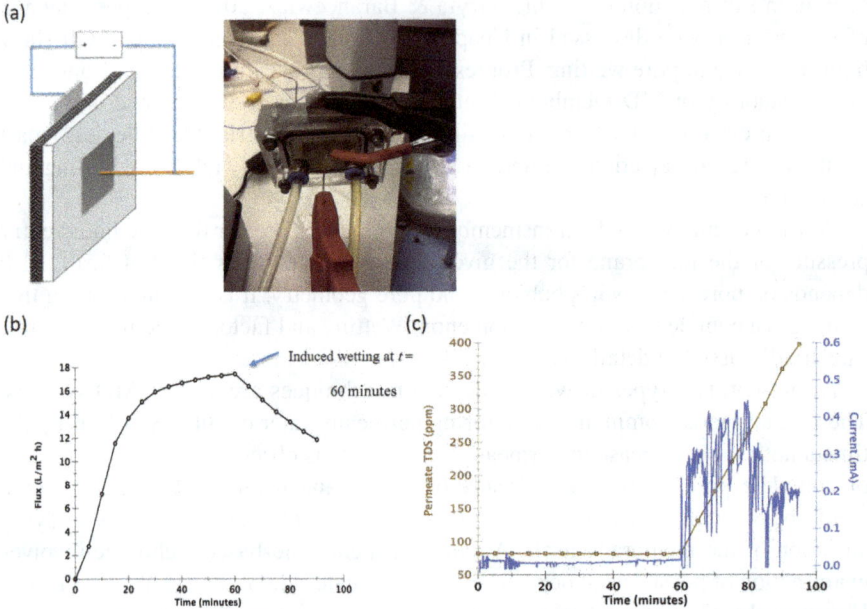

FIGURE 7.2 (a) MD-electrochemical setup for real-time wetting detection using electrically conducting membrane; (b) permeate flux and (c) permeate TDS during DCMD with real-time electrolytic wetting detection using electrically conductive membrane.

Source: Ahmed, F. E., Lalia, B. S., & Hashaikeh, R. (2017). Membrane-based detection of wetting phenomenon in direct contact membrane distillation. *Journal of Membrane Science*, *535*, 89–93. Reprinted with permission.

2014; Lalia et al., 2013). Our study found that when the carbon-cloth-electrospun PVDF-co-HFP (CC-ES PH) was subjected to an anodic potential of 1 V, wetting via ethanol injection at 60 minutes was detected instantaneously, as shown in Figure 7.2. At the point of wetting, a sharp increase in current is observed when ions in the feed pass through the membrane and complete the circuit (Ahmed et al., 2017). This method provides instantaneous detection of changes in electrical signal corresponding to membrane pore wetting.

7.2.1.2 Electrically Conducting Spacers for Wetting Detection

Shifting the focus to electrically conducting spacers, Alpatova et al. coated 3D-printed acrylic spacers with Pt to achieve early pore wetting detection (Ahmed et al., 2017) using a similar approach (Alpatova et al., 2022). A 0.6 μm thin Pt layer was coated using physical vapor deposition (PVD).

The role of spacers in enhancing shear stress at the membrane spacer and limiting the buildup of foulants can be further expanded by using electrically conductive membranes for early detection of wetting. Spacers also lower temperature polarization and enhance both mass and heat transfer to drive vapor permeation across the membrane (Phattaranawik et al., 2003). The work of Phattaranawik

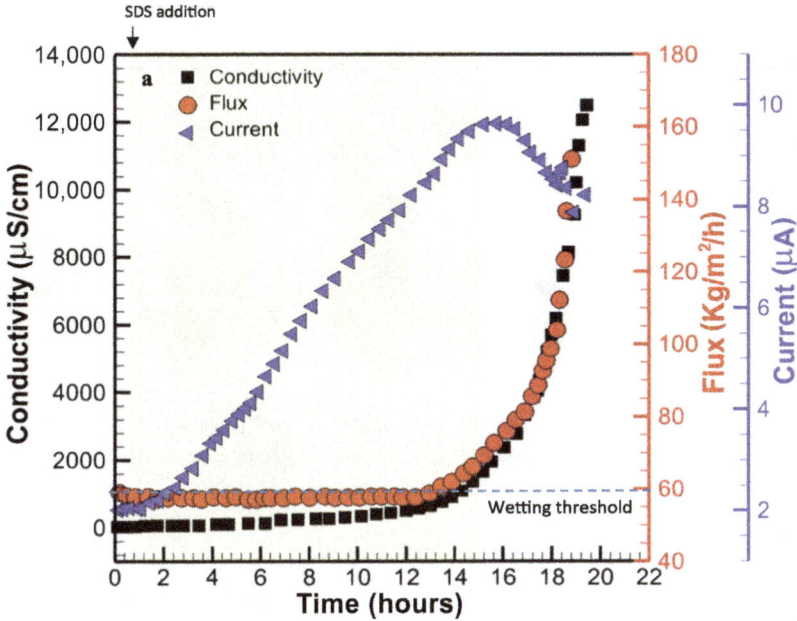

FIGURE 7.3 Changes in DCMD process parameters vs. DCMD time with pore wetting induced by 200 mg L^{-1} of sodium dodecyl sulfate.

Source: Alpatova, A., Qamar, A., Alhaddad, M., Kerdi, S., Soo Son, H., Amin, N., & Ghaffour, N. (2022). In situ conductive spacers for early pore wetting detection in membrane distillation. *Separation and Purification Technology*, 294, 121162. Reprinted with permission.

et al. introduced an electrochemical approach to tailor membrane spacers to detect electrical current. The spacers were inserted in both feed and coolant channels and connected to a power supply. Electric current was measured during DCMD as wetting was induced by either 200 mg L^{-1} sodium dodecyl sulfate (SDS) or ethanol in 15 g L^{-1} NaCl. They explored the effect of applied electrical potential on the intensity of the generated current and noted that the electrical current amplifies when the potential is 1–2 V.

In Figure 7.3, three distinct phases of electrical current can be observed from the onset of wetting: increase, short plateau, and decrease (Alpatova et al., 2022). The increase in electric current is due to increased salt penetration into the permeate side.

They also investigated spacer design by comparing the performance of a 1-helical spacer to a plain cylindrical filament, with and without Pt coating. Their findings suggest that the spacer with helical element generated higher permeate flux due to enhanced channel porosity. They also confirmed that, at a thickness of 0.6 μm, the coating had no effect on permeate flux (Figure 7.4), although the presence of spacers did significantly improve flow through the cell compared with the no-spacer condition. NaCl rejection also remained unaffected by surface coating.

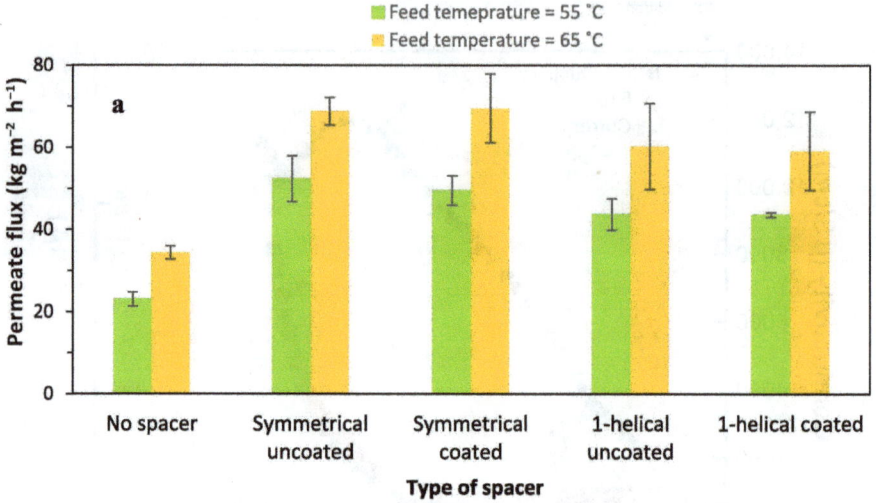

FIGURE 7.4   Effect of spacer type and Pt coating on permeate flux.

Source: Alpatova, A., Qamar, A., Alhaddad, M., Kerdi, S., Soo Son, H., Amin, N., & Ghaffour, N. (2022). In situ conductive spacers for early pore wetting detection in membrane distillation. *Separation and Purification Technology*, *294*, 121162. Reprinted with permission.

7.2.2 FOULING

As we saw in Section 2.4, interactions between the membrane and feed constituents can cause various types of fouling in MD systems. Although the extent of fouling is more severe in pressure-driven processes, it can still block pores, reduce vapor flux, and even induce failure in the form of pore wetting. While hydrophobic interactions are considered responsible for organic fouling in MD, there are little real-time characterization tools to detect and understand how fouling occurs in these systems.

The focus on MD membranes and configurations means that little attention has been paid to real-time in situ fouling detection mechanisms, especially those that do not involve optical techniques and thus transparent or semi-transparent modules. Currently, the most common methods of detecting fouling include monitoring the transmembrane flux and salt rejection, both of which are somewhat delayed and do not give an insight into temporal or spatial evolution of fouling of MD membranes.

The use of electrochemical impedance spectroscopy (EIS) for fouling characterization has been highlighted in Section 6.1.3. It is a powerful non-invasive tool used to study electrochemical systems and coatings and has recently emerged as an online monitoring tool for membrane processes in which evolution of the foulant layer can be characterized through impedance measurements. In RO, EIS has been applied to monitor organic and inorganic fouling of non-conducting polymeric membranes using a four-terminal system in RO (Gaedt et al., 2002), MF, UF (Gaedt et al., 2002), and NF (Xu et al., 2011). These studies have found that EIS can be used as a method for early detection of fouling as compared to the conventional methods of observing changes in flux and salt rejection, yet it has not been extensively applied to emerging processes such as MD. Chen et al. probed early detection of surfactant-induced pore wetting using impedance measurements (Chen et al., 2017).

In MD, the use of EIS on a conductive MD membrane for early detection of fouling was first reported by (Ahmed et al., 2018). The electrically conductive membranes were prepared by incorporating silica gel with an additive, either polyvinyl alcohol (PVA) or networked cellulose (NC) to CNT membranes, followed by calcination and fluorination to increase hydrophobicity (Ahmed et al., 2018). Networked cellulose is a unique networked form of cellulose prepared by controlled dissolution and regeneration of microcrystalline cellulose in sulfuric acid (Hashaikeh & Abushammala, 2011). Using colloidal silica as foulant, the electrical conducting membrane was used as an electrode in an MD-electrochemical cell. EIS measurements were taken every hour for 15 hours to monitor changes with an applied alternating voltage of 100 mV at 50 frequencies for each run over the range of 10^{-2} to 10^5 Hz.

7.2.2.1 Electrical Impedance Spectroscopy (EIS)

EIS can be used to study the intrinsic structure or dynamic processes that influence conductance, resistance, or capacitance of an electrochemical system.

EIS is a powerful electroanalytical technique used to analyze interfacial properties in an electrochemical system. In the case of membrane distillation, EIS measures features that change at the electrode interface due to fouling. In EIS, the impedance in an electrochemical circuit is determined when a small applied signal potential $v = v_0 \sin(\omega t)$ yields a linear current response with the same frequency but different amplitude.

The current response is described as:

$$i = i_0(\omega t + \sin \theta) \qquad\qquad \text{Equation 7.1}$$

where i_0 is the current amplitude, and θ is the phase difference.

The impedance, calculated digitally, is:

$$Z(\omega) = \frac{v_0}{i_0}(\cos\theta + j\sin\theta) \qquad\qquad \text{Equation 7.2}$$

where j is the imaginary constant defined by $j^2 = -1$. The impedance thus has a magnitude $|Z| = \frac{v_0}{i_0}$ and phase θ. The impedance is measured over a range of frequencies.

The reciprocal of the impedance is referred to as the admittance and is represented by:

$$Y(\omega) \equiv \frac{1}{Z(\omega)} \equiv G(\omega) + j\omega C(w) \qquad\qquad \text{Equation 7.3}$$

where the conductance G is the ability of the system to conduct electric charge, and the capacitance C is the ability of the system to store electric charge (Antony et al., 2013). Each layer has a conductive and capacitive component such that overall G and C are frequency dependent. In terms of area of measurement, the admittance can be rewritten as:

$$y(\omega) \equiv \frac{Y(\omega)}{A} \equiv \frac{G(\omega)}{A} + j\omega\frac{C(\omega)}{A} \equiv g(\omega) + j\omega c(w) \qquad \text{Equation 7.4}$$

Changes in the system are studied by observing impedance, conductance, and capacitance spectra over a range of frequencies at given time intervals throughout a dynamic process.

When EIS is used for online fouling characterization with non-conducting membranes, a four-terminal system EIS system is employed to differentiate between the impedance of the membrane and that of the electrode–electrolyte interface. On the other hand, the use of a conducting membrane significantly simplifies the EIS setup needed as the electrically conducting surface acts as working electrode and enables clearer differentiation between electrode and electrode–electrolyte interface. This is because the impedance of the electrically conducting membrane is much lower than that of the electrode–electrolyte interface.

EIS is a non-invasive technique in which a small amplitude sinusoidal potential $v = v_0 \sin(\omega t)$ is applied to an electrochemical system (two-electrode, three-electrode, or four-electrode) at a range of frequencies. The measured current response is linear in that it has the same frequency as the input signal but different amplitude i_0 and phase.

EIS could observe fouling earlier than the two conventional methods of flux decline and salt rejection. Membrane capacitance increases with fouling as the foulant layer accumulates on the membrane surface (Figure 7.5). Overall conductance and capacitance depend on frequency, as shown in Figure 7.5a. Capacitance increases as the foulant layer accumulates on the membrane surface, as shown by the schematic in Figure 7.5b. Conductance decreases drastically at later stages (>12 hours of operation). Initially, capacitance is a result of the double layers formed at the surfaces of the two electrodes during EIS, as well as membrane porosity. Over time, the accumulation of a diffuse oxide layer on the membrane surface separating the two electrodes forms a dielectric space that causes the capacitance to increase (Ahmed et al., 2018). This dielectric space densifies as fouling progresses.

Low membrane porosity (47%) and relatively high thickness (1.3mm) contributed to low permeate flux (4.3 L m^{-2} h^{-1}); nevertheless, this groundbreaking work provided an insight into the direct use of electrically conducting membranes for online monitoring of MD systems (Ahmed et al., 2018).

EIS has been applied to pilot-scale RO desalination plants in the past (Ho et al., 2017); however, these involve the use of canary cells fitted with electrodes that use a bypass of the feed stream as intake, and analysis is carried out under operating conditions similar to the actual module. In situ monitoring using EIS on electrically conductive membranes is a drastically more realistic approach to the described use of canary cells demonstrating in situ fouling evolution of the actual membrane under

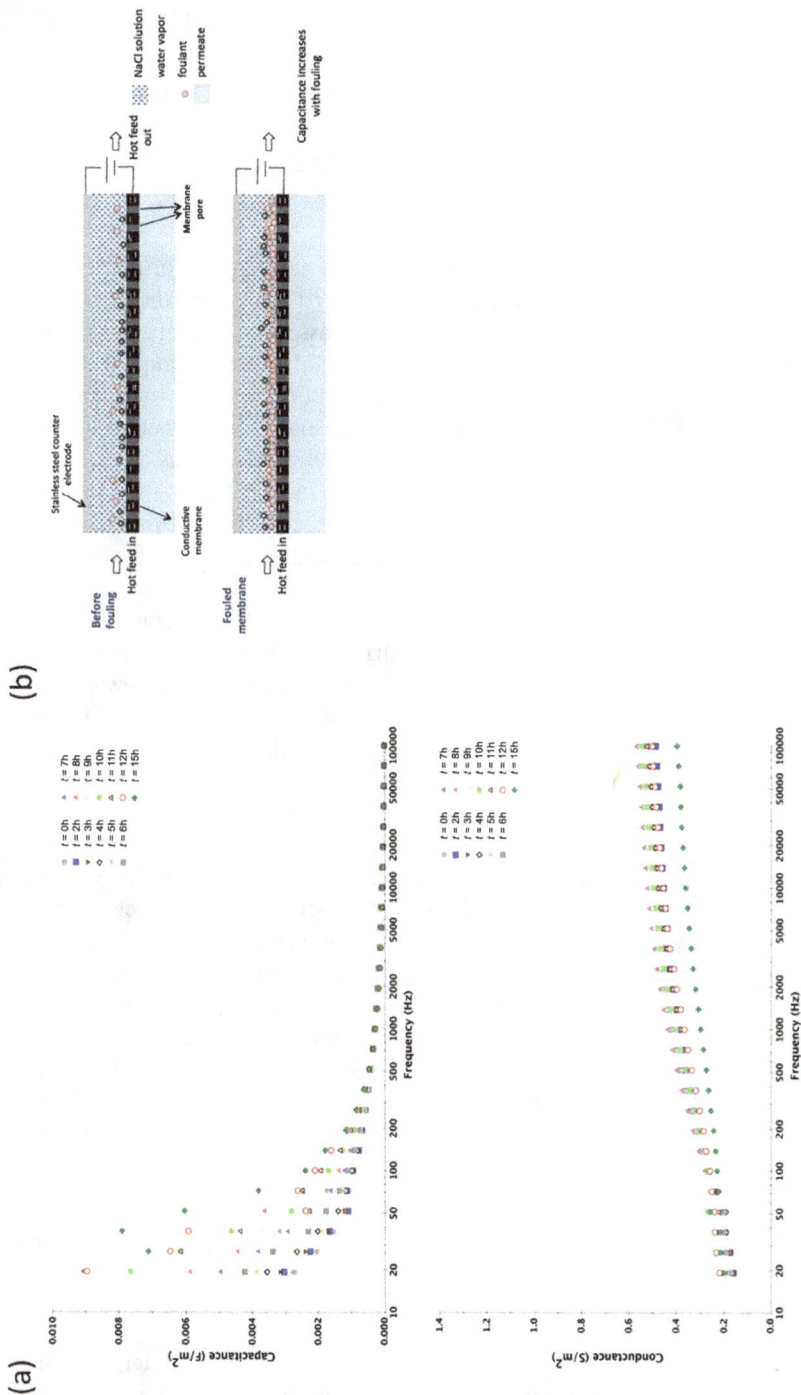

FIGURE 7.5 (a) Capacitance and conductance spectra obtained from electrochemical impedance measurements; (b) schematic demonstrating increase in membrane capacitance with fouling.

Source: Ahmed, F. E., Hilal, N., & Hashaikeh, R. (2018). Electrically conductive membranes for in situ fouling detection in membrane distillation using impedance spectroscopy. *Journal of Membrane Science, 556,* 66–72. Reprinted with permission.

real conditions. Furthermore, since the membrane serves as an electrode, a simplified two-electrode system is sufficient compared to the typically employed four-terminal method for fouling monitoring using EIS to account for effects of voltage electrode–solution interfaces.

7.3 CHALLENGES IN TECHNOLOGY DEVELOPMENT

7.3.1 BARRIERS TO MD COMMERCIALIZATION

Technology readiness level (TRL) indicates the stage of development of new clean energy and clean water technologies. The term was first coined by the National Aeronautics and Space Administration (NASA) to assess space-related R&D. Measured on a scale from 1 to 9 (Mankins, 1995), a TRL of 9 indicates high technological maturity and implementation of a system in operational environment (Ahmed et al., 2021). In 2016, MD classified as a low TRL technology (Lienhard et al., 2016)

The Committee on Determinants of Market Adoption of Advanced Energy Efficiency and Clean Energy Technologies, appointed by the National Research Council, cites the need for significant improvements in permeate flux by at least an order of magnitude in existing MD systems, along with field demonstrations to prove feasibility (National Academies of Sciences et al., 2016).

As a phase change process, the energy consumption in MD systems remains orders of magnitude higher than that for pressure-driven desalination processes. Yet, as previously mentioned, niche applications, such as treatment of oily wastewater or treatment of hypersaline feeds, are areas where MD may compete with other technologies, especially if waste heat or renewable energy is available at low cost.

With the aim of increasing technological maturity, the years between 2010 and 2020 saw 12 MD pilot-scale investigations in various locations using technology from different developers (Hussain et al., 2021). A few pilots have incorporated solar power with MD systems (Mohamed et al., 2017; Ruiz-Aguirre et al., 2018; Schwantes et al., 2013). Additionally, the pilot-scale MD development has been strongly supported by research projects such as SolMed, "Desalination by Solar Powered Membrane Distillation: Material and Process Optimization" (Schwantes et al., 2013), funded by the European Regional Development Fund. Further improvements in MD systems are expected through implementing design improvements that would lower energy requirements such as module stacking and multistage MD (Carrero-Parreño et al., 2019; Chung et al., 2016), as well as utilization of condensation energy to heat incoming feed (Ghim et al., 2021). Currently, there is limited research on the latter (Ma et al., 2021).

Table 7.3 shows existing technology developers for MD, as reviewed by Hussain et al. (Hussain et al., 2021).

7.3.2 UPSCALING ELECTRICALLY CONDUCTING SYSTEMS FOR MD

We argue that electrically conducting surfaces can be facilitated in large-scale plants, as little design modification is required beyond electrical connections. Since MD is still developing as a technology, it would be possible to build a pilot-scale

TABLE 7.3

MD Technology Developers.

Developer	Membrane Configuration	MD system Configuration	Country
Fraunhofer	Spiral wound	Air gap	Germany
SolarSpring	Spiral wound	Air gap	Germany
Scarab/Xzero	Flat-sheet	Air gap	Sweden
TNO/Memstill	Proprietary	Air gap	Netherlands
Memsys GmbH	Flat-sheet	Vacuum air gap	Germany, China
Keppel/Memstill	Flat-sheet	Air gap and DCMD	Singapore
Aquatech	Flat-sheet	Vacuum	USA
KmX corporation	Hollow fiber	Vacuum	Canada, USA
Econity	Hollyow fiber	Vacuum	Korea
Memsift Innovation	Hollow fiber	Vacuum (Joule Thomson)	Singapore

Source: Hussain, A., Janson, A., Matar, J. M., & Adham, S. (2022). Membrane distillation: Recent technological developments and advancements in membrane materials. *Emergent Materials*, 5(2), 347–367.

plant around conductive membranes. Future work should include long-term studies of several months at both lab and pilot scales.

A major drawback of applying electrically conductive membranes in MD, however, is that many materials with high electrical conductivity also have high thermal conductivity. In metals, this relationship is known as the Wiedemann–Franz law. While this may not be an issue in other processes, in MD particularly, a highly thermal conducting membrane translates into greater thermal losses through the membrane. One strategy to deal with this is to focus on the spacer rather than on the membrane itself. Furthermore, introducing electrical conductivity without compensating membrane selectivity and permeability is a challenge, especially when high porosity and hydrophobicity need to be maintained, as in with MD. CNTs are widely used to prepare electrically conducting surfaces for MD. As we see in Chapter 9, the toxicity of CNTs is under debate, especially when it is in contact with the permeate as it can deteriorate permeate quality and raise potential environmental and health concerns with toxic leaching into drinking water, etc.

7.4 CONCLUSION

In the next few decades, membrane distillation is set to build on its current momentum and reach commercial-scale success. In MD, in which a porous membrane separates the heated feed from the lower temperature permeate. As a desalination process, MD requires substantial improvements to compete with RO and other commercial processes. Often in the form of membrane development, these improvements to reduce energy consumption can also be prompted by early characterization of operational challenges and timely control. In particular, MD membranes succumb to pore wetting, which is a failure mode that increases permeate conductivity. Early detection of wetting allows reversal methods, such as dryout and backwashing with pressurized air, to be applied in a timely manner, increasing membrane lifetime and keeping operation costs low. MD-electrochemical cells, in which an electrically conducting surface is applied as membrane or spacer, can help address challenges associated with wetting and fouling through early online detection. In particular, electrochemical impedance spectroscopy is a powerful non-invasive tool that can be used to characterize fouling in MD systems. Additionally, an electrically conducting surface can aid in early wetting detection such that the circuit is complete when pore wetting occurs; this method is faster and just as reliable as conventional methods for wetting detection. Additionally, these methods can be applied directly to membrane modules, provided that electrical connections are made, and are thus a stronger and more reliable indicator of operational challenges compared to the use of bypass streams in canary cells. In situ monitoring tools involving electrically conducting surfaces remain an important component of emerging desalination technologies such as MD and may provide the push needed to drive these technologies toward commercial use.

BIBLIOGRAPHY

Adham, S., Hussain, A., Matar, J. M., Dores, R., & Janson, A. (2013). Application of membrane distillation for desalting brines from thermal desalination plants. *Desalination, 314*, 101–108. https://doi.org/10.1016/j.desal.2013.01.003

Ahmed, F. E., Hilal, N., & Hashaikeh, R. (2018). Electrically conductive membranes for in situ fouling detection in membrane distillation using impedance spectroscopy. *Journal of Membrane Science, 556*, 66–72. https://doi.org/10.1016/j.memsci.2018.03.069

Ahmed, F. E., Khalil, A., & Hilal, N. (2021). Emerging desalination technologies: Current status, challenges and future trends. *Desalination, 517*, 115183. https://doi.org/10.1016/j.desal.2021.115183

Ahmed, F. E., Lalia, B. S., & Hashaikeh, R. (2015). A review on electrospinning for membrane fabrication: Challenges and applications. *Desalination, 356*, 15–30.

Ahmed, F. E., Lalia, B. S., & Hashaikeh, R. (2017). Membrane-based detection of wetting phenomenon in direct contact membrane distillation. *Journal of Membrane Science, 535*, 89–93.

Ali, M. I., Summers, E. K., Arafat, H. A., & Lienhard, J. H. V. (2012). Effects of membrane properties on water production cost in small scale membrane distillation systems. *Desalination, 306*, 60–71. https://doi.org/10.1016/j.desal.2012.07.043

Alklaibi, A. M., & Lior, N. (2005). Membrane-distillation desalination: Status and potential. *Desalination, 171*(2), 111–131. https://doi.org/10.1016/j.desal.2004.03.024

Alpatova, A., Qamar, A., Alhaddad, M., Kerdi, S., Soo Son, H., Amin, N., & Ghaffour, N. (2022). In situ conductive spacers for early pore wetting detection in membrane distillation. *Separation and Purification Technology, 294*, 121162. https://doi.org/10.1016/j.seppur.2022.121162

Amy, G., Ghaffour, N., Li, Z., Francis, L., Linares, R. V., Missimer, T., & Lattemann, S. (2017). Membrane-based seawater desalination: Present and future prospects. *Desalination, 401*, 16–21. https://doi.org/10.1016/j.desal.2016.10.002

Antony, A., Chilcott, T., Coster, H., & Leslie, G. (2013). In situ structural and functional characterization of reverse osmosis membranes using electrical impedance spectroscopy. *Journal of Membrane Science, 425–426*, 89–97. http://doi.org/10.1016/j.memsci.2012.09.028

Banat, F., Jumah, R., & Garaibeh, M. (2002). Exploitation of solar energy collected by solar stills for desalination by membrane distillation. *Renewable Energy, 25*(2), 293–305. https://doi.org/10.1016/S0960-1481(01)00058-1

Banat, F., Jwaied, N., Rommel, M., Koschikowski, J., & Wieghaus, M. (2007). Performance evaluation of the "large SMADES" autonomous desalination solar-driven membrane distillation plant in Aqaba, Jordan. *Desalination, 217*(1), 17–28. https://doi.org/10.1016/j.desal.2006.11.027

Benyahia, F. (2019). *Membrane-distillation in desalination.* CRC Press.

Bouguecha, S., Hamrouni, B., & Dhahbi, M. (2005). Small scale desalination pilots powered by renewable energy sources: Case studies. *Desalination, 183*(1), 151–165. https://doi.org/10.1016/j.desal.2005.03.032

Carrero-Parreño, A., Onishi, V. C., Ruiz-Femenia, R., Salcedo-Díaz, R., Caballero, J. A., & Reyes-Labarta, J. A. (2019). Optimization of multistage membrane distillation system for treating shale gas produced water. *Desalination, 460*, 15–27.

Chen, Y., Wang, Z., Jennings, G. K., & Lin, S. (2017). Probing pore wetting in membrane distillation using impedance: Early detection and mechanism of surfactant-induced wetting. *Environmental Science & Technology Letters, 4*(11), 505–510. doi:10.1021/acs.estlett.7b00372

Chung, H. W., Swaminathan, J., & Warsinger, D. M. (2016). Multistage vacuum membrane distillation (MSVMD) systems for high salinity applications. *Journal of Membrane Science, 497*, 128–141.

Criscuoli, A., Carnevale, M. C., & Drioli, E. (2008). Evaluation of energy requirements in membrane distillation. *Chemical Engineering and Processing: Process Intensification, 47*(7), 1098–1105.

Criscuoli, A., Carnevale, M. C., & Drioli, E. (2013). Modeling the performance of flat and capillary membrane modules in vacuum membrane distillation. *Journal of Membrane Science, 447*, 369–375. https://doi.org/10.1016/j.memsci.2013.07.044

Curcio, E., & Drioli, E. (2005). Membrane distillation and related operations—a review. *Separation and Purification Reviews, 34*(1), 35–86.

Dow, N., Gray, S., Zhang, J., Ostarcevic, E., Liubinas, A., Atherton, P., . . . Duke, M. (2016). Pilot trial of membrane distillation driven by low grade waste heat: Membrane fouling and energy assessment. *Desalination, 391*, 30–42.

Drioli, E., Ali, A., & Macedonio, F. (2015). Membrane distillation: Recent developments and perspectives. *Desalination, 356*, 56–84. https://doi.org/10.1016/j.desal.2014.10.028

Drioli, E., Criscuoli, A., & Curcio, E. (2011). *Membrane contactors: Fundamentals, applications and potentialities.* Elsevier.

Duong, H. C., Chuai, D., Woo, Y. C., Shon, H. K., Nghiem, L. D., & Sencadas, V. (2018). A novel electrospun, hydrophobic, and elastomeric styrene-butadiene-styrene membrane for membrane distillation applications. *Journal of Membrane Science, 549*, 420–427. https://doi.org/10.1016/j.memsci.2017.12.024

Duong, H. C., Cooper, P., Nelemans, B., Cath, T. Y., & Nghiem, L. D. (2016). Evaluating energy consumption of air gap membrane distillation for seawater desalination at pilot scale level. *Separation and Purification Technology, 166*, 55–62. https://doi.org/10.1016/j.seppur.2016.04.014

Ejaz Ahmed, F., Lalia, B. S., Hilal, N., & Hashaikeh, R. (2014). Underwater superoleophobic cellulose/electrospun PVDF–HFP membranes for efficient oil/water separation. *Desalination, 344*, 48–54. https://doi.org/10.1016/j.desal.2014.03.010

Francis, L., Ahmed, F. E., & Hilal, N. (2022). Electrospun membranes for membrane distillation: The state of play and recent advances. *Desalination, 526*, 115511. https://doi.org/10.1016/j.desal.2021.115511

Gaedt, L., Chilcott, T. C., Chan, M., Nantawisarakul, T., Fane, A. G., & Coster, H. G. L. (2002). Electrical impedance spectroscopy characterisation of conducting membranes. *Journal of Membrane Science, 195*(2), 169–180. http://dx.doi.org/10.1016/S0376-7388(01)00542-7

Ghim, D., Wu, X., Suazo, M., & Jun, Y.-S. (2021). Achieving maximum recovery of latent heat in photothermally driven multi-layer stacked membrane distillation. *Nano Energy*, *80*, 105444.

Gryta, M. (2005). Long-term performance of membrane distillation process. *Journal of Membrane Science*, *265*(1), 153–159. https://doi.org/10.1016/j.memsci.2005.04.049

Gryta, M., & Barancewicz, M. (2010). Influence of morphology of PVDF capillary membranes on the performance of direct contact membrane distillation. *Journal of Membrane Science*, *358*(1–2), 158–167. http://dx.doi.org/10.1016/j.memsci.2010.04.044

Guan, G., Yang, X., Wang, R., Field, R., & Fane, A. G. (2014). Evaluation of hollow fiber-based direct contact and vacuum membrane distillation systems using aspen process simulation. *Journal of Membrane Science*, *464*, 127–139.

Guo, F., Servi, A., Liu, A., Gleason, K. K., & Rutledge, G. C. (2015). Desalination by membrane distillation using electrospun polyamide fiber membranes with surface fluorination by chemical vapor deposition. *ACS Applied Materials & Interfaces*, *7*(15), 8225–8232. doi:10.1021/acsami.5b01197

Hashaikeh, R., & Abushammala, H. (2011). Acid mediated networked cellulose: Preparation and characterization. *Carbohydrate Polymers*, *83*(3), 1088–1094. https://doi.org/10.1016/j.carbpol.2010.08.081

Ho, J. S., Sim, L. N., Webster, R. D., Viswanath, B., Coster, H. G. L., & Fane, A. G. (2017). Monitoring fouling behavior of reverse osmosis membranes using electrical impedance spectroscopy: A field trial study. *Desalination*, *407*, 75–84. https://doi.org/10.1016/j.desal.2016.12.012

Huang, Y.-X., Wang, Z., Jin, J., & Lin, S. (2017). Novel janus membrane for membrane distillation with simultaneous fouling and wetting resistance. *Environmental Science & Technology*, *51*(22), 13304–13310. doi:10.1021/acs.est.7b02848

Hussain, A., Janson, A., Matar, J. M., & Adham, S. (2021). Membrane distillation: Recent technological developments and advancements in membrane materials. *Emergent Materials*. doi:10.1007/s42247-020-00152-8

Kesieme, U. K., Milne, N., Aral, H., Cheng, C. Y., & Duke, M. (2013). Economic analysis of desalination technologies in the context of carbon pricing, and opportunities for membrane distillation. *Desalination*, *323*, 66–74. https://doi.org/10.1016/j.desal.2013.03.033

Koschikowski, J., Wieghaus, M., & Rommel, M. (2003). Solar thermal-driven desalination plants based on membrane distillation. *Desalination*, *156*(1), 295–304. https://doi.org/10.1016/S0011-9164(03)00360-6

Lalia, B. S., Guillen-Burrieza, E., Arafat, H. A., & Hashaikeh, R. (2013). Fabrication and characterization of polyvinylidenefluoride-co-hexafluoropropylene (PVDF-HFP) electrospun membranes for direct contact membrane distillation. *Journal of Membrane Science*, *428*, 104–115. http://dx.doi.org/10.1016/j.memsci.2012.10.061

Leitch, M. E., Li, C., Ikkala, O., Mauter, M. S., & Lowry, G. V. (2016). Bacterial nanocellulose aerogel membranes: Novel high-porosity materials for membrane distillation. *Environmental Science & Technology Letters*, *3*(3), 85–91. doi:10.1021/acs.estlett.6b00030

Li, X., Deng, L., Yu, X., Wang, M., Wang, X., García-Payo, C., & Khayet, M. (2016). A novel profiled core–shell nanofibrous membrane for wastewater treatment by direct contact membrane distillation. *Journal of Materials Chemistry A*, *4*(37), 14453–14463.

Liao, Y., Loh, C.-H., Wang, R., & Fane, A. G. (2014). Electrospun superhydrophobic membranes with unique structures for membrane distillation. *ACS Applied Materials & Interfaces*, *6*(18), 16035–16048. doi:10.1021/am503968n

Lienhard, J. H., Thiel, G. P., Warsinger, D. M., & Banchik, L. D. (2016). *Low carbon desalination: Status and research, development, and demonstration needs*. Report of a workshop conducted at the Massachusetts institute of technology in association with the Global Clean Water Desalination Alliance.

Lu, K.-J., Zuo, J., & Chung, T.-S. (2017). Novel PVDF membranes comprising n-butylamine functionalized graphene oxide for direct contact membrane distillation. *Journal of Membrane Science, 539*, 34–42. https://doi.org/10.1016/j.memsci.2017.05.064

Ma, Q., Xu, Z., & Wang, R. (2021). Distributed solar desalination by membrane distillation: Current status and future perspectives. *Water Research, 198*, 117154. https://doi.org/10.1016/j.watres.2021.117154

Macedonio, F., Curcio, E., & Drioli, E. (2007). Integrated membrane systems for seawater desalination: Energetic and exergetic analysis, economic evaluation, experimental study. *Desalination, 203*(1), 260–276. https://doi.org/10.1016/j.desal.2006.02.021

Mankins, J. C. (1995, April). Technology readiness levels. *White Paper, 6*(1995), 1995.

Minier-Matar, J., Hussain, A., Janson, A., Benyahia, F., & Adham, S. (2014). Field evaluation of membrane distillation technologies for desalination of highly saline brines. *Desalination, 351*, 101–108. https://doi.org/10.1016/j.desal.2014.07.027

Mohamed, E. S., Boutikos, P., Mathioulakis, E., & Belessiotis, V. (2017). Experimental evaluation of the performance and energy efficiency of a vacuum multi-effect membrane distillation system. *Desalination, 408*, 70–80. https://doi.org/10.1016/j.desal.2016.12.020

National Academies of Sciences, E., Medicine Board on Science, T., Economic Policy, A., National Academies of Sciences, E., Medicine Board on, E., & Environmental Systems, A. (2016). *The power of change: Innovation for development and deployment of increasingly clean electric power technologies*. The National Academies Press.

Phattaranawik, J., Jiraratananon, R., & Fane, A. (2003). Effects of net-type spacers on heat and mass transfer in direct contact membrane distillation and comparison with ultrafiltration studies. *Journal of Membrane Science, 217*(1–2), 193–206.

Ruiz-Aguirre, A., Andrés-Mañas, J. A., Fernández-Sevilla, J. M., & Zaragoza, G. (2018). Experimental characterization and optimization of multi-channel spiral wound air gap membrane distillation modules for seawater desalination. *Separation and Purification Technology, 205*, 212–222. https://doi.org/10.1016/j.seppur.2018.05.044

Schneider, K., Hölz, W., Wollbeck, R., & Ripperger, S. (1988). Membranes and modules for transmembrane distillation. *Journal of Membrane Science, 39*(1), 25–42. https://doi.org/10.1016/S0376-7388(00)80992-8

Schwantes, R., Cipollina, A., Gross, F., Koschikowski, J., Pfeifle, D., Rolletschek, M., & Subiela, V. (2013). Membrane distillation: Solar and waste heat driven demonstration plants for desalination. *Desalination, 323*, 93–106. https://doi.org/10.1016/j.desal.2013.04.011

Su, C.-I., Chang, J., Tang, K., Gao, F., Li, Y., & Cao, H. (2017). Novel three-dimensional superhydrophobic and strength-enhanced electrospun membranes for long-term membrane distillation. *Separation and Purification Technology, 178*, 279–287. https://doi.org/10.1016/j.seppur.2017.01.050

Su, C.-I., Shih, J.-H., Huang, M.-S., Wang, C.-M., Shih, W.-C., & Liu, Y.-S. (2012). A study of hydrophobic electrospun membrane applied in seawater desalination by membrane distillation. *Fibers and Polymers, 13*(6), 698–702. doi:10.1007/s12221-012-0698-3

Summers, E. K., Arafat, H. A., & Lienhard, J. H. (2012). Energy efficiency comparison of single-stage membrane distillation (MD) desalination cycles in different configurations. *Desalination, 290*, 54–66. https://doi.org/10.1016/j.desal.2012.01.004

Swaminathan, J., Chung, H. W., Warsinger, D. M., AlMarzooqi, F. A., Arafat, H. A., & Lienhard V, J. H. (2016). Energy efficiency of permeate gap and novel conductive gap membrane distillation. *Journal of Membrane Science, 502*, 171–178. https://doi.org/10.1016/j.memsci.2015.12.017

Tijing, L. D., Choi, J.-S., Lee, S., Kim, S.-H., & Shon, H. K. (2014). Recent progress of membrane distillation using electrospun nanofibrous membrane. *Journal of Membrane Science, 453*, 435–462. https://doi.org/10.1016/j.memsci.2013.11.022

Wang, X., Zhang, L., Yang, H., & Chen, H. (2009). Feasibility research of potable water production via solar-heated hollow fiber membrane distillation system. *Desalination*, *247*(1–3), 403–411.

Wang, Z., & Lin, S. (2017). Membrane fouling and wetting in membrane distillation and their mitigation by novel membranes with special wettability. *Water Research*, *112*, 38–47. https://doi.org/10.1016/j.watres.2017.01.022

Xu, Y., Wang, M., Ma, Z., & Gao, C. (2011). Electrochemical impedance spectroscopy analysis of sulfonated polyethersulfone nanofiltration membrane. *Desalination*, *271*(1), 29–33. https://doi.org/10.1016/j.desal.2010.12.007

Yao, M., Tijing, L. D., Naidu, G., Kim, S.-H., Matsuyama, H., Fane, A. G., & Shon, H. K. (2020). A review of membrane wettability for the treatment of saline water deploying membrane distillation. *Desalination*, *479*, 114312. https://doi.org/10.1016/j.desal.2020.114312

Zakrzewska-Trznadel, G., Harasimowicz, M., & Chmielewski, A. G. (1999). Concentration of radioactive components in liquid low-level radioactive waste by membrane distillation. *Journal of Membrane Science*, *163*(2), 257–264. https://doi.org/10.1016/S0376-7388(99)00171-4

Zaragoza, G., Andrés-Mañas, J. A., & Ruiz-Aguirre, A. (2018). Commercial scale membrane distillation for solar desalination. *NPJ Clean Water*, *1*(1), 20. doi:10.1038/s41545-018-0020-z

Zheng, Y. (2019). 3—Fabrication on bioinspired surfaces. In Y. Zheng (Ed.), *Bioinspired design of materials surfaces* (pp. 99–146). Elsevier.

Zuo, G., & Wang, R. (2013). Novel membrane surface modification to enhance anti-oil fouling property for membrane distillation application. *Journal of Membrane Science*, *447*, 26–35. https://doi.org/10.1016/j.memsci.2013.06.053

8 Electrically Tunable Membrane Systems

8.1 INTRODUCTION

The membrane industry has undergone a period of unprecedented transformation in recent decades, with the synthesis of various membrane materials, modification of existing membranes, and module configurations. Novel synthesis and fabrication techniques have expanded the range of polymers and ceramics available to develop membranes and related components. Advanced techniques for chemical modification of existing membranes have helped researchers attain specific surface characteristics and separation behavior. Fouling control has been a significant part of membrane development as fouling deters performance and increases energy costs.

So far, we have only considered electrically conductive materials in terms of fabrication techniques and novel membranes/spacers to minimize the effect of particulate, organic, inorganic, and biofouling. Highlighted mechanisms for fouling mitigation using electrically conducting membranes or spacers include electrochemical destruction of toxic contaminants at the membrane–feedwater or spacer–feedwater interface, localized changes in pH, and electrostatic forces. The use of electrically conductive surfaces in membrane separation, however, goes far beyond fouling mitigation. Increasing the presence of complex pollutants and foulants have risen the demand for stimuli-responsive membranes with in situ control of permselectivity (ability to discriminate between anions and cations) and the ability to differentiate between similar ions. The performance of a membrane separation process depends on membrane selectivity and permeability. Selectivity determines the extent to which desired molecules are separated from the rest, while permeability is the rate at which molecules pass through a membrane material. Membrane selectivity typically depends on membrane affinity to the substance and pore size distribution. Permeability depends on pore size and surface properties. It would be of interest if pore size could be tuned through an external stimulus during operation and not only during membrane fabrication, allowing electronic circuits to control performance. Such membranes, inspired by the gating function in cell membranes, can be designed by incorporating stimuli-responsive materials in various forms into non-responsive membrane substrates to regulate substance transport and allow the membranes to operate as functional gates (Kaner et al., 2016b; Luo et al., 2015; Wandera et al., 2010).

In molecular biology, ion channels are *gated* pores that demonstrate ion selectivity and open and close in response to a specific stimulus (Alberts et al., 2002). Ion channels are part of all biological membranes used in cell physiology. Inspired

DOI: 10.1201/9781003144991-8

by the biological function of cell membranes in nature, biomimetic stimuli-responsive membranes have the ability to switch their permselectivity in response to an external stimuli such as temperature, pH, UV/light, ionic strength, or electric and magnetic fields. Responsive membranes essentially serve as smart "valves", allowing flux and rejection to be controlled dynamically as an external stimulus modifies membrane structure and transport properties (Formoso et al., 2017) and can find application in sensing, drug delivery (García-Fernández et al., 2020), and separation. One main type of stimuli that causes ion channels to open is a change in voltage across the membrane, also known as voltage-gated channels. Several cell membrane transport mechanisms are voltage dependent (Bezanilla & Stefani, 1998). In fact, electroporation is a well-known biological technique to drive ionic and molecular transport through pores. Electroporation is the structural rearrangement of the cell membrane under the application of strong electric field pulses that create aqueous pathways or pores (Chang & Reese, 1990; Tieleman, 2004). Electroporation refers to electrically induced pore formation in biological membranes (Kevin Li et al., 2013). In cell membranes, the electric field serves two purposes: (1) It causes pore formation, and (2) it provides a local driving force for molecular transport (Weaver, 1995).

Synthetic electroresponsive membranes that mimic nature have the potential to tune ion selectivity in water treatment and desalination, further expanding the applicability of membrane technology. Smart gated membranes can be used in different configurations such as flat-sheet or hollow fiber, and they can be applied to various desalination and water treatment processes (Bandehali et al., 2021). Not only do such membranes demonstrate switchable performance, but they can also enable fouling mitigation and tunable self-cleaning. The challenge lies in developing membranes that simultaneously exhibit high flux, significant response to environmental stimuli, adequate mechanical strength, and ease of fabrication. Given the scope of this book, we are most interested in electroresponsive surfaces and voltage gating using electrically conductive membranes during operation. Voltage gating allows surface properties and/or pore size to be tuned in response to applied electric fields. In fact, electroactive surfaces that use electrical simulation to tune pore size, charge, and/or temperature, which in turn controls membrane permeability and/or selectivity, may prove to be the next generation of materials for separation applications. Electroresponsive surfaces can be made of electroactive polymers such that the application of an electric field induces changes in pore structure. Zhu and Jassby have reviewed the energy considerations and challenges of electroactive membranes in water treatment; however, the focus of their review is limited to fouling prevention and control and does not draw attention to electrically tunable membrane properties (Zhu & Jassby, 2019). Considerable efforts have been made in design and application of electroresponsive smart gating membranes, but a thorough review in this area remains lacking. In this chapter, we take a closer look at electrically tuned permselectivity, which allows optimization of ion, solute or particle transport via application of an electric field, as well as electrostatically charged surfaces. Later, we include selected studies on electroresponsive materials for MD, which includes electrothermal heaters embedded or coated in membranes and spacers for enhanced performance (MD).

8.2 ELECTRICALLY TUNABLE PERFORMANCE OF PRESSURE-DRIVEN PROCESSES

Electroresponsiveness can be incorporated through the use of electroresponsive polymers, copolymers, mixtures of polymers, and additives, during membrane fabrication or through post-treatment after membrane formation through coating a thin layer, etc. (Kaner et al., 2016a). Electrically conductive polymers can be used to functionalize membranes to make them electroresponsive. When compared to other stimuli such as pH or ionic strength, electric fields do not require changes in the solution conditions and can be applied externally, providing simple and effective control (Durmaz et al., 2021). Electrical switching allows control over the conformation of the electroresponsive layer with simple electronic circuits. Although promising, electric control of water transport is far from its full potential. To be used in membrane separation processes, these conductive polymers need to be tailored with specific reactivity, polarity, conformation, and often functionality (Baker, 2012). Chemical grafting and physical coating are commonly used methods of introducing electroresponsive domains into membrane materials after membrane formation. On the other hand, electroresponsive materials can be introduced during membrane formation via physical blending with the membrane base material or chemical grafting prior to membrane formation. In this section, we discuss some materials that have been applied as voltage-gated membranes for desalination and water treatment.

8.2.1 Polyelectrolyte Gels

The concept of electrically controlled permeation was introduced by Yamauchi et al. in 1993 (Yamauchi et al., 1993), who reported the switchable permeation rate of a high-molecular-weight protein, trypsin, through a polyelectrolyte gel membrane composed of polyvinyl alcohol (PVA) and polyacrylic acid (PAA). This permeation rate could be increased upon application of an electric field as trypsin was transported toward the oppositely charged electrode. The applied electric field caused isometric contraction, which widened the pore channels enough to allow diffusion of the protein. When considering membrane permeability as a function of electric field strength in 150 mM phosphate buffer using trypsin with a molecular weight of 23 kDa, they found that the membrane was only permeable under an electric field and that the permeation rate increased with an increasing electric field. Polyelectrolyte hydrogels experience reversible expansion and contraction in response to an electric field, allowing solute permeation to be controlled in membranes fabricated using polyelectrolyte hydrogels (Shiga & Kurauchi, 1990). The control of structure in polyelectrolyte brush systems in response to applied electric field is caused by the presence of charged monomers and counter-ions (Weir et al., 2011). Studies have shown that the application of an electric field allows for reversible swelling and de-swelling of polyelectrolyte hydrogels (Ouyang et al., 2009; F. Zhou et al., 2008). The conformational response along polyelectrolyte chains may be a result of the balancing forces between chain dynamics, electric polarization, counter-ion binding, and hydrodynamic pressure (Mahinthichaichan et al., 2020). It has also been shown that polyelectrolytes such as PAA undergo repetitive switching between low conductance and high conductance depending on the applied voltage (Zhitenev et al., 2007). Moghaddasi designed an artificial gate using polyelectrolyte brushes in a cylindrical nanochannel, which can be opened by applying a tangent electric field and closed by removing the electric field (Moghaddasi Fereidani, 2015). Bliznyuk and coworkers found that the in-plane electrical conductivity of grafted polymer layers of PAA increases by at least 2 orders of magnitude when exposed to water compared to its insulating state under ambient conditions (Bliznyuk et al., 2014). Khor et al. also grew PAA brushes from the surface of a percolating network of CNTs coating a UF membrane (Khor et al., 2019). Positive potentials increased flux by 85%, while negative potentials reduced flux by 30% compared to when no potential was applied as the grafted charged polymer brush expanded and collapsed under the effect of an electric field. Sun and coworkers iterate that incorporating electrically conducting membranes for salt rejection requires a shift in module design (M. Sun et al., 2021).

8.2.2 CARBON NANOTUBES

Chen and coworkers developed a conducting nanoporous membrane from CNT-chitosan-gelatin for the separation of proteins with electrically tuned selectivity (using ±5 V) (Chen & Chuang, 2013). CNT was used to introduce electrical conductivity. A positive potential allows the membrane to prevent negatively charged molecules to pass while allowing diffusion of positively charged molecules. The selectivity is reversed when the polarity of the applied potential is changed. Electrically conductive membranes made with PVA and CNT thin films on UF support can be used for electrostatic repulsion of alginate found in foulant solutions (Alexander V. Dudchenko et al., 2014). When the voltage is increased, the membrane becomes more repulsive to AA in the feed solution. As an antifouling strategy, Dudchenko et al. found that applying PVA-CNT membranes to prevent fouling electrostatically reduces operation costs (Alexander V. Dudchenko et al., 2014). They found that diffusion of negatively charged molecules is faster when a positive potential is applied to the membrane as compared to negative. Wei et al. used anodic potentials to regulate the transport of gold nanoparticles through CNT hollow fiber membranes with hydrophobic pore channels (Wei et al., 2015). They chose gold nanoparticles with well-defined diameters due to their charged surface and easily regulated sizes. The CNT membrane was fabricated using electrophoretic deposition with Cu wire as a template (Wei et al., 2014). The resulting membrane has an average pore size of 120 nm, while the gold nanoparticles have a diameter of 10 nm. When no voltage is applied, GNPs penetrate the CNT membrane pores freely due to the much smaller particle diameter compared to average membrane pore size. In the presence of anodic potential, the membrane rejects up to 98% of the gold nanoparticles at 1.2 V. Increasing applied voltage from 0.4 to 1.2 V increases GNP rejection (see Figure 8.1a). Figure 8.1b shows the reversible dynamic switch between open and closed states when a voltage of 1.2 V is applied and removed cyclically. The transport of gold particles is regulated across the pore channels of CNT hollow fiber membranes by opening and closing, allowing dynamic control of membrane selectivity. The authors attribute the reversible switching to CNT pore channels with different polarities, which can cause or prevent the formation of noncovalent interactions with charged nanoparticles (Wei et al., 2015).

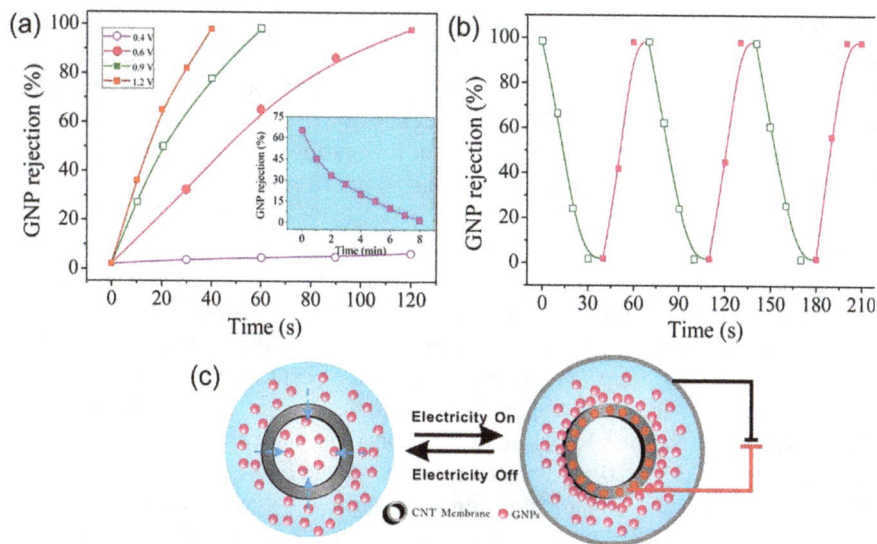

FIGURE 8.1 (a) Rejection of gold nanoparticles at different voltages as a function of time. (b) Dynamic switching between permeation and rejection of gold nanoparticles by CNT membranes (solid squares: electricity on; hollow squares: electricity off); flow rate = 10 mL cm^{-2} h^{-1}; transmembrane pressure = 0.002 MPa. (c) Schematic illustration for switch between opening and closing of membrane pores using 10 nm gold nanoparticles.

Source: Adapted with permission from Wei, G., Quan, X., Chen, S., Fan, X., Yu, H., & Zhao, H. Voltage-gated transport of nanoparticles across free-standing all-carbon-nanotube-based hollow-fiber membranes. *ACS Appl. Mater. Interfaces* 2015, 7(27), 14620–14627. Copyright © 2015 American Chemical Society.

Zhang used polystyrenesulfonate (PSS) as a dopant during the polymerization of PANI to obtain electrically active PANI and then incorporated CNTs, preparing PANI-PSS/CNT NF membranes (H. Zhang et al., 2019). They used the resulting membrane as cathode to evaluate ion separation with and without electrical assistance. Increasing externally applied voltage from 0 to 2.5 V increased surface charge density of the membrane from 11.9 to 73 mC m^{-2} to 6.1 times its value without applied potential. This increase in surface charge density increases membrane rejection for Na$_2$SO$_4$ from 82 to 93% and for NaCl from 54 to 82%, while retaining high water flux ~ 14 L m^{-2} h^{-1} bar^{-1}. According to the Donnan steric pore model commonly applied to NF, an applied potential increases the Donnan potential difference between membrane and bulk solution, which leads to increased ion transfer resistance and enhances ion rejection.

8.2.3 GRAPHENE

Graphene has been proposed as a promising material for a solid state–ion channel for sensing and separation due to its unique properties, which include atomic-level thickness, mechanical strength, chemical stability, and controllable formation of nanopores. Gated graphene pores have been previously applied to the sensing of DNA and proteins (Traversi et al., 2013; Venkatesan et al., 2012), but only limited studies exist in gated nanopores for selective control of ionic flux as desalination and water treatment membranes. Cantley et al. showed that inter-cation selectivity of graphene nanopores can be controlled at low gating voltages of under 500 mV (Cantley et al., 2019). Nanopores as large as 50 nm in graphene membranes reveal inter-cation selectivity with 20 times preference for K^+ over divalent cations. The authors used molecular dynamics to demonstrate that ion selectivity is due to ion transport occurring across thin water films along the edge of the graphene pore and that the transport of ions is strongly dependent on an externally applied electric field. Other authors have also used molecular dynamics simulations to show that bioinspired nanopores in graphene sheets can be exploited through electrically tunable ion selectivity (Ni et al., 2018) and also to discriminate between two similar ions, namely Na^+ and K^+, through functionalization of the pores with carboxylate groups (He et al., 2013). Transmembrane voltage bias causes the nanopore to mimic a voltage-gated biological channel. Under low voltage bias, the membrane demonstrates Na^+ selectivity, and under high voltage bias, the nanopore becomes K^+-selective as blockage by Na^+ is destabilized and the stronger affinity for carboxylate groups slows their passage. The authors point to the importance of their work in biomimetic membrane design for nanofiltration for N^+/K^+ separation. Another group mimicked the structures and mechanisms of a biological K^+ channel by constructing a multilayer modified graphene stacked nanochannel. Their study demonstrated a voltage-gated K^+/Na^+ separation at the edges of each pore, with a separation ratio between K^+ and Na^+ of close to 4 (K. Gong et al., 2020). Widakdo and coworkers experimentally demonstrated reversed electrical-switchable permselectivity in a graphene-organic framework membrane for CO_2 separation (Widakdo et al., 2021). Polydopamine (PDA) was added to the PVDF/graphene membranes to facilitate gas permselectivity under applied voltage. An increase in applied potential caused permeation and MWCO to decrease during cross-flow filtration.

Although the ultrafast water permeation of water and molecular sieving was previously reported (Abraham et al., 2017; Joshi et al., 2014; Nair et al., 2012; P. Sun et al., 2016), Zhang et al. were the first to introduce conductive filaments in the graphene oxide for electrically controlled water transport through micrometer-thick graphene oxide membranes. They concentrated an electric field around these conductive filaments to ionize water molecules inside the capillaries and impede water transport (K. G. Zhou et al., 2018). Figure 8.2 shows water permeation switching through graphene oxide membrane under the effect of an electric field.

Hu et al. fabricated an electrically conducting 3D hybrid membrane using reduced GO and CNTs (Hu et al., 2018). The intercalation of CNTs in the GO matrix created water transport channels and improved permeability, while simultaneously increasing active adsorption sites for salt ions. Applying a bias increased the number of capacitive counter-ions adsorbed, which resulted in stronger electrostatic repulsion between capacitive ions and ions in the solution and improved NaCl salt rejection to 71%, three times greater than without bias when a GO-CNT membrane with 15 wt% CNTs was employed.

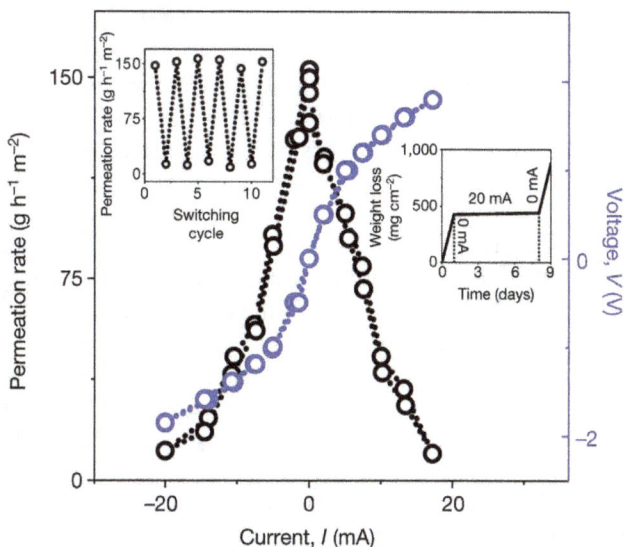

FIGURE 8.2 Water permeation through graphene oxide membrane after conductive filament formation and corresponding I–V characteristic curve. Inset on left shows continuous switching of water permeation rate between 0 and 1.8 V. Inset on right shows effect of long-term switching off on weight loss. All weight loss measurements were performed inside a dry chamber with 10% relative humidity.

Source: Zhou, K. G., Vasu, K. S., Cherian, C. T., Neek-Amal, M., Zhang, J. C., Ghorbanfekr-Kalashami, H., . . . Nair, R. R. (2018). Electrically controlled water permeation through graphene oxide membranes. *Nature*, 559(7713), 236–240. Reprinted with permission.

8.2.4 Conducting Polymers

The ion selectivity of membranes modified with nanoparticles or zwitterionic groups can be tuned using an electric field. With electrostatic charging of the membrane surface, oppositely charged ions will be attracted to the membrane surface, and similarly charged ions will be rejected. Alternatively, a diffusion-limited oxidation approach consists of coating a polymeric membrane with a conductive layer. Weidlich and Mangold coated PVDF and polyethersulfone (PES) membranes with PPy, varying polymerization time and counter-ions incorporated into the PPy coating during polymerization (Weidlich & Mangold, 2011). Using a Pt mesh as a counter-electrode and 0.005 M $CaCl_2$, they studied the rejection of Ca^{2+} ions when ±0.8 V is applied to the conducting membrane. Compared to 0 V, the rejection of Ca^{2+} ions increased when −0.8 V was applied and decreased when +0.8 V was applied. In the case of positive potential, desorption of Ca^{2+} ions from the membrane into the permeate leads to lower rejection. However, they noticed that PPy is not suitable for repeated cycles due to anodic oxidation, and thus other materials should be sought. Xu et al. studied the membrane flux and molecular weight cut-off (MWCO) through conductive polyaniline (PANI) membranes prepared via non-solvent-induced phase separation (NIPS) using neutral polyethylene glycol (PEG) feed solutions under the effect of an applied electrical potential (L. Xu et al., 2018). High potentials of 9 and 30 V were applied to study the electrical tunability of these membranes, which necessitates further studies to evaluate the energy consumption and feasibility of applying PANI membranes to membrane separation. The same group applied another synthesis method to prevent leaching of small-acid-doped PANI by using 2-acrylamido-2-methyl-1-propane sulfonic acid (PAMPSA) as a polymer acid template for polymerization (L. L. Xu et al., 2019). The PANI-PAMPSA membranes demonstrated greater tensile strength and were acid leach resistant, and their electrical conductivity was three orders of magnitude greater than that of PANI membrane with post cast doping with small acid. Rohani and Yusoff used in situ chemical oxidative interfacial polymerization to synthesize PANI-coated PVDF nanofiltration membranes with tunable separation selectivity under external electrical potential (Rohani & Yusoff, 2019). Of solution and diffusion cell polymerization, the latter created a greater thin PANI layer after 48 hours of reaction time. The electrical conductivity of the resulting PANI coating was 6.7 S cm^{-1} with continuous electrical connectivity. The hydrophilicity of the membranes is electrically tunable as the water contact angle decreased by 30–40% when an external electrical potential was applied. Electrically controlled membranes can show tunable selectivity for non-ionic and non-polar molecules; however, in this specific study, the permeation tunability was not noticeable due to the small acid dopant used during membrane fabrication.

Polypyrrole (PPy) is another electrically conductive polymer that has been used to fabricate pore-size-tunable membranes for selective separation (X. Tan et al., 2019). PPy is known for its reversible volume change during redox processes (Jiang et al., 2016). When a cathodic potential is applied, ions are embedded in the polymer, which leads to volume expansion. On the other hand, an anodic potential causes the ions in the membrane to de-intercalate and enter the bulk solution, which causes volume contraction. Applying an external potential tunes membrane pores as the volume of the membrane itself is altered. Tan et al. discovered that pores become enlarged under oxidation voltage, which can be reversed by applying a reduction voltage. This enhanced switchable sieving can alleviate membrane fouling by adjusting pore size in situ along with cleaning. The applicability of an n-doped silicon membrane with nanopores for electrically tunable separation has also been investigated via modeling (Jou et al., 2014).

8.2.5 MXENES

MXenes are a relatively new family of 2D transition metal carbides, nitrides, and carbonitrides that offer the combination of high electronic conductivity, hydrophilicity, and mechanical and chemical stability (Hart et al., 2019; Ling et al., 2014). Their regular structure and subnanometer channels make stacked MXene membranes suitable for desalination via molecular sieving (Gao et al., 2020). Recently, there have been some studies on electrical tunability of MXene membranes. For instance, Fan et al. demonstrated ion sieving of mixed K$^+$/Pb^{2+} pairs and high rejection of heavy metal Pb^{2+} ions through a thermal cross-linked 2D MXene membrane by applying an external voltage (Fan et al., 2020). During the drying process, dehydration and cross-linking of hydroxyls caused the interlayer spacing to be approximately 6.7 ~ 6.92 Å. Under external voltage, the rejection of hydrated Pb^{2+} increases up to 99%,

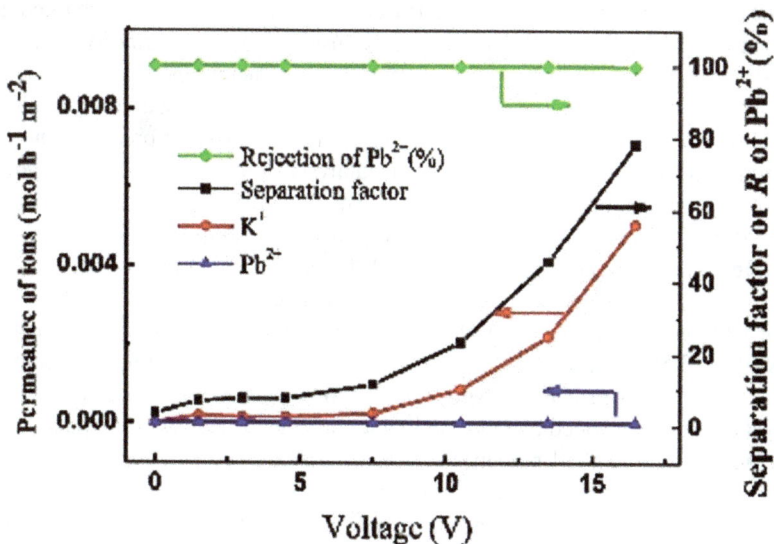

FIGURE 8.3 Separation of mixed ions (K[+]/Pb[2+]) and rejection rate of Pb[2+] for MXene membrane under different applied voltage.

Source: Fan, Y., Li, J., Wang, S., Meng, X., Zhang, W., Jin, Y., ... Liu, S. (2020). Voltage-enhanced ion sieving and rejection of Pb[2+] through a thermally cross-linked two-dimensional MXene membrane. *Chemical Engineering Journal, 401*, 126073. Reprinted with permission.

while hydrated K[+] transports through membrane channels. The separation factor of mixed K[+] and Pb[2+] reaches 78 when a voltage of 16.5 V is applied to a 365 nm thick membrane. Additionally, the membrane retains its mechanical integrity and does not swell under applied potential. Figure 8.3 shows that the separation factor increases with increasing voltage for the thermal cross-linked MXene membrane. Recently, Zhang et al. demonstrated electric field modulated water permeation through laminar $Ti_3C_2T_x$ (L. Zhang et al., 2022). Water permeation was enhanced by a factor of 70 when −3 V was applied, while dye rejection rates remained >93%; the authors attributed this increase to enhanced water/MXene interaction under electric field, i.e. enhanced hydrophilicity of the nanosheets.

8.2.6 Other

Sb-containing compounds generated in industrial effluents from mining, electronics, and textile industries carry health risks as determined by the Environmental Protection Agency. Liu and coworkers developed an electroactive TiO_2-modified CNT filter that, under the effect of an electric field, transformed Sb(III) to less toxic Sb(V), which is then sequestered by TiO_2. (Liu et al., 2019) Increasing applied voltage and feed flow rate enhanced removal of Sb(III) due to electrochemical reactivity, small pore size, and greater number of exposed sorption sites. They also showed that used filters can be regenerated using aqueous NaOH.

As compared to electroresponsive permselectivity, ion exchange and charged nanofiltration membranes can also demonstrate permselectivity, or the ability to discriminate between anions and cations (Millet, 2015). These membranes have fixed charges on the surface and are widely used for water treatment and desalination. Nanofiltration is used for water softening by removal of large ions, as well as for pretreatment to RO. Ion exchange membranes are used in electrodialysis to concentrate or dilute aqueous electrolyte solutions such as saline feedwater. These membranes require low potentials, in the range of mV, as compared to electroresponsive membranes.

Guo et al. used tubular asymmetric TiO_2-based UF membranes with a stainless steel counter-electrode applying cell potentials between 0 and 10 V for the separation of NO_3^- using 1 and 10 mM $NaNO_3$ solutions (Guo et al., 2016). Increasing the potential to 10 V increased NO_3^- rejection to 68%. A similar trend for the removal of ClO_4^- from solutions of $NaClO_4$ with 95% rejection at 10 V confirmed that oxyanion removal could be enhanced by increasing potential due to electrostatic repulsion. Both permeate flux and applied potential affected removal rate. The absence of reduced species indicated that the ions were not electrochemically reduced. Shi and coworkers recently reported electroenhanced separation of oil droplets using electroresponsive copper metallic membranes (Shi et al., 2022). They prepared copper flat-sheet membranes via dry pressing of copper powder and subsequent heat treatment in H_2 atmosphere. Applying a negative potential to a Cu membrane with large pores (~ 2.1 μm) helped reject even small oil droplets in the range of 1 μm. The applied potential also helped redistribute surfactants around oil droplets due to electrostatic repulsion, improving the steric hindrance effect between neighboring droplets, which prevented oil droplets from coalescing and forming a foulant layer on the membrane surface. The ionic strength affects electrically induced fouling control of the membranes as under high ionic strengths, there is reduced repulsion between the membrane and oil droplets, which leads to complete blocking. Although there is generally a trade-off between permeation and selectivity, electroenhanced filtration through the copper membranes led to a simultaneous increase in both as the permeation increased from ~1026 to 2516 $L\ m^{-2}\ h^{-1}\ bar^{-1}$, and rejection increased from ~87 to 98%.

Ceramic membranes are another class of materials that offer many advantages over traditionally used polymeric materials such as narrow pore size distribution, durability, high chemical, mechanical and thermal stability, and strong resistance to erosion and fouling (De Napoli et al., 2011). They may also be promising as electroresponsive materials as they are more stable than polymers, especially under oxidative conditions (Purkait et al., 2018). Currently, not many studies have focused on electroresponsive ceramic membranes. The pore diameter of ceramic anodic alumina oxide membranes can be reduced through adjustment of voltage during anodization (Miedema, 2012).

8.3 ELECTRICALLY TUNABLE PERFORMANCE OF MEMBRANE DISTILLATION

Thus far, we have looked at various applications for wastewater treatment and ion removal through MF, UF, NF, and RO membranes. Permselectivity of membranes in these processes can be controlled using an applied potential with choice of material, filler, and responsiveness to electric field all determining the end result. Electrically tunable performance may be extended to other emerging technologies. In particular, electrothermal materials whose temperature changes under the effect of an applied electric field can be incorporated into membrane modules for thermal processes such as MD.

MD is a thermally driven separation process used to separate volatile compounds from a heated feed solution. The feed, which is in contact with a porous, hydrophobic membrane, is heated, while the other side of the membrane is kept at a cooler temperature. The temperature difference across the membrane induces a vapor pressure gradient that causes pure water to vaporize on the feed side and transport through membrane pores, where it is condensed and collected according to the selected configuration. MD can be applied to a range of feed types from brackish water to highly saline brine. Compared to the more conventional RO process used for desalination, MD membranes are less prone to fouling and can be readily applied to hypersaline feed solutions such as desalination brine or produced water from the oil and gas industry. In these aspects, MD holds tremendous promise and may surpass existing technologies. One aspect that has prevented widespread commercial use of MD is its low energy efficiency. To tackle energy efficiency, much work has been carried out on devising new MD configurations. Since MD relies on thermal distillation, heat transfer plays a crucial role in the performance of MD systems including energy efficiency. Typically, the feed solution to the MD module is heated externally, causing heat losses outside the module to contribute to the low energy efficiency of the system. Additionally, the energy efficiency is also lowered by a phenomenon called temperature polarization, which refers to the bulk temperature of the feed solution being greater than the temperature at the membrane surface, due to conductive heat losses through the membrane. This difference in temperature between bulk and surface lowers the driving force needed for mass transfer through the membrane. To combat these issues, researchers have directed their focus to self-heating membranes for direct heating of the feed at the feed–membrane interface where vaporization occurs. Internal heat generation from the membrane can be exploited for this purpose. As reviewed by our group recently, many such alternate techniques exist for heating in MD systems, including microwave heating, induction heating, joule heating, and photothermal heating (Ahmed et al., 2020a). Some researchers have embedded photothermal particles in membranes and coatings (Dongare et al., 2017; Huang et al., 2019; Politano et al., 2017; Said et al., 2019; Y. Z. Tan et al., 2018; Wu et al., 2018; Ye et al., 2019) to induce heating via the thermoplasmonic effect (B. Gong et al., 2019; Wang, 2018). Tan et al. employed metallic spacers for localized induction heating of the feed near the membrane surface (Y. Z. Tan et al., 2020). Alsaati and Marconnet used an electric heater to heat the region of the fluid near the surface of a silver membrane for air gap MD (Alsaati & Marconnet, 2018). Although the high thermal conductivity of the membrane is suitable for uniform heating, it also poses the risk of large conductive losses through the membrane. Direct surface Joule heating of the membrane using appropriate electrically conductive materials may increase vapor flux by raising the feed temperature at the membrane surface and keeping the driving force for mass transfer high. Direct surface heating reduces the unwanted effects of temperature polarization (Deshmukh et al., 2018).

These multifunctional membrane heaters have the potential to revolutionize niche technologies such as MD.

8.3.1 Joule Heating

Electrothermal heaters provide a simple form of direct heating in which a current passing through the film causes a rise in temperature. This phenomenon is also known as Joule heating, resistive heating, or ohmic heating. Suitable materials must be electrically conducting so that current may pass through them continuously, but the conductivity must be relatively low in order to ensure sufficiently high resistance for this type of heating, which is also known as resistive or ohmic heating. Graphene (Sui et al., 2011) and CNT (Jia et al., 2018) film heaters have been reported as efficient electric heaters in air. However, few studies report the use of these materials for electric heating in applications where ionizable media such as water are involved, owing to instability. In fact, Dudchenko et al. demonstrate that a DC voltage cannot be used to heat CNT films in water due to eventual oxidative degradation (Alexander V Dudchenko et al., 2017). Most of the work carried out in electrically enhanced MD performance via electrical heating is done on very small cells with millimeter-range dimensions, far from practical use. Additionally, while electrothermal heating is promising, relying on it alone may take away from the touted benefits of MD, which include the ability to use waste heat. Using two heat sources together may break the barrier faced by direct surface heating (Deshmukh et al., 2018) and allow it to be used in large-scale systems.

As a consequence, our group developed CNT-based membrane heaters for both DCMD and AGMD in conjunction with a low heated feed in order to reduce the amount of high grade electrical energy required. Today, four conventional configurations exist: direct contact membrane distillation (DCMD), air gap membrane distillation (AGMD), vacuum membrane distillation (VMD), and sweeping gas membrane distillation (SGMD). Of these, DCMD is most commonly used in lab-scale studies as it is simple and easy to use. In DCMD, the heated feed flows on one side of the membrane, while the cooler permeate flows on the other side; both streams are, as the name suggests, in direct contact with the membrane.

8.3.2 Direct Electric Heating during DCMD for Brackish Water Desalination

We have modified a commercial polypropylene (PP) membrane with a layer of CNTs using two different methods: tape casting and vacuum filtration, applying each to enhance the performance of MD for brackish and seawater desalination using separate configurations. In this first part, we discuss coating CNTs on the PP membrane by first forming an ink-like suspension of CNS in ethanol and water through rigorous homogenization using PVDF as a binder (Ahmed et al., 2020b) and then using this membrane for electrically assisted MD with low temperature heated feed solutions. The SEM image in Figure 8.4a shows a layer of CNTs coated on the PP membrane. The advantage of using the doctor blade method is control over thickness. Although the porosity of the CNT-PP membrane was slightly lower than that of PP alone (Figure 8.4b), this helped in keeping the liquid entry pressure (LEP) high, which translates into better non-wetting under MD conditions. When a current is passed through the conductive CNS layer, it acts as an electrothermal heater; the rate of heating is proportional to the amplitude of the applied AC potential.

A lab-scale pilot was used to evaluate the performance of the membrane for brackish water desalination (TDS 10,000 mg L^{-1}); both feed and permeate streams had a flow rate of 20 L h^{-1}, and permeate mass was recorded every 60 seconds. Electrothermal heating was employed in conjunction with heating of the circulating feed, with and without AC bias of 24 V at 50 Hz at three feed temperatures: 40, 50, or 60 °C. Inlet and outlet temperatures, permeate mass, and permeate conductivity were recorded every 60 seconds. Permeate flux was calculated from the permeate mass according to the following equation:

$$J\left(\text{kg}\,\text{m}^{-2}\,\text{h}^{-1}\right) = \frac{\left(\text{Change in permeate mass}\right)}{t * A_m} \qquad \text{Equation 8.1}$$

where t is the time, and A_m is the active membrane area.

Salt rejection (SR) was calculated from feed and permeate conductivities according to the following equation:

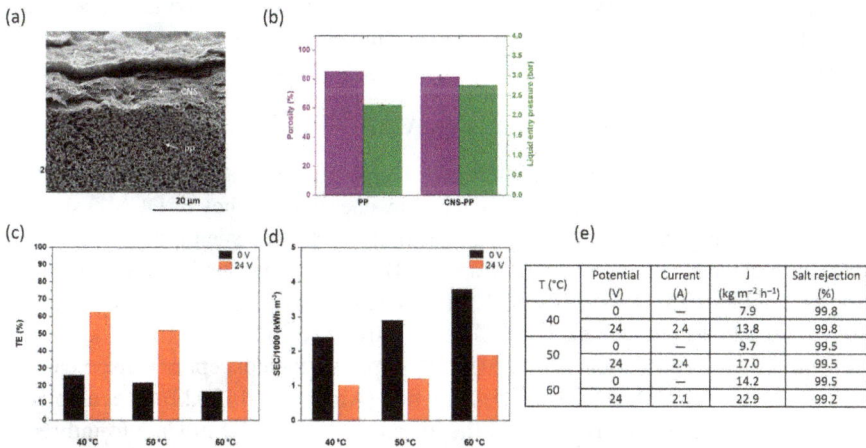

T (°C)	Potential (V)	Current (A)	J (kg m^{-2} h^{-1})	Salt rejection (%)
40	0	—	7.9	99.8
	24	2.4	13.8	99.8
50	0	—	9.7	99.5
	24	2.4	17.0	99.5
60	0	—	14.2	99.5
	24	2.1	22.9	99.2

FIGURE 8.4 (a) SEM image of CNS-PP membrane applied to electrothermal heating in DCMD; (b) porosity of polypropylene membrane before and after coating with CNS; (c–d) thermal efficiency and specific energy consumption with and without application of 24 V at initial feed temperatures of 40, 50, and 60 °C; (e) water flux and salt rejection through CNS membrane with and without electric field at different initial feed temperatures.

Source: Ahmed, F. E., Lalia, B. S., Hashaikeh, R., & Hilal, N. (2020). Enhanced performance of direct contact membrane distillation via selected electrothermal heating of membrane surface. *Journal of Membrane Science, 610*, 118224. https://doi.org/10.1016/j.memsci.2020.118224. Reprinted with permission.

$$SR(\%) = \frac{k_{feed} - k_{permeate}}{k_{feed}} \cdot 100 \qquad \text{Equation 8.2}$$

where k_{feed} and $k_{permeate}$ are the conductivities of the feed and permeate, respectively.

Thermal efficiency and specific energy consumption were also calculated. Figure 8.4d–f shows the performance parameters and the effect of electrically assisted heating on the performance of a DCMD system for brackish water desalination at different initial feed temperatures.

Applying AC potential across the CNS layer increased permeate flux at all three temperatures by 75, 76, and 61% (Ahmed et al., 2020b), and salt rejection is maintained at >99%. One may argue that increasing feed temperature also enhanced permeate flux, but we showed that at higher feed temperatures, the greater difference between the feed and ambient temperatures caused greater heat losses and decreased the thermal energy efficiency; thus incorporating electrically assisted heating at lower temperatures is a more efficient way of increasing permeate flux. This result is emphasized by Elmarghany et al., who carried out a thermal analysis of DCMD systems (Elmarghany et al., 2019). The increase in flux is surprisingly complemented by a reduction in specific energy consumption of >50% at all three temperatures. Although SEC values are high, they are comparable to other reported values for lab-scale DCMD systems. Other ways to lower SEC that could be incorporated include using higher flow rates, upscaling, and using heat recovery devices (Alexander V. Dudchenko et al., 2017; Khayet & Matsuura, 2011). This work demonstrated that combining electrothermal surface heating with MD should be considered in the upscaling of MD systems.

8.3.3 Direct Electric Heating during AGMD for Seawater Desalination

Assisted Joule heating improves thermal energy efficiency and increases the productivity of MD systems. Nevertheless, the low energy efficiency of DCMD systems makes this configuration particularly undesirable for large-scale units. Studies show that the greater efficiency of AGMD and VMD systems has made them the configuration of choice for larger pilot-scale MD systems (Alkhudhiri et al., 2012, 2013a, 2013b; Guillén-Burrieza et al., 2011; Guillén-Burrieza et al., 2012). To further reduce the high grade electrical energy needed, we explored the concept of pulsed electric heating during AGMD of seawater desalination (Ahmed et al., 2022). Li et al. placed a graphene oxide conductive layer at the air gap of an AGMD module to indirectly heat the feed through reverse Joule heating, but our group was the first to apply an electrically conductive surface for the direct heating of AGMD feed at the membrane surface (Ahmed et al., 2022). We used vacuum filtration to produce CNT-PP films with varying CNT mass loading and pressed these CNT films onto commercial MD membranes. The electrothermal heating effect of CNT films was investigated in air experimentally and numerically using a 3D thermal-electric simulation, and the membrane with the optimum heating effect was applied to interval electrothermal heating in conjunction with low-temperature bulk heating during seawater desalination by AGMD, and we investigated the effect of intermittent electrical heating on vapor flux and thermal efficiency.

8.3.3.1 Specific Thermal Energy Consumption

Specific thermal energy consumption (STEC) was calculated as a means of determining operating costs. STEC is made up of the energy input needed to heat the bulk feed solution and the additional thermal energy input to heat the feed at the membrane surface.

By energy conservation, heat flux through the membrane is equivalent to the heat lost by the feed as it is circulated.

Q_m, heat flux through the membrane in kW, is expressed as:

$$Q_m = \dot{m}_f C_{p_w} \left(T_{f,in} - T_{f,out} \right) \qquad \text{Equation 8.3}$$

where $\dot{m}_f$ is the mass flow rate in kg s^{-1}, C_{p_w} is the feedwater specific heat in kJ kg^{-1} °C^{-1}, $T_{f,in}$, and $T_{f,out}$ are the inlet and outlet feed temperatures.

The energy consumed for heating of the feed Q_{in}, when no electrical assistance is applied, is:

$$Q_{in} = Q_m \qquad \text{Equation 8.4}$$

When electrical heating is applied, the overall heat flux Q_m, as calculated from Equation 8.4, now considers both energy input to heat bulk feed (Q_{in}) as well as energy input via membrane surface heating. The additional energy gained by the feed from Joule heating Q_j in kJ is calculated by subtracting heat flux without Joule heating Q_{in} from heat flux with intermittent Joule heating Q_m.

The specific thermal energy consumption (STEC) can then be calculated in kWh m^{-3} from the following equation (Elmarghany et al., 2019; Soomro & Kim, 2018):

$$\text{STEC} = \frac{(Q_{in} + Q_j)\rho}{JA_m} \qquad \text{Equation 8.5}$$

where ρ is the density of water in kg m^{-3}, J is the permeate flux through the membrane in kg m^{-2} h^{-1}, and A_m is the active area of the membrane in m^2.

8.3.3.2 Gained Output Ratio (GOR)

GOR is an indicator of energy efficiency commonly used in thermal systems including MD. It is the ratio of latent heat required to evaporate the produced water to the amount of heat supplied to the system and is determined by:

$$\text{GOR} = \frac{JA\Delta H_v}{Q_m} \qquad \text{Equation 8.6}$$

where ΔH_v is the latent heat of vaporization in J kg^{-1} (Alkhudhiri & Hilal, 2018; Alexander V Dudchenko et al., 2017).

The effect of electric field on the heating of CNT films depends on the material's electrical and thermal properties. The specific heat capacity c_p determines how much energy a material requires to raise its temperature by 1 K. The CNS films have a

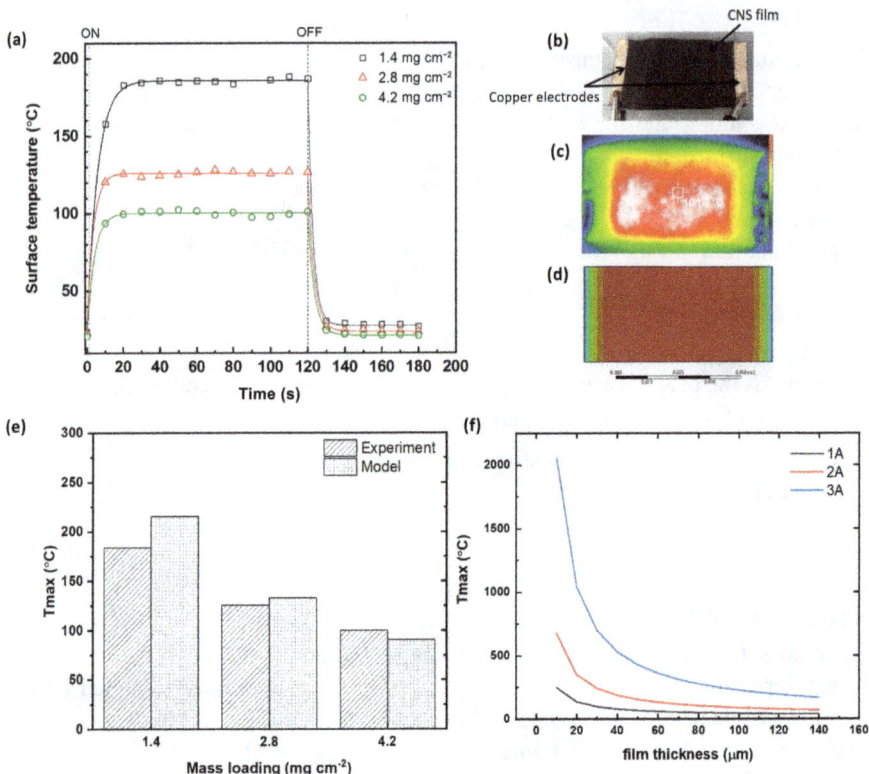

FIGURE 8.5 Joule heating of CNS films in air: (a) time-dependent CNS surface temperature with varying CNS loading under the effect of 2 A current; solid lines represent curves fitted to exponential growth and Newton's law of cooling; (b) photograph of CNS film during Joule heating; (c) steady-state temperature distribution on Joule heated CNS as measured by infrared camera; (d) simulation of temperature distribution on Joule heated membrane using thermal electric analysis; (e) maximum surface temperature of films under 2 A current with different CNS mass loadings; (f) theoretical maximum surface temperature of CNS films in air as a function of film thickness and electric current.

Source: Ahmed, F. E., Lalia, B. S., Hashaikeh, R., & Hilal, N. (2022). Intermittent direct joule heating of membrane surface for seawater desalination by air gap membrane distillation. *Journal of Membrane Science, 648*, 120390. Reprinted with permission.

specific heat capacity of 340 J g^{-1} K^{-1} at room temperature. This is lower than the reported values for CNT sheets as thermal transport depends on factors such as number of walls, defects, aspect ratio, etc. It is also lower than that of nichrome, the most widely used material for electric heating. Therefore, less energy is needed to raise the temperature of CNS films compared to nichrome. The CNS films demonstrated rapid and reversible heating under an applied current (Figure 8.5a). The maximum surface temperature reached for CNT mass loadings of 1.4 mg cm^{-2}, 2.8 mg cm^{-2}, and 4.2 mg cm^{-2} were 184 °C, 126 °C, and 100 °C, respectively (Ahmed et al., 2022). Increasing

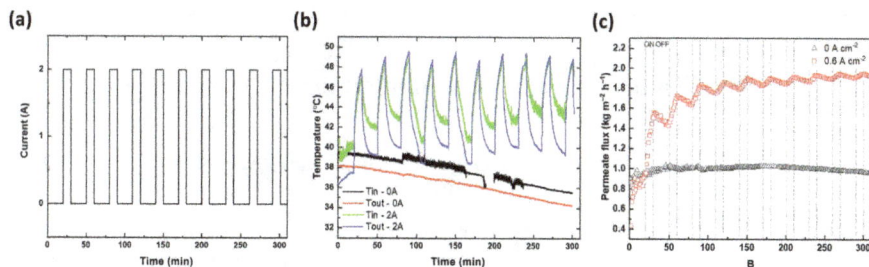

FIGURE 8.6 (a) Applied current profile; (b) temperature profile of CNS-PP membrane with and without applied current; (c) permeate flux with and without applied current, during AGMD using 35,000 ppm NaCl as feed solution.

Source: Ahmed, F. E., Lalia, B. S., Hashaikeh, R., & Hilal, N. (2022). Intermittent direct Joule heating of membrane surface for seawater desalination by air gap membrane distillation. *Journal of Membrane Science*, 648, 120390. Reprinted with permission.

CNT mass loading reduces heating effect due to better contact between conducting nanotubes and lower resistance. Figure 8.5b–f compares temperature distribution and maximum temperature reached through electric heating for the experiment and model. One limitation in this fabrication technique is that standalone vacuum filtered CNT films were first prepared and then pressed onto PP, which limits the minimum achievable thickness of the CNT layer.

During AGMD, a current of 0.6 A cm^{-2} was passed through the CNT-PP membrane at 20 minute intervals for 10 minutes each, as shown in Figure 8.6a. Figure 8.6b–c shows the temperature profile of the feed inlet and outlet and the permeate flux with and without applied current. After the first cycle of applied potential, the temperature at the feed outlet increases by 10 °C, which is accompanied by a 51% increase in flux. During heating and cooling cycles, the electrically induced temperature profile follows exponential growth and decay.

Table 8.1 shows the average permeate flux, STEC, and GOR with and without Joule heating. Average permeate flux increases by 78% for the entire run when current is applied, due to internal heat generation and greater feed temperature at the membrane surface (Ahmed et al., 2022). This increase in flux is not without compromising STEC, which increases by 34% due to the energy needed for localized Joule heating. However, GOR, which indicates how efficiently thermal energy is utilized, increases by 25% with electrically assisted heating. Potential future work could include studying the effect of module length and air gap width on MD performance.

This work demonstrated that intermittent electrically assisted MD is an energy efficient method of enhancing MD performance. By combining electrothermal surface heating with MD, these studies pave the way for smart, low-energy MD systems.

8.4 CONCLUSION

Electric fields can be exploited to control the performance of membrane systems. In this chapter, we discussed electrically enhanced performance of both

TABLE 8.1

Performance Indicators with and without Intermittent Surface Heating for AGMD of 35,000 ppm NaCl.

	Average Permeate Flux (kg m^{-2} h^{-1})	STEC (kWh m^{-3})	GOR
Without intermittent Joule heating	0.65 ± 0.042	4600 ± 56	0.059
With Joule heating (0.6 A cm^{-2})	1.15 ± 0.055	6185 ± 78	0.074

Source: Ahmed, F. E., Lalia, B. S., Hashaikeh, R., & Hilal, N. (2022). Intermittent direct joule heating of membrane surface for seawater desalination by air gap membrane distillation. *Journal of Membrane Science, 648,* 120390. Reprinted with permission.

pressure-driven and temperature-driven membrane processes including microfiltration, ultrafiltration, nanofiltration, and membrane distillation. In this first part, we have dealt with smart electroresponsive membranes with electrically tunable permselectivity.

Smart gate membranes aim to mimic the behavior of cell membranes through the design of voltage-controlled nanopores. In recent years, many fabrication techniques and materials have been employed for electrically driven permselectivity. Some common materials that have been studied include polyelectrolyte gels, carbon nanotubes, graphene, and conducting polymers such as polyaniline and polypyrrole and MXenes. Voltage-gated membranes can be formed into flat-sheet or hollow fiber membranes and applied to desalination of seawater, brine treatment, mineral recovery, and water treatment. More complex configurations such as spiral wound have not yet been demonstrated in modeling or experimental studies. Functional electroresponsive materials can be used to form the membrane or be introduced into its pores to provide control over pore size, flux, and solute rejection ratio through the application of anodic or cathodic potentials. In this chapter, we have reviewed advances in voltage-gated membranes in desalination and water treatment. However, voltage-gated synthetic membranes is a relatively new field with many challenges to address. Incorporating an electroresponsive material by grafting, coating, or pore-filling may improve membrane response to electric field, but it may also negatively impact other membrane properties such as flux, mechanical strength, etc. The application of voltage-gated membranes is also limited by fabrication methods, many of which are difficult and/or expensive to scale up. More work is needed on the careful design of responsive domains without compromising membrane performance and adopting scalable fabrication techniques. Still, exciting advances in this area are certain to inspire future studies to optimize the use of electroresponsive membranes and maneuver their performance using electric fields.

Under suitable applied electric potential, electroresponsive gating membranes alter the mobility of charged ions into the membrane, which causes pore size to change. Reversible swelling of the polymer affects ion transport and consequently membrane selectivity. They are used not only in sensors, drug delivery, space, and energy applications but can be beneficial to separation processes where ion transport needs to be tuned. Electroactive polymers can be classified as (1) redox polymers, (2) electronically conducting polymers, or (3) loaded ionomers (Lyons, 2013). The scope of the present section is to discuss high conductivity electroactive polymers, including properties, preparation, and use in membrane separation.

In the latter part of the chapter, we considered the electrically tunable surface temperature of hydrophobic MD membranes to increase driving force and reduce temperature polarization in order to obtain greater flux at better thermal energy efficiency. We have focused on electrically assisted Joule heating for direct surface heating for two configurations—the simpler, lab-scale appropriate DCMD and the more commercially apt AGMD—to show that electrothermal heating may pave the way for energy-efficient MD and provide the boost this technology needs to go from lab- and pilot-scale to commercial use. The applicability of electrically assisted self-heating should be considered in future MD studies as more electrically conductive materials evolve as membrane heaters.

BIBLIOGRAPHY

Abraham, J., Vasu, K. S., Williams, C. D., Gopinadhan, K., Su, Y., Cherian, C. T., . . . Grigorieva, I. V. (2017). Tunable sieving of ions using graphene oxide membranes. *Nature Nanotechnology, 12*(6), 546–550.

Ahmed, F. E., Lalia, B. S., Hashaikeh, R., & Hilal, N. (2020a). Alternative heating techniques in membrane distillation: A review. *Desalination, 496*, 114713. https://doi.org/10.1016/j.desal.2020.114713

Ahmed, F. E., Lalia, B. S., Hashaikeh, R., & Hilal, N. (2020b). Enhanced performance of direct contact membrane distillation via selected electrothermal heating of membrane surface. *Journal of Membrane Science, 610*, 118224. https://doi.org/10.1016/j.memsci.2020.118224

Ahmed, F. E., Lalia, B. S., Hashaikeh, R., & Hilal, N. (2022). Intermittent direct joule heating of membrane surface for seawater desalination by air gap membrane distillation. *Journal of Membrane Science, 648*, 120390. https://doi.org/10.1016/j.memsci.2022.120390

Alberts, B., Johnson, A., Lewis, J., Raff, M., Roberts, K., & Walter, P. (2002). *Ion channels and the electrical properties of membranes molecular biology of the cell* (4th ed.). Garland Science.

Alkhudhiri, A., Darwish, N., & Hilal, N. (2012). Treatment of high salinity solutions: Application of air gap membrane distillation. *Desalination, 287*, 55–60. https://doi.org/10.1016/j.desal.2011.08.056

Alkhudhiri, A., Darwish, N., & Hilal, N. (2013a). Produced water treatment: Application of air gap membrane distillation. *Desalination, 309*, 46–51. https://doi.org/10.1016/j.desal.2012.09.017

Alkhudhiri, A., Darwish, N., & Hilal, N. (2013b). Treatment of saline solutions using air gap membrane distillation: Experimental study. *Desalination, 323*, 2–7. https://doi.org/10.1016/j.desal.2012.09.010

Alkhudhiri, A., & Hilal, N. (2018). 3—Membrane distillation—principles, applications, configurations, design, and implementation. In V. G. Gude (Ed.), *Emerging technologies for sustainable desalination handbook* (pp. 55–106). Butterworth-Heinemann.

Alsaati, A., & Marconnet, A. M. (2018). Energy efficient membrane distillation through localized heating. *Desalination, 442*, 99–107. https://doi.org/10.1016/j.desal.2018.05.009

Baker, R. W. (2012). *Membrane technology and applications.* Wiley.

Bandehali, S., Parvizian, F., Hosseini, S. M., Matsuura, T., Drioli, E., Shen, J., . . . Adeleye, A. S. (2021). Planning of smart gating membranes for water treatment. *Chemosphere, 283*, 131207. https://doi.org/10.1016/j.chemosphere.2021.131207

Bezanilla, F., & Stefani, E. (1998). *Gating currents methods in enzymology* (Vol. 293, pp. 331–352). Academic Press.

Bliznyuk, V., Galabura, Y., Burtovyy, R., Karagani, P., Lavrik, N., & Luzinov, I. (2014). Electrical conductivity of insulating polymer nanoscale layers: Environmental effects. *Physical Chemistry Chemical Physics, 16*(5), 1977–1986. doi:10.1039/C3CP54020K

Cantley, L., Swett, J. L., Lloyd, D., Cullen, D. A., Zhou, K., Bedworth, P. V., . . . Bunch, J. S. (2019). Voltage gated inter-cation selective ion channels from graphene nanopores. *Nanoscale, 11*(20), 9856–9861. doi:10.1039/C8NR10360G

Chang, D. C., & Reese, T. S. (1990). Changes in membrane structure induced by electroporation as revealed by rapid-freezing electron microscopy. *Biophysical Journal, 58*(1), 1–12. doi:10.1016/S0006-3495(90)82348-1

Chen, P.-R., & Chuang, Y.-J. (2013). The development of conductive nanoporous chitosan polymer membrane for selective transport of charged molecules. *Journal of Nanomaterials, 2013*, 980857. https://doi.org/10.1155/2013/980857

De Napoli, I. E., Catapano, G., Ebrahimi, M., & Czermak, P. (2011). 5.56—Novel and current techniques to produce endotoxin-free dialysate in dialysis centers. In M. Moo-Young (Ed.), *Comprehensive biotechnology* (2nd ed., pp. 741–752). Academic Press.

Deshmukh, A., Boo, C., Karanikola, V., Lin, S., Straub, A. P., Tong, T., . . . Elimelech, M. (2018). Membrane distillation at the water-energy nexus: Limits, opportunities, and challenges. *Energy & Environmental Science, 11*(5), 1177–1196. doi:10.1039/C8EE00291F

Dongare, P. D., Alabastri, A., Pedersen, S., Zodrow, K. R., Hogan, N. J., Neumann, O., . . . Halas, N. J. (2017). Nanophotonics-enabled solar membrane distillation for off-grid water purification. *Proceedings of the National Academy of Sciences, 114*(27), 6936. doi:10.1073/pnas.1701835114

Dudchenko, A. V., Chen, C., Cardenas, A., Rolf, J., & Jassby, D. (2017). Frequency-dependent stability of CNT Joule heaters in ionizable media and desalination processes. *Nature Nanotechnology, 12*(6), 557.

Dudchenko, A. V., Rolf, J., Russell, K., Duan, W., & Jassby, D. (2014). Organic fouling inhibition on electrically conducting carbon nanotube–polyvinyl alcohol composite ultrafiltration membranes. *Journal of Membrane Science, 468*, 1–10. https://doi.org/10.1016/j.memsci.2014.05.041

Durmaz, E. N., Sahin, S., Virga, E., de Beer, S., de Smet, L. C. P. M., & de Vos, W. M. (2021). Polyelectrolytes as building blocks for next-generation membranes with advanced functionalities. *ACS Applied Polymer Materials, 3*(9), 4347–4374. doi:10.1021/acsapm.1c00654

Elmarghany, M. R., El-Shazly, A. H., Salem, M. S., Sabry, M. N., & Nady, N. (2019). Thermal analysis evaluation of direct contact membrane distillation system. *Case Studies in Thermal Engineering, 13*, 100377. https://doi.org/10.1016/j.csite.2018.100377

Fan, Y., Li, J., Wang, S., Meng, X., Zhang, W., Jin, Y., . . . Liu, S. (2020). Voltage-enhanced ion sieving and rejection of Pb2+ through a thermally cross-linked two-dimensional MXene membrane. *Chemical Engineering Journal, 401*, 126073. https://doi.org/10.1016/j.cej.2020.126073

Formoso, P., Pantuso, E., De Filpo, G., & Nicoletta, F. P. (2017). Electro-conductive membranes for permeation enhancement and fouling mitigation: A short review. *Membranes, 7*(3), 39. doi:10.3390/membranes7030039

Gao, L., Li, C., Huang, W., Mei, S., Lin, H., Ou, Q., . . . Zhang, H. (2020). MXene/polymer membranes: Synthesis, properties, and emerging applications. *Chemistry of Materials, 32*(5), 1703–1747. doi:10.1021/acs.chemmater.9b04408

García-Fernández, A., Lozano-Torres, B., Blandez, J. F., Monreal-Trigo, J., Soto, J., Collazos-Castro, J. E., . . . Martínez-Máñez, R. (2020). Electro-responsive films containing voltage responsive gated mesoporous silica nanoparticles grafted onto PEDOT-based conducting polymer. *Journal of Controlled Release, 323*, 421–430. https://doi.org/10.1016/j.jconrel.2020.04.048

Gong, B., Yang, H., Wu, S., Yan, J., Cen, K., Bo, Z., & Ostrikov, K. K. (2019). Superstructure-enabled anti-fouling membrane for efficient photothermal distillation. *ACS Sustainable Chemistry & Engineering.* doi:10.1021/acssuschemeng.9b06160

Gong, K., Fang, T., Wan, T., Yan, Y., Li, W., & Zhang, J. (2020). Voltage-gated multilayer graphene nanochannel for K+/Na+ separation: A molecular dynamics study. *Journal of Molecular Liquids, 317*, 114025. https://doi.org/10.1016/j.molliq.2020.114025

Guillén-Burrieza, E., Blanco, J., Zaragoza, G., Alarcón, D.-C., Palenzuela, P., Ibarra, M., & Gernjak, W. (2011). Experimental analysis of an air gap membrane distillation solar desalination pilot system. *Journal of Membrane Science, 379*(1), 386–396. https://doi.org/10.1016/j.memsci.2011.06.009

Guillén-Burrieza, E., Zaragoza, G., Miralles-Cuevas, S., & Blanco, J. (2012). Experimental evaluation of two pilot-scale membrane distillation modules used for solar desalination. *Journal of Membrane Science, 409–410*, 264–275. https://doi.org/10.1016/j.memsci.2012.03.063

Guo, L., Jing, Y., & Chaplin, B. P. (2016). Development and characterization of ultrafiltration TiO2 magnéli phase reactive electrochemical membranes. *Environmental Science & Technology, 50*(3), 1428–1436. doi:10.1021/acs.est.5b04366

Hart, J. L., Hantanasirisakul, K., Lang, A. C., Anasori, B., Pinto, D., Pivak, Y., . . . Taheri, M. L. (2019). Control of MXenes' electronic properties through termination and intercalation. *Nature Communications, 10*(1), 522. doi:10.1038/s41467-018-08169-8

He, Z., Zhou, J., Lu, X., & Corry, B. (2013). Bioinspired graphene nanopores with voltage-tunable ion selectivity for Na+ and K+. *ACS Nano, 7*(11), 10148–10157. doi:10.1021/nn4043628

Hu, C., Liu, Z., Lu, X., Sun, J., Liu, H., & Qu, J. (2018). Enhancement of the Donnan effect through capacitive ion increase using an electroconductive rGO-CNT nanofiltration membrane. *Journal of Materials Chemistry A, 6*(11), 4737–4745.

Huang, Q., Gao, S., Huang, Y., Zhang, M., & Xiao, C. (2019). Study on photothermal PVDF/ATO nanofiber membrane and its membrane distillation performance. *Journal of Membrane Science, 582*, 203–210. https://doi.org/10.1016/j.memsci.2019.04.019

Jia, S.-L., Geng, H.-Z., Wang, L., Tian, Y., Xu, C.-X., Shi, P.-P., . . . Kong, J. (2018). Carbon nanotube-based flexible electrothermal film heaters with a high heating rate. *Royal Society Open Science, 5*(6), 172072. doi:10.1098/rsos.172072

Jiang, Y., Hu, C., Cheng, H., Li, C., Xu, T., Zhao, Y., . . . Qu, L. (2016). Spontaneous, straightforward fabrication of partially reduced graphene oxide–polypyrrole composite films for versatile actuators. *ACS Nano, 10*(4), 4735–4741.

Joshi, R., Carbone, P., Wang, F.-C., Kravets, V. G., Su, Y., Grigorieva, I. V., . . . Nair, R. R. (2014). Precise and ultrafast molecular sieving through graphene oxide membranes. *Science, 343*(6172), 752–754.

Jou, I. A., Melnikov, D. V., Nadtochiy, A., & Gracheva, M. E. (2014). Charged particle separation by an electrically tunable nanoporous membrane. *Nanotechnology, 25*(14), 145201.

Kaner, P., Bengani-Lutz, P., Sadeghi, I., & Asatekin, A. (2016a). Responsive filtration membranes by polymer self-assembly. *Technology, 4*(04), 217–228.

Kaner, P., Bengani-Lutz, P., Sadeghi, I., & Asatekin, A. (2016b). Responsive filtration membranes by polymer self-assembly. *Technology, 04*(04), 217–228. doi:10.1142/S2339547816500096

Kevin Li, S., Hao, J., & Liddell, M. (2013). Chapter 11—electrotransport across membranes in biological media: Electrokinetic theories and applications in drug delivery. In S. M. Becker & A. V. Kuznetsov (Eds.), *Transport in biological media* (pp. 417–454). Elsevier.

Khayet, M., & Matsuura, T. (2011). *Membrane distillation: Principles and applications*. Elsevier Science.

Khor, C. M., Zhu, X., Messina, M. S., Poon, S., Lew, X. Y., Maynard, H. D., & Jassby, D. (2019). Electrically mediated membrane pore gating via grafted polymer brushes. *ACS Materials Letters, 1*(6), 647–654. doi:10.1021/acsmaterialslett.9b00298

Ling, Z., Ren, C. E., Zhao, M.-Q., Yang, J., Giammarco, J. M., Qiu, J., . . . Gogotsi, Y. (2014). Flexible and conductive MXene films and nanocomposites with high capacitance. *Proceedings of the National Academy of Sciences, 111*(47), 16676–16681. doi:10.1073/pnas.1414215111

Liu, Y., Wu, P., Liu, F., Li, F., An, X., Liu, J., . . . Sand, W. (2019). Electroactive modified carbon nanotube filter for simultaneous detoxification and sequestration of Sb(III). *Environmental Science & Technology, 53*(3), 1527–1535. doi:10.1021/acs.est.8b05936

Luo, F., Xie, R., Liu, Z., Ju, X.-J., Wang, W., Lin, S., & Chu, L.-Y. (2015). Smart gating membranes with in situ self-assembled responsive nanogels as functional gates. *Scientific Reports, 5*(1), 14708. doi:10.1038/srep14708

Lyons, M. E. G. (2013). *Electroactive polymer electrochemistry: Part 1: Fundamentals*. Springer US.

Mahinthichaichan, P., Tsai, C.-C., Payne, G. F., & Shen, J. (2020). Polyelectrolyte in electric field: Disparate conformational behavior along an aminopolysaccharide chain. *ACS Omega, 5*(21), 12016–12026. doi:10.1021/acsomega.0c00164

Miedema, H. (2012). Ion-selective biomimetic membranes. In C. Hélix-Nielsen (Ed.), *Biomimetic membranes for sensor and separation applications* (pp. 63–86). Springer Netherlands.

Millet, P. (2015). 9—Hydrogen production by polymer electrolyte membrane water electrolysis. In V. Subramani, A. Basile, & T. N. Veziroğlu (Eds.), *Compendium of hydrogen energy* (pp. 255–286). Woodhead Publishing.

Moghaddasi Fereidani, R. (2015). *Electrical gating of nanochannels of polyelectrolyte brushes*. Sharif University of Technology. http://repository.sharif.edu/resource/419718/-/&-from=search&&query=polyelectrolytes&field=subjectkeyword&count=20&execute=true (47425 (04))

Nair, R., Wu, H., Jayaram, P. N., Grigorieva, I. V., & Geim, A. (2012). Unimpeded permeation of water through helium-leak–tight graphene-based membranes. *Science, 335*(6067), 442–444.

Ni, Z., Qiu, H., & Guo, W. (2018). Electrically tunable ion selectivity of charged nanopores. *The Journal of Physical Chemistry C, 122*(51), 29380–29385. doi:10.1021/acs.jpcc.8b10191

Ouyang, H., Xia, Z., & Zhe, J. (2009). Static and dynamic responses of polyelectrolyte brushes under external electric field. *Nanotechnology, 20*(19), 195703.

Politano, A., Argurio, P., Di Profio, G., Sanna, V., Cupolillo, A., Chakraborty, S., . . . Curcio, E. (2017). Photothermal membrane distillation for seawater desalination. *Advanced Materials, 29*(2), 1603504. doi:10.1002/adma.201603504

Purkait, M. K., Sinha, M. K., Mondal, P., & Singh, R. (2018). Chapter 6—electric field-responsive membranes. In M. K. Purkait, M. K. Sinha, P. Mondal, & R. Singh (Eds.), *Interface science and technology* (Vol. 25, pp. 173–191). Elsevier.

Rohani, R., & Yusoff, I. I. (2019). Towards electrically tunable nanofiltration membranes: Polyaniline-coated polyvinylidene fluoride membranes with tunable permeation properties. *Iranian Polymer Journal, 28*(9), 789–800. doi:10.1007/s13726-019-00744-0

Said, I. A., Wang, S., & Li, Q. (2019). Field demonstration of a nanophotonics-enabled solar membrane distillation reactor for desalination. *Industrial & Engineering Chemistry Research, 58*(40), 18829–18835. doi:10.1021/acs.iecr.9b03246

Shi, Y., Zheng, Q., Ding, L., Yang, F., Jin, W., Tang, C. Y., & Dong, Y. (2022). Electro-enhanced separation of microsized oil-in-water emulsions via metallic membranes: Performance and mechanistic insights. *Environmental Science & Technology, 56*(7), 4518–4530.

Shiga, T., & Kurauchi, T. (1990). Deformation of polyelectrolyte gels under the influence of electric field. *Journal of Applied Polymer Science, 39*(11–12), 2305–2320.

Soomro, M. I., & Kim, W.-S. (2018). Parabolic-trough plant integrated with direct-contact membrane distillation system: Concept, simulation, performance, and economic evaluation. *Solar Energy, 173*, 348–361. https://doi.org/10.1016/j.solener.2018.07.086

Sui, D., Huang, Y., Huang, L., Liang, J., Ma, Y., & Chen, Y. (2011). Flexible and transparent electrothermal film heaters based on graphene materials. *Small, 7*(22), 3186–3192.

Sun, M., Wang, X., Winter, L. R., Zhao, Y., Ma, W., Hedtke, T., . . . Elimelech, M. (2021). Electrified membranes for water treatment applications. *ACS ES&T Engineering, 1*(4), 725–752. doi:10.1021/acsestengg.1c00015

Sun, P., Wang, K., & Zhu, H. (2016). Recent developments in graphene-based membranes: Structure, mass-transport mechanism and potential applications. *Advanced Materials, 28*(12), 2287–2310.

Tan, X., Hu, C., Zhu, Z., Liu, H., & Qu, J. (2019). Electrically pore-size-tunable polypyrrole membrane for antifouling and selective separation. *Advanced Functional Materials, 29*(35), 1903081. https://doi.org/10.1002/adfm.201903081

Tan, Y. Z., Chandrakant, S. P., Ting Ang, J. S., & Chew, J. W. (2020). Localized induction heating of metallic spacers for energy-efficient membrane distillation. *Journal of Membrane Science*, 118150. https://doi.org/10.1016/j.memsci.2020.118150

Tan, Y. Z., Wang, H., Han, L., Tanis-Kanbur, M. B., Pranav, M. V., & Chew, J. W. (2018). Photothermal-enhanced and fouling-resistant membrane for solar-assisted membrane distillation. *Journal of Membrane Science, 565*, 254–265. https://doi.org/10.1016/j.memsci.2018.08.032

Tieleman, D. P. (2004). The molecular basis of electroporation. *BMC Biochemistry, 5*(1), 10. doi:10.1186/1471-2091-5-10

Traversi, F., Raillon, C., Benameur, S. M., Liu, K., Khlybov, S., Tosun, M., . . . Radenovic, A. (2013). Detecting the translocation of DNA through a nanopore using graphene nanoribbons. *Nature Nanotechnology, 8*(12), 939–945. doi:10.1038/nnano.2013.240

Venkatesan, B. M., Estrada, D., Banerjee, S., Jin, X., Dorgan, V. E., Bae, M.-H., . . . Bashir, R. (2012). Stacked graphene-Al2O3 nanopore sensors for sensitive detection of DNA and DNA–protein complexes. *ACS Nano, 6*(1), 441–450. doi:10.1021/nn203769e

Wandera, D., Wickramasinghe, S. R., & Husson, S. M. (2010). Stimuli-responsive membranes. *Journal of Membrane Science, 357*(1), 6–35. https://doi.org/10.1016/j.memsci.2010.03.046

Wang, P. (2018). Emerging investigator series: The rise of nano-enabled photothermal materials for water evaporation and clean water production by sunlight. *Environmental Science: Nano, 5*(5), 1078–1089. doi:10.1039/C8EN00156A

Weaver, J. C. (1995). Electroporation theory. In J. A. Nickoloff (Ed.), *Plant cell electroporation and electrofusion protocols* (pp. 3–28). Springer.

Wei, G., Quan, X., Chen, S., Fan, X., Yu, H., & Zhao, H. (2015). Voltage-gated transport of nanoparticles across free-standing all-carbon-nanotube-based hollow-fiber membranes. *ACS Applied Materials & Interfaces, 7*(27), 14620–14627. doi:10.1021/acsami.5b01183

Wei, G., Yu, H., Quan, X., Chen, S., Zhao, H., & Fan, X. (2014). Constructing all carbon nanotube hollow fiber membranes with improved performance in separation and antifouling for water treatment. *Environmental Science & Technology, 48*(14), 8062–8068. doi:10.1021/es500506w

Weidlich, C., & Mangold, K.-M. (2011). Electrochemically switchable polypyrrole coated membranes. *Electrochimica Acta, 56*(10), 3481–3484.

Weir, M. P., Heriot, S. Y., Martin, S. J., Parnell, A. J., Holt, S. A., Webster, J. R. P., & Jones, R. A. L. (2011). Voltage-induced swelling and deswelling of weak polybase brushes. *Langmuir, 27*(17), 11000–11007. doi:10.1021/la201343w

Widakdo, J., Huang, T.-H., Subrahmanya, T. M., Austria, H. F. M., Hung, W.-S., Wang, C.-F., . . . Lai, J.-Y. (2021). Tailoring of graphene–organic frameworks membrane to enable reversed electrical-switchable permselectivity in CO2 separation. *Carbon, 182*, 545–558. https://doi.org/10.1016/j.carbon.2021.06.047

Wu, X., Jiang, Q., Ghim, D., Singamaneni, S., & Jun, Y.-S. (2018). Localized heating with a photothermal polydopamine coating facilitates a novel membrane distillation process. *Journal of Materials Chemistry A, 6*(39), 18799–18807.

Xu, L. L., Shahid, S., Holda, A. K., Emanuelsson, E. A. C., & Patterson, D. A. (2018). Stimuli responsive conductive polyaniline membrane: In-filtration electrical tuneability of flux and MWCO. *Journal of Membrane Science, 552*, 153–166. https://doi.org/10.1016/j.memsci.2018.01.070

Xu, L. L., Shahid, S., Patterson, D. A., & Emanuelsson, E. A. C. (2019). Flexible electro-responsive in-situ polymer acid doped polyaniline membranes for permeation enhancement and membrane fouling removal. *Journal of Membrane Science, 578*, 263–272. https://doi.org/10.1016/j.memsci.2018.09.070

Yamauchi, T., Kokufuta, E., & Osada, Y. (1993). Electrically controlled protein permeation through a poly(vinyl alcohol)/poly(acrylic acid) composite membrane. *Polymer Gels and Networks, 1*(4), 247–255. https://doi.org/10.1016/0966-7822(93)90003-Z

Ye, H., Li, X., Deng, L., Li, P., Zhang, T., Wang, X., & Hsiao, B. S. (2019). Silver nanoparticle-enabled photothermal nanofibrous membrane for light-driven membrane distillation. *Industrial & Engineering Chemistry Research, 58*(8), 3269–3281. doi:10.1021/acs.iecr.8b04708

Zhang, H., Quan, X., Fan, X., Yi, G., Chen, S., Yu, H., & Chen, Y. (2019). Improving ion rejection of conductive nanofiltration membrane through electrically enhanced surface charge density. *Environmental Science & Technology, 53*(2), 868–877. doi:10.1021/acs.est.8b04268

Zhang, L., Kan, X., Huang, T., Lao, J., Luo, K., Gao, J., . . . Jiang, L. (2022). Electric field modulated water permeation through laminar Ti3C2Tx MXene membrane. *Water Research, 219*, 118598. https://doi.org/10.1016/j.watres.2022.118598

Zhitenev, N. B., Sidorenko, A., Tennant, D. M., & Cirelli, R. A. (2007). Chemical modification of the electronic conducting states in polymer nanodevices. *Nature Nanotechnology, 2*(4), 237–242. doi:10.1038/nnano.2007.75

Zhou, F., Biesheuvel, P. M., Choi, E.-Y., Shu, W., Poetes, R., Steiner, U., & Huck, W. T. (2008). Polyelectrolyte brush amplified electroactuation of microcantilevers. *Nano Letters, 8*(2), 725–730.

Zhou, K. G., Vasu, K. S., Cherian, C. T., Neek-Amal, M., Zhang, J. C., Ghorbanfekr-Kalashami, H., . . . Nair, R. R. (2018). Electrically controlled water permeation through graphene oxide membranes. *Nature, 559*(7713), 236–240. doi:10.1038/s41586-018-0292-y

Zhu, X., & Jassby, D. (2019). Electroactive membranes for water treatment: Enhanced treatment functionalities, energy considerations, and future challenges. *Accounts of Chemical Research, 52*(5), 1177–1186. doi:10.1021/acs.accounts.8b00558

9 Future Prospects

9.1 INTRODUCTION

The preceding chapters have conveyed the contributions of electrically conductive membrane systems for fouling mitigation and overall performance enhancement in membrane-based separation processes. The extensive literature available in this still young, thriving field shows its crucial role in future membrane module development for low energy desalination and water treatment. The present book did not aim to provide a comprehensive overview but has instead delved into specific topics and examples to help the reader grasp the status of desalination and water treatment, challenges related to fouling and energy, and the role of electrically conducting materials in overcoming those challenges. The book also discussed recent advancements in both electrically conducting membranes and spacers as smart devices with several benefits on membrane transport behavior and fouling mitigation. We hope that this book has provided an account of the principles, achievements, and possibilities of electrically conducting membrane systems in desalination and water treatment that is adequately informative for experts in membrane process engineering to foresee how these transformative materials, fabrication techniques, and application will guide future research.

Without simultaneous growth in material synthesis and fabrication techniques to scalable methods, electrically conductive membrane systems are likely to be limited, for a while, to small-scale systems suitable only for remote water treatment applications. Upscaling depends on the commercial viability of existing fabrication processes and then on implementing conductive membranes in modules. We conclude that the applicability of electrically conductive membranes relies on five aspects: (1) simplified desalination; (2) low cost, scalable, and sustainable fabrication; (3) implementation of electrically conducting membranes and spacers in existing modules; (4) strategies to develop new processes and configurations that encompass electrically conducting materials; and (5) systematic evaluation of the toxicology of nanomaterials (Figure 9.1).

9.2 SIMPLIFIED DESALINATION PRETREATMENT

Pretreatment is essential to most membrane desalination plants. Pretreatment aims to remove contaminants or potential foulants in the source water to prevent membrane fouling. Pretreatment involves a series of unit operations, each of which is designed to tackle foulant-induced risks during the desalination process. While pretreatment is used in all desalination processes, membrane-based processes often have more rigorous pretreatment requirements due to the propensity of membranes to foul. Pretreatment removes feed constituents such as sediment and microbes, which may

DOI: 10.1201/9781003144991-9

FIGURE 9.1 Facets that will determine the future outlook of electrically conductive membrane systems for fouling mitigation.

cause membrane fouling and deter the desalination process. The extent of pretreatment depends very strongly on feedwater constituents and quality: Certain source waters may require negligible pretreatment, while some more challenging feeds may need rigorous pretreated to guard the RO operation. The type and location of intake influence the nature and composition of foulants in the source water; thus pretreatment must be adjusted to the characteristics of the specific source water. Table 9.1 shows pretreatment considerations and the resulting propensity of fouling for seawater RO plants (Voutchkov, 2010).

Pretreatment adds to the costs of the plant in the form of both infrastructure and operating expenses, depending on the type of unit operations that constitute pretreatment. Lack of appropriate pretreatment measures can be detrimental for the desalination plant and can in turn cause delays in deadlines being met as well as severe financial losses. Insights on fouling mechanisms and types, as well as an example of the consequences of insufficient pretreatment, were discussed in Chapter 2. Many conventional and non-conventional pretreatment technologies have been applied to RO plants. In conventional pretreatment, the source water goes through conditioning via coagulation, flocculation, and pH adjustment, followed by granular media filtration such as sand filtration to remove particulates. Other conventional technologies include acidification, UV radiation, use of antiscalants, and dissolved air flotation (DAF). However, due to greater separation efficiency, lower space requirements, and modularity, membrane-based processes have increasingly been used for pretreatment. Common processes include MF, UF, and NF that can remove dissolved and suspended matter of different sizes prior to RO desalination. In Chapter 1, we observed characteristics of each of these membrane processes, including operating conditions, size of particle removed, energy, etc. These are all considerations when designing appropriate

TABLE 9.1

Seawater Quality Characterization for Pretreatment.

Parameter	Pretreatment Consideration
Turbidity (NTU)	High levels > 0.1 mg L^{-1} may lead to fouled membranes. Values > 50 NTU usually requires DAF and filtration.
TOC (mg L^{-1})	High contents > 2 mg L^{-1} may lead to organic or biofouling.
SDI_{15}	Pretreatment is a must for SDI > 4.
TSS (mg L^{-1})	The parameter assesses the amount of residuals. It does not correlate well with turbidity > 5 NTU.
Iron (mg L^{-1})	State of the iron is important. In reduced forms, ≤2 mg L^{-1} is tolerable for RO membranes, while in oxidized forms, >0.05 mg L^{-1} is detrimental to performance.
Manganese (mg L^{-1})	State of the manganese is important. In reduced forms, ≤0.1 mg L^{-1} is tolerable for RO membranes, while in oxidized forms, >0.02 mg L^{-1} is detrimental to performance.
Silica (mg L^{-1})	Concentrations > 20 mg L^{-1} cause accelerated fouling.
Chlorine (mg L^{-1})	Concentrations > 0.01 mg L^{-1} cause RO membrane damage.
Temperature	Intake temperature is critical. $T \leq 12$ °C causes increase in unit energy use. $T \geq 35$ °C can lead to enhanced mineral scaling and biofouling. $T > 45$ °C may cause permanent damage to RO membranes.
Oil	Concentrations > 0.02 mg L^{-1} causes accelerated organic fouling.
pH	For pH < 4 and pH > 11, long-term exposure will cause RO membrane damage.

Source: Voutchkov, N. (2010). Considerations for selection of seawater filtration pretreatment system. *Desalination, 261*(3), 354–364. Reprinted with permission.

membrane-based pretreatment. Conventional processes are still widely used in pretreatment systems due to built infrastructure and other social barriers that are not in the scope of this book.

The potential of electrically conducting membrane systems in desalination can be fulfilled via two routes. The first involves replacing conventional pretreatment and incorporating electrically conducting materials in membrane-based pretreatment systems (MF, UF, and NF), while the second vastly eliminates costs associated with pretreatment by using electrically conducting RO membranes to remove and degrade various feed contaminants, clean membrane surface, and reject salts with high separation efficiencies. We are most interested in estimating the reduction in pretreatment costs brought about by using electrically conducting membrane systems.

As large-scale studies are still lacking in electrically conducting membrane systems, a fair cost comparison is possible only for small-scale decentralized systems. A common five-stage small-scale desalination system consists of a sediment filter, carbon filters, RO membrane, and carbon filter for post-treatment. Decentralized desalination systems are simply small-scale desalination plants close to a source of saline water supply and freshwater demand. As large plants are multiyear projects with financing, building, and operating considerations, small and medium-sized decentralized plants offer a competitive solution for densely populated urban areas, as well as remote areas. Since these smaller systems cannot benefit from economies of scale, their cost-effectiveness and environmental sustainability must be balanced (Younos & Lee, 2020). Compared to large-scale desalination systems, small-scale systems have lower capital construction costs, can be operated in situ by keeping the distance between a raw water source and the point of demand to a minimum, and are flexible with respect to available solutions due to decentralization (Delgado & Moreno, 2008). However, small-scale RO systems typically have a lower life expectancy.

Pretreatment is commonly used as a means to prevent fouling in NF and RO systems. As explored in previous chapters, electrically conducting membrane systems have been shown to reduce fouling and tackle feed contaminants at the membrane surface. This reduces the need for rigorous and expensive pretreatment, as challenging material can be dealt with by the smart membranes or spacers. This greatly simplifies the design and complexity of the desalination plant and lowers capital costs associated with aggressive pretreatment without compromising the frequency of membrane replacement.

Conventional pretreatment includes sediment filters and activated carbon filters, the latter of which are also used in the post-treatment stage. Decentralized systems in urban areas are often used to purify tap water, and these additional filters last about 6–18 months depending on usage and feedwater quality (Peter-Varbanets et al., 2009). UF membranes, when used in lieu of conventional pretreatment, remove most pathogens such as bacteria and also particles and colloidal material. However, the more complex system translates into high investment costs and maintenance by qualified workers. Although fouling can be due to many constituents in the feed, it is widely associated with natural organic matter (NOM) fractions and inorganic compounds. Pretreatment can be reduced or eliminated through clever design to lower the cost and maintenance of the system. Eliminating pretreatment reduces product water cost by 6–9% (Anis et al., 2019). Table 9.2 compares various aspects pertaining to conventional pretreatment, MF/UF pretreatment, and electrically conducting RO membrane system without pretreatment. The feedwater source is a dominant factor in determining the extent to which pretreatment is reduced. Removal of chlorine via chemical dosing is no longer required when an electrically conductive membrane system is used (Chapter 5), as chlorine can be removed electrochemically through anodic reactions at the surface of the conductive membrane or spacer. Similarly, scale-forming ions may also be removed through electrostatic repulsion prior to RO, potentially eliminating the need for antiscalants. We foresee that exciting progress in the types of components that can be electrochemically removed or inactivated by the desalination membrane/spacer will further elucidate the cost reductions enabled by reducing the capital and energy required for pretreatment.

TABLE 9.2

Comparison and Analysis of Various Aspects between Conventional Pretreatment, MF/UF Pretreatment and Electrically Conducting RO Membrane System.

	Conventional Pretreatment	MF/UF Pretreatment	Electrically Conducting RO Membrane System	Comments
Capital costs	Competitive with MF/UF	Competitive for small-scale systems (≤ 2000 m³ day⁻¹); more expensive for large-scale systems	Low	Depends on water production capacity.
Energy requirements	Low compared to MF/UF	Higher than conventional	Low; depends on feed characteristics and material of membrane electrode or spacer electrode.	Application of voltage needed, but no or low pretreatment when electrically conducting membranes are used as many pathogens and other foulants can be removed at membrane surface; depends on feed characteristics.
Chemical costs	High due to coagulants and process chemicals needed for optimization	Depends on raw water quality; relatively low chemical use; due to low chlorine resistance of many RO membranes, chemical dosing is employed to remove chlorine during pretreatment; antiscalants are used to delay onset of scaling.	Low; electrical potential is used to remove chlorine and sealant ions, drastically lowering need for chemicals.	
RO capital cost	Low separation efficiency limits RO flux; hence capital cost is higher than membrane pretreatment.	20% higher RO flux possible due to lower SDI of feed; hence RO capital cost is lower.	Low/moderate; electrical connections and power supply needed only; RO capital cost depends on feed SDI and specific removal rates via applied potential.	
RO operating costs	Higher fouling potential translates to higher cleaning frequency and membrane replacement costs.	MF/UF pretreatment lowers net driving pressure and fouling potential, which prolongs membrane life and reduces operating costs. Cleaning frequency is reduced by 10–100% compared to when conventional pretreatment is used.	Low; many foulants are removed from feed via electric potential; electric potential is also used to inactivate bacteria, causing biofouling; dependent on feed characteristics and nature of constituents to be removed electrically.	
Footprint	Large	Relatively low; compact and modular; 30–60% lower than conventional pretreatment.	Low; reduction of pretreatment requirements decreases footprint.	

9.3 SCALABLE FABRICATION

First, it is vital to investigate and demonstrate the upscaling of existing membrane and spacer fabrication techniques including CNT synthesis by chemical vapor deposition (CVD), electrospinning of nanofibers, and 3D printing of spacer materials. As an example, modified geometries have been introduced to the simple electrospinning setup to increase production capacity. These include multineedle, co-axial, and tri-axial approaches. Industrial upscaling of electrospinning for membrane fabrication has been reviewed by Ahmed and coworkers (Ahmed et al., 2015).

Similarly, the intrinsically unique properties of CNTs and graphene are often not extended to the macroscale due to practical difficulty in fabricating aligned, defect-free membranes or, in the case of graphene, monolayer-thick membranes. Today, CNT membranes can be prepared with ease, but their poor mechanical strength can be an issue with respect to filtration applications. There have been continuous efforts in scalable methods for producing graphene-based membranes recently (Gupta et al., 2021; Hong et al., 2017; Kidambi et al., 2018; Kim et al., 2020; Li et al., 2018; Liu et al., 2021; Vlassiouk, 2017; Yang et al., 2019). Hence, translating much of the work done on molecular dynamics simulations to full-scale studies necessitates the ability to manufacture CNT and graphene membranes on a large scale without loss of properties. With electrically conducting nanocomposites specifically, filler content and dispersion need more attention with regard to the specific application. Furthermore, the environmental and health hazards pertaining to nanoparticle release or leakage remain unclear. For more details on strategies toward scalable manufacturing of CNT-based membranes, the reader is referred to a review by Rashed et al. (Rashed et al., 2021). Other electrically conducting membranes, demonstrated only as flat-sheet in lab-scale studies, can be fabricated in larger dimensions and other configurations using advanced manufacturing techniques such as extrusion paste processing (Shi et al., 2022).

Today, innovation in conducting materials must be targeted to facilitate sustainable and low-cost fabrication. Shortcomings of commonly used materials include corrosion, oxidation, and high cost. At the fabrication stage, efforts need to be directed at achieving controlled porosity conductivity, stability, and reactivity for long-term stable and reliable operation. Additionally, other functional nanomaterials and catalysts may be incorporated into the conductive layer to achieve selective removal of contaminants especially when complex feed solutions are concerned.

With respect to membrane spacer fabrication, previous chapters have shown the promise of 3D printing as a potential technique to develop electrically conducting spacers of desired geometry for fouling mitigation. Based on the authors' own experience in this area, preparing a 3D printing dope solution with appropriate electrical conductivity and rheology is a perplexing task. In addition, while 3D printing has emerged as an exciting technique for spacer fabrication, not enough studies focus on electrically conducting 3D-printed spacers for desalination and water treatment. Certain 3D printing modes may be advantageous over others due to higher resolution and continuous production capabilities, the latter of which is particularly significant for production scalability. Another hindrance to the use of 3D printing is its cost, which is estimated to drop by between 50–75% within the next decade (Soo et al., 2021). With accessibility to 3D printing, electrically conducting spacers may not be too far off.

9.4 MODULE INTEGRATION

In reverse osmosis desalination plants, sensing is carried out through sensors installed directly on the spiral wound module or by using an ex situ canary cell that simulates plant operation and uses bypassed feed stream. Electrochemical techniques such as impedance spectroscopy have been incorporated in canary cells using four-electrode systems to characterize fouling through impedance measurements (Ho et al., 2017). As the use of electrochemical techniques for real-time fouling characterization grows, we are forced to transform the configuration of sensors and enable simpler designs. Fortunately, electrically conducting surfaces permit this as (1) a simple two-electrode system can be used instead of three-electrodes since the surface itself is conductive, and (2) if a conductive membrane module is used, there is no need to replicate plant conditions in a canary cell, and the sensor can be installed directly within the membrane module. This extends to all membrane processes including microfiltration, ultrafiltration, nanofiltration, and membrane distillation.

Figure 9.2a shows the schematic of a spiral wound module with electrically conducting membranes and counter-electrodes. This is one example of how electrically conductive materials can be incorporated into existing large-scale modules with compact spacing. However, many variations are possible, for example, 3D printing of both cathodic and anodic feed spacers on the membrane surface. The scaling up of electrically conducting membranes is dependent on progress in flexible electrode materials that can be integrated into existing spiral wound modules or, ideally, by embedding both cathode and anode into the module without additional layers. Figure 9.2a is a pressure vessel containing several spiral wound modules as used in large-scale RO systems today.

Another concern that necessitates further research into electrically conducting materials is the long-term stability of existing materials under the effect of an electric field, particularly when the applied potential is anodic. Commonly used materials are susceptible to electrochemical degradation via oxidation, for which voltage and time need to be carefully optimized. Smart designs such as integration of anode and cathode directly on the membrane surface offer the benefit of

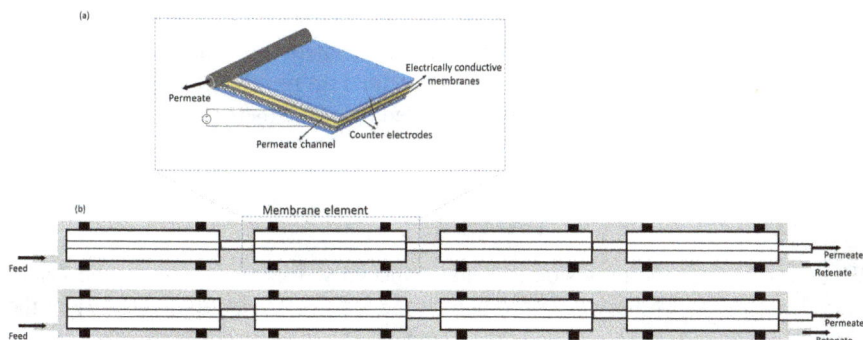

FIGURE 9.2   Schematic of (a) spiral wound membrane module using electrically conductive membranes and (b) pressure vessel that includes several such membrane elements.

integrated spacers, simple design, and cleaning via hydrogen generation, as well as simultaneous chlorine disinfection (Khalil et al., 2022). Industrial adoption with real-world applications requires the development of membranes with higher electrochemical stability (Halali et al., 2021).

9.5 PROCESS OPTIMIZATION

Chemical-free removal of foulants could make electrically conducting membranes and spacers a viable solution for desalination and water treatment. Exploitation of electrically conducting surfaces requires thorough studies of the electrochemical properties and optimization of electric field amplitude, frequency, and duration for a given application. Depending on the feed solution, undesirable competing reactions may take away from the overall efficiency of the separation process. Toxic organic compounds can also be formed depending on feedwater constituents and applied electric field.

It has been argued that electrically conducting membranes for NF and RO will be difficult without a paradigm shift in module design (Sun et al., 2021). Electrically conducting membranes and spacers can combine fouling control and contaminant degradation with filtration through direct and free-radical-mediated electro-oxidation, electroreduction, electrostatic repulsion, and electroporation, as discussed in previous chapters. Tunable pore sizes can also reduce or eliminate the need for pre-treatment. Future work should involve achieving 100% contaminant removal efficiency together with low energy consumption and high current efficiency, as well as improved performance for sensing and resource recovery. The prospect of a single membrane module to carry out all of these functions is transformative in the field of membrane technology.

Currently, the lack of integration into large-scale modules has limited electrically conducting materials to off-grid distributed applications. More effort is needed in the design of electrically conducting modules to enable their use in traditional membrane processes. This includes integrating these systems into compact spiral wound modules suitable for centralized desalination and water treatment. Fouling control via electrically conducting surfaces should adopt energy-saving strategies such as intermittent as opposed to continuous self-cleaning. Energy analysis of applying electrically conductive surfaces to membrane systems is still lacking as many studies focus only on the effect of applying a potential on membrane flux and/or rejection. Further work on energy analysis is needed to ensure their competitiveness with mature membranes.

9.6 TOXICITY OF NANOMATERIALS

When it comes to the provision of water, especially drinking water, public health takes priority. As we have seen throughout the text, nanoscale materials are at the forefront of science and technology development, and the area of electrically conductive membrane systems is no different in this regard. Current and future developments in benchtop science and applied technology continue to be propelled by advances in nanomaterials. Their nanoscale dimensions yield unique morphological and surface properties that enable electrochemical reactivity and other desirable behavior. Yet there is a charging debate on the environmental risks and hazards posed by nano-materials. Studies on nanoscale particles have prompted the emergence of a new discipline, nanotoxicology, that deals with the safety evaluation of nanostructures and nanodevices (Oberdörster et al., 2005). Unfortunately, the number of studies on their toxicity remains very limited given their domination in various fields. In water treatment specifically, studying the safety risks associated with nanomaterials is a relatively new area (Bharati et al., 2017).

Nanoparticles are known to be more chemically reactive than their bulk counter-parts. Their high surface area also makes the human body more accessible to them with greater bioavailability, which may lead to greater toxicity (Bharati et al., 2017). The surface area of nanomaterials may facilitate the binding and transport of chem-ical pollutants. They may compromise the human immune system response (Bharati et al., 2017).

The toxicity of nanomaterials arises not just from their composition, which is sim-ilar to that of the parent bulk materials, but also from physicochemical properties such as size, surface chemistry, and roughness. Chellappa et al. reveal that the cytotoxic-ity of TiO_2 nanoparticles is dependent on size and crystalline structure (Chellappa et al., 2015). The same is true for silver nanoparticles. Fortunately, like other aspects of nanoparticles, toxicity can also be tuned by manipulating their physicochemical properties. However, there is inadequate data on which properties control toxicity and to what extent. Currently, literature on the correlation between exposure time, dosage, and health impact is also lacking. Moore argues that the harmful effects of nanoscale particulates released into the aquatic environment is even more poorly understood (Moore, 2006). Researchers agree that in-depth systematic ecotoxico-logical assessments and life cycle analyses still need to be carried out to identify suitable forms of CNTs and graphene-based nanomaterials with minimal health and environmental impacts (Ali et al., 2019; Perreault et al., 2015).

This enormous lack of awareness carries forward to companies involved in the manufacture of nanomaterials and nanodevices, which are simply unable to provide consumers with accurate risk assessment because the scientific literature is so lim-ited. We expect initiatives undertaken especially in the EU and the United States to produce relevant data on the toxicological aspects of (exposure to) engineering nanoparticles in the near future.

As part of the scientific community, it is our role to devise appropriate testing pro-tocols and predictive tools to assess and address the risk of nanoparticles throughout product life cycle. Environmental regulators and decision makers rely on scientists to provide them with data on not only technical but also toxicological aspects of the materials and processes we advocate. This is necessary in order for the water sector to avoid drastically negative consequences in the long term, as was the case with plant biotechnology (Moore, 2006). Nel et al. press for urgent implementation of principles and test procedures for the safe manufacture and use of nanomateri-als, especially for commercial manufacture (Nel et al., 2006). The future outlook of nanomaterials must include studies of the physiological effects of exposure to nanoparticles during membrane and spacer element manufacture and a deeper under-standing of their interaction in aquatic environments.

9.7 CONCLUSION

Global research efforts have driven technological innovation in materials especially with respect to membrane separation. As we have previously seen, the research community is actively involved with optimizing membrane technology for low energy desalination and water treatment. Electrically conducting membranes and spacers can mitigate fouling through early detection, self-cleaning, and tunable permselectivity to provide demonstrated reduction in energy consumption. Applicability to real systems, however, will depend on fine-tuning membrane electrodes and integrating them with electronic circuits and machine learning algorithms. While an enormous amount of progress has been made, efforts are needed to facilitate commercial application of electrically conductive membranes and spacers for fouling mitigation. This includes further manipulation of electrically conductive materials through continuous progress in fabrication techniques that would introduce opportunities for advanced separation. A smart electrically conducting RO membrane or spacer is capable of simplifying the desalination process by reducing or eliminating stages of pretreatment, which in turn lowers capital and operating costs while retaining high RO membrane lifetime and low downtime. Compared to conventional and MF/UF pretreatment, electrically conducting membrane systems use fewer chemicals, have a smaller footprint, and require a lower capital cost while their energy cost depends on the materials involved and feed characteristics.

The integration of electrochemical processes with membrane separation through the use of electrically conducting materials is a promising process intensification strategy. This is partly due to the progress in electrochemical sensing and fouling control, which has thus far been limitedly applied to membrane separation in desalination and water treatment. Strides in battery and energy storage research would be beneficial when considering membrane electrodes for separation to deploy these technologies at a large scale. In future work, the selection of appropriate electrochemical configurations to couple with separation will be significant in applying large-scale processes. What makes the marriage of electrochemical processes with membrane separation so promising is that both are governed by colloidal and interfacial chemistry. New processes may become available as the intersection of interfacial design, molecular engineering, and electrochemical engineering rapidly advances. However, one must proceed with caution when incorporating nanomaterials as the toxicology of nanomaterials through product lifestyle is poorly understood.

BIBLIOGRAPHY

Ahmed, F. E., Lalia, B. S., & Hashaikeh, R. (2015). A review on electrospinning for membrane fabrication: Challenges and applications. *Desalination*, *356*, 15–30.

Ali, S., Rehman, S. A. U., Luan, H.-Y., Farid, M. U., & Huang, H. (2019). Challenges and opportunities in functional carbon nanotubes for membrane-based water treatment and desalination. *Science of the Total Environment*, *646*, 1126–1139. https://doi.org/10.1016/j.scitotenv.2018.07.348

Anis, S. F., Hashaikeh, R., & Hilal, N. (2019). Reverse osmosis pretreatment technologies and future trends: A comprehensive review. *Desalination*, *452*, 159–195. https://doi.org/10.1016/j.desal.2018.11.006

Bharati, R., Sundaramurthy, S., & Thakur, C. (2017). 14—Nanomaterials and food-processing wastewater. In A. M. Grumezescu (Ed.), *Water purification* (pp. 479–516). Academic Press.

Chellappa, M., Anjaneyulu, U., Manivasagam, G., & Vijayalakshmi, U. (2015). Preparation and evaluation of the cytotoxic nature of TiO2 nanoparticles by direct contact method. *International Journal of Nanomedicine*, *10*(Suppl 1), 31.

Delgado, D. J., & Moreno, P. (2008). *Desalination research progress*. Nova Science Pub Incorporated.

Gupta, A., Sharma, C. P., & Arnusch, C. J. (2021). Simple scalable fabrication of laser-induced graphene composite membranes for water treatment. *ACS ES&T Water*, *1*(4), 881–887.

Halali, M. A., Larocque, M., & de Lannoy, C.-F. (2021). Investigating the stability of electrically conductive membranes. *Journal of Membrane Science*, *627*, 119181. https://doi.org/10.1016/j.memsci.2021.119181

Ho, J. S., Sim, L. N., Webster, R. D., Viswanath, B., Coster, H. G. L., & Fane, A. G. (2017). Monitoring fouling behavior of reverse osmosis membranes using electrical impedance spectroscopy: A field trial study. *Desalination*, *407*, 75–84. https://doi.org/10.1016/j.desal.2016.12.012

Hong, S., Constans, C., Surmani Martins, M. V., Seow, Y. C., Guevara Carrio, J. A., & Garaj, S. (2017). Scalable graphene-based membranes for ionic sieving with ultrahigh charge selectivity. *Nano Letters*, *17*(2), 728–732.

Khalil, A., Ahmed, F. E., Hashaikeh, R., & Hilal, N. (2022). 3D printed electrically conductive interdigitated spacer on ultrafiltration membrane for electrolytic cleaning and chlorination. *Journal of Applied Polymer Science, n/a*(n/a), 52292. https://doi.org/10.1002/app.52292

Kidambi, P. R., Mariappan, D. D., Dee, N. T., Vyatskikh, A., Zhang, S., Karnik, R., & Hart, A. J. (2018). A scalable route to nanoporous large-area atomically thin graphene membranes by roll-to-roll chemical vapor deposition and polymer support casting. *ACS Applied Materials & Interfaces*, *10*(12), 10369–10378.

Kim, J. H., Choi, Y., Kang, J., Choi, E., Choi, S. E., Kwon, O., & Kim, D. W. (2020). Scalable fabrication of deoxygenated graphene oxide nanofiltration membrane by continuous slot-die coating. *Journal of Membrane Science*, *612*, 118454.

Li, G., Law, W.-C., & Chan, K. C. (2018). Floating, highly efficient, and scalable graphene membranes for seawater desalination using solar energy. *Green Chemistry*, *20*(16), 3689–3695.

Liu, Z., Ma, Z., Qian, B., Chan, A. Y., Wang, X., Liu, Y., & Xin, J. H. (2021). A facile and scalable method of fabrication of large-area ultrathin graphene oxide nanofiltration membrane. *ACS Nano*, *15*(9), 15294–15305.

Moore, M. N. (2006). Do nanoparticles present ecotoxicological risks for the health of the aquatic environment? *Environment International*, *32*(8), 967–976. https://doi.org/10.1016/j.envint.2006.06.014

Nel, A., Xia, T., Mädler, L., & Li, N. (2006). Toxic potential of materials at the nanolevel. *Science*, *311*(5761), 622–627.

Oberdörster, G., Oberdörster, E., & Oberdörster, J. (2005). Nanotoxicology: An emerging discipline evolving from studies of ultrafine particles. *Environmental Health Perspectives*, *113*(7), 823–839.

Perreault, F., De Faria, A. F., & Elimelech, M. (2015). Environmental applications of graphene-based nanomaterials. *Chemical Society Reviews*, *44*(16), 5861–5896.

Peter-Varbanets, M., Zurbrügg, C., Swartz, C., & Pronk, W. (2009). Decentralized systems for potable water and the potential of membrane technology. *Water Research*, *43*(2), 245–265. https://doi.org/10.1016/j.watres.2008.10.030

Rashed, A. O., Merenda, A., Kondo, T., Lima, M., Razal, J., Kong, L., . . . Dumée, L. F. (2021). Carbon nanotube membranes—Strategies and challenges towards scalable manufacturing and practical separation applications. *Separation and Purification Technology*, *257*, 117929. https://doi.org/10.1016/j.seppur.2020.117929

Shi, Y., Zheng, Q., Ding, L., Yang, F., Jin, W., Tang, C. Y., & Dong, Y. (2022). Electro-enhanced separation of microsized oil-in-water emulsions via metallic membranes: Performance and mechanistic insights. *Environmental Science & Technology*, *56*(7), 4518–4530.

Soo, A., Ali, S. M., & Shon, H. K. (2021). 3D printing for membrane desalination: Challenges and future prospects. *Desalination*, *520*, 115366. https://doi.org/10.1016/j.desal.2021.115366

Sun, M., Wang, X., Winter, L. R., Zhao, Y., Ma, W., Hedtke, T., . . . Elimelech, M. (2021). Electrified membranes for water treatment applications. *ACS ES&T Engineering*, *1*(4), 725–752. doi:10.1021/acsestengg.1c00015

Vlassiouk, I. V. (2017). A scalable graphene-based membrane. *Nature Nanotechnology*, *12*(11), 1022–1023.

Voutchkov, N. (2010). Considerations for selection of seawater filtration pretreatment system. *Desalination*, *261*(3), 354–364. https://doi.org/10.1016/j.desal.2010.07.002

Yang, E., Karahan, H. E., Goh, K., Chuah, C. Y., Wang, R., & Bae, T.-H. (2019). Scalable fabrication of graphene-based laminate membranes for liquid and gas separations by crosslinking-induced gelation and doctor-blade casting. *Carbon*, *155*, 129–137.

Younos, T., & Lee, J. (2020). Desalination: Concept and system components. In V. S. Saji, A. A. Meroufel, & A. A. Sorour (Eds.), *Corrosion and fouling control in desalination industry* (pp. 3–27). Springer International Publishing.

For Product Safety Concerns and Information please contact our EU
representative GPSR@taylorandfrancis.com
Taylor & Francis Verlag GmbH, Kaufingerstraße 24, 80331 München, Germany

www.ingramcontent.com/pod-product-compliance
Lightning Source LLC
Chambersburg PA
CBHW060332220326
41598CB00023B/2682